Functional Molecules in Tracing and Cancer Therapeutics

Functional Molecules in Tracing and Cancer Therapeutics

Kang-Nan Wang

MDPI

Basel • Beijing • Wuhan • Barcelona • Belgrade • Novi Sad • Cluj • Manchester

Editor
Kang-Nan Wang
State Key Laboratory of
Crystal Materials
Shandong University
Jinan
China

Editorial Office
MDPI AG
Grosspeteranlage 5
4052 Basel, Switzerland

This is a reprint of articles from the Special Issue published online in the open access journal *International Journal of Molecular Sciences* (ISSN 1422-0067) (available at: www.mdpi.com/journal/ijms/special_issues/45RFG12T1D).

For citation purposes, cite each article independently as indicated on the article page online and as indicated below:

Lastname, A.A.; Lastname, B.B. Article Title. *Journal Name* **Year**, *Volume Number*, Page Range.

ISBN 978-3-7258-2492-2 (Hbk)
ISBN 978-3-7258-2491-5 (PDF)
doi.org/10.3390/books978-3-7258-2491-5

Contents

About the Editor

Kang-Nan Wang

Kang-Nan Wang received his Ph.D. in Chemical Biology from Sun Yat-sen University in 2019 and subsequently conducted postdoctoral research at the National University of Singapore. He is currently a Professor at the State Key Laboratory of Crystal Materials, Shandong University, where he serves as a Master's/PhD Supervisor. His research interests focus on "Molecular Probe Imaging and Theranostics". In recent years, he has concentrated on functional fluorescent molecules, with a particular emphasis on achieving high-resolution analysis and the dynamic intervention of active molecules and organelles under pathological conditions. His key research areas include (1) developing organelle-targeted imaging probes to elucidate the functions and interactions of key organelles; (2) creating novel ratiometric fluorescent probes to uncover how active molecules regulate organelle homeostasis; and (3) proposing light-activated, organelle-based synergistic theranostic strategies to overcome challenges such as low therapeutic efficiency and uncontrolled treatment processes.

Preface

In recent years, research on functional targeting molecules has made remarkable progress in the medical field. These molecules are not only used for sensing and imaging but also show significant potential in anti-inflammatory and antitumor applications. Some research teams have achieved groundbreaking advancements by integrating multiple functions into a single molecule, thereby expanding its scope of application. In the context of disease treatment, functional molecules such as targeted drug delivery systems, imaging agents for tumor and inflammation characterization, and molecular probes for detecting biomarkers associated with cancer and other diseases are gradually transforming traditional diagnostic and therapeutic approaches. The rapid development of functional targeting molecules has brought new hope to cancer treatment.

Since joining the State Key Laboratory of Crystal Materials at Shandong University, I have been devoted to the study of light-functional fluorescent molecules and diagnostic probes, witnessing firsthand the rapid progress in this field. Given the significance and broad application prospects of this research area, I am honored to have been invited to serve as a Guest Editor for the *International Journal of Molecular Sciences* (IJMS), where I have collaborated with leading scholars from around the world to share and explore the latest applications and future developments of functional molecules in fluorescent imaging and cancer-targeted therapy.

IJMS plays a crucial role on the global academic stage, and I look forward to seeing it provide more opportunities for researchers to showcase their work. I also wish *IJMS* continued success in its future endeavors.

Kang-Nan Wang
Editor

International Journal of *Molecular Sciences*

MDPI

Review

Assessment of the Impact of Trace Essential Metals on Cancer Development

Aleksandra Górska [1], Agnieszka Markiewicz-Gospodarek [1,*], Mateusz Trubalski [2], Marta Żerebiec [2], Julia Poleszak [2] and Renata Markiewicz [3]

[1] Department of Normal, Clinical and Imaging Anatomy, Medical University of Lublin, 4 Jaczewskiego St., 20-090 Lublin, Poland; aleksandragorska@umlub.pl

[2] Students Scientific Association, Department of Normal, Clinical and Imaging Anatomy, Medical University of Lublin, 4 Jaczewskiego St., 20-090 Lublin, Poland; mateusztrub@gmail.com (M.T.); zerebiecm22@gmail.com (M.Ż.); juliapoleszak3@gmail.com (J.P.)

[3] Occupational Therapy Laboratory, Chair of Nursing Development, Medical University of Lublin, 4 Staszica St., 20-081 Lublin, Poland; renatamarkiewicz@umlub.pl

* Correspondence: agnieszkamarkiewiczgospodarek@umlub.pl

Abstract: This study examines the impact of zinc, copper, cobalt, iron, and manganese on cancer development, considering their dual roles as potential promoters or inhibitors within tumorigenesis. A comprehensive analysis of existing literature and experimental data is conducted to elucidate the intricate relationship between these trace elements and cancer progression. The findings highlight the multifaceted effects of zinc, copper, cobalt, iron, and manganese on various aspects of cancer development, including cell proliferation, angiogenesis, and metastasis. Understanding the nuanced interactions between these trace elements and cancer could offer crucial insights into tumorigenesis mechanisms and facilitate the identification of novel biomarkers and therapeutic targets for cancer prevention and treatment strategies. This research underscores the importance of considering the roles of essential trace elements in cancer biology and may ultimately contribute to advancements in precision medicine approaches for combating cancer.

Keywords: heavy metals; zinc; copper; cobalt; iron; manganese; carcinogenesis

check for updates

Citation: Górska, A.; Markiewicz-Gospodarek, A.; Trubalski, M.; Żerebiec, M.; Poleszak, J.; Markiewicz, R. Assessment of the Impact of Trace Essential Metals on Cancer Development. *Int. J. Mol. Sci.* **2024**, *25*, 6842. https://doi.org/10.3390/ijms25136842

Academic Editor: Kang-Nan Wang

Received: 13 May 2024
Revised: 14 June 2024
Accepted: 19 June 2024
Published: 21 June 2024

1. Introduction

Carcinogenesis is a complex and protracted progression characterized by a series of stages. The conventional understanding posits that the development of cancer comprises a triphasic process, encompassing initiation, promotion, and progression stages. It encompasses genetic mutations, genomic instability, heightened activity of oncogenes, suppression of tumour suppressor genes, alterations in genetic material, and aberrations in cellular metabolism [1,2]. The initiation and progression of carcinogenesis are influenced by exogenous and endogenous factors, coupled with individual elements such as genetic predisposition [3,4]. It can be inferred that the amalgamation of diverse risk factors exerts the most significant impact on the evolution of cancer [4].

An extremely important group of exogenous carcinogenesis factors is the group of heavy metals (HMs). In biological systems, heavy metals have been documented to impact various cellular organelles and constituents, including the cell membrane, mitochondria, lysosomes, endoplasmic reticulum, nuclei, and certain enzymes associated with metabolic processes, detoxification mechanisms, and damage repair pathways. Metal ions have been identified to engage with cellular components like DNA and nuclear proteins, inducing DNA damage and structural alterations that can result in carcinogenesis [5,6].

Heavy metals such as arsenic, cadmium, chromium, and nickel are well-documented carcinogens, but the focus of this article will be on the tumorigenic effects of copper, cobalt, iron, zinc, and manganese. These metals are not only essential for various biological functions but also pose a risk of toxicity and carcinogenesis when present in abnormal concentrations. Copper is known for its role in angiogenesis and metastasis [7,8], while cobalt has been implicated in hypoxia signalling pathways that can promote tumour growth [9]. Iron, through its participation in the Fenton reaction, can generate reactive oxygen species (ROS) leading to oxidative stress and DNA damage [10,11]. Zinc plays a crucial role in DNA synthesis and repair but can also influence cancer progression depending on its concentration and cellular context [12]. Manganese, while essential for enzymatic functions, has been associated with neurotoxicity and potential carcinogenic effects through mechanisms that are still being elucidated [13].

The aim of this article is to compare the impact of selected heavy metals on carcinogenesis. We will focus on tumorigenesis induced by copper, cobalt, iron, zinc, and manganese. Our goal is to gather and summarize information on the role of these metals in cancer development, as they have not been collectively discussed in this context before. This review seeks to summarize how these essential, yet potentially harmful elements contribute to the complex process of carcinogenesis.

2. Methods and Search Criteria

To conduct a literature review on carcinogenesis and the relationship with heavy metals, the following steps were undertaken.

2.1. *Defining the Scope and Objectives*

The review aimed to interstrand the mechanisms of carcinogenesis, with a particular focus on the role of heavy metals. Specific heavy metals of interest included: zinc, copper, cobalt, iron, and manganese.

2.2. *Search Strategy*

Comprehensive searches were conducted in multiple scientific databases, including PubMed, Scopus, and Web of Science. Keywords used for the search included combinations of terms such as: carcinogenesis, heavy metals, cancer, metal toxicity, zinc, copper, cobalt, iron and manganese.

2.3. *Inclusion Criteria*

Peer-reviewed articles, studies published within the last 20 years, research focusing on the biological mechanisms of heavy metals in carcinogenesis, and reviews or meta-analyses on the topic.

2.4. *Exclusion Criteria*

Articles not available in English, studies with insufficient data on heavy metals, and research focused on non-cancerous outcomes.

Each study was assessed for methodological quality including sample size, study design, statistical analysis, and potential biases. High-quality studies were given more weight in the synthesis of findings.

Data from the selected studies were synthesized to identify common terms and patterns regarding the role of heavy metals in carcinogenesis. Mechanisms of action, such as DNA damage, oxidative stress, and disruption of cellular signaling pathways, were highlighted. Differences in the carcinogenic potential of different heavy metals were also discussed.

3. The Influence of Specific Heavy Metals on the Carcinogenic Process

3.1. *Zinc*

The indispensability of zinc (Zn) for human physiology was conclusively determined in 1963. Over the last five decades, remarkable progress has been witnessed in both the clinical and fundamental aspects of understanding zinc metabolism in humans [8]. The human body's mass contains approximately 2–3 g of zinc, with skeletal muscle and bone accounting for around 57% and 29% of the total zinc content, respectively. The heart and blood plasma are recognized to contain 0.4% and 0.1% of the body's zinc, respectively. Insufficient dietary intake, reduced absorption, or elevated zinc loss can lead to a state of deficiency [10]. Reduced serum zinc levels have been observed in various cancer patients, encompassing those with breast and prostate cancers [11].

Zinc, a micronutrient essential for all living organisms, plays a crucial role in various biochemical pathways within human cells. One of these roles is that Zn forms associations with more than 2500 proteins, representing approximately 10% of the total human proteome [10] and assumes a vital role as a structural constituent within structural motifs termed "zinc fingers", present in diverse RNA and DNA binding proteins. In this capacity, zinc contributes to the preservation of structural integrity for a substantial portion of these proteins [10,12]. Zinc is also indispensable for diverse enzyme activities, gene expression, and critical cellular functions, including cellular proliferation. A noteworthy transcription factor activated by zinc is the metal response element-binding transcription factor-1 (MTF-1), a protein featuring six zinc fingers and multiple domains. Metal regulatory transcription factor-1 (MTF-1) functions as a zinc sensor, regulating the expression of genes pivotal for zinc homeostasis and providing protection against metal toxicity and oxidative stress (OS) [12]. Oxidative stress and heightened inflammatory cytokines are acknowledged as significant contributing factors in numerous age-associated chronic diseases, mutagenesis, and cancer [8].

Oxidative stress is characterized by an imbalance between the generation of free radicals and reactive metabolites, commonly known as oxidants or reactive oxygen species (ROS), and their removal through protective mechanisms termed antioxidants. Reactive oxygen species are natural by-products of cellular metabolism, serving essential roles in activating signaling pathways within animal cells in response to alterations in intra- and extracellular environmental conditions [13]. In cells experiencing chronic inflammation, the substantial release of ROS and reactive nitrogen species (RNS) attracts an increased number of activated immune cells, thereby amplifying dysregulated processes, and culminating in a preneoplastic state. If the production of cellular ROS/RNS surpasses the endogenous antioxidant response, irreversible oxidative damage to nucleic acids, lipids, and proteins may occur, inducing genetic and/or epigenetic alterations that disrupt the regulation of oncogenes and tumour suppressor genes. The processes of oxidative stress and chronic inflammation are intricately linked, and the inability to inhibit these processes may result in genetic/epigenetic changes that initiate carcinogenesis (Figure 1) [14]. This illustrates how important it is to maintain the correct level of zinc in the body. Physiological levels of zinc demonstrate an inhibitory effect on the generation of reactive oxygen species, encompassing superoxide anion ($\cdot O^-$), hydrogen peroxide (H_2O_2), and hydroxyl radical ($OH\cdot$), as well as reactive nitrogen species, including peroxynitrite. The direct antioxidant function of zinc ions is associated with their interaction with thiol groups, thereby shielding them from oxidative processes. Zn serves as a cofactor for the antioxidant enzyme Cu, Zn-superoxide dismutase (SOD1), and its activity is attenuated in conditions of zinc deficiency. Furthermore, studies indicate that zinc may indirectly modulate the functionality of other antioxidant enzymes [15].

As mentioned earlier, diminished serum zinc levels have been observed in various cancer patients, including those with breast, prostate, and endometrial cancers (Table 1) [16].

Due to the increased incidence of cancer in recent years, an attempt was made to conduct in vitro tests. In the case of gastric cancer, a high frequency of mutations has been associated with PI3K-Akt-mTOR signaling pathways, and genetic changes in this pathway have had a direct impact on the progression of many cancers [17]. In the above study, the created piperine-loaded ZnO nanocomposite (ZnO-Pip-NC) was found to have anticancer activity in the case of gastric cancer [17].

Table 1. Characteristics of breast, prostate, and endometrial cancers.

Type of Cancer	Characterization	Ref.
Breast cancer (BC)	Breast cancer, the most common cancer in women worldwide, has diverse types based on hormone and human epithelial growth factor receptor 2 (HER2) status: luminal A/B, HER2-positive, and triple-negative (TNBC). Zinc imbalance is linked to breast cancer, with low serum zinc but high zinc in cancer tissues. Zinc plays a crucial role in cancer progression, affecting cell transformation and tumor aggressiveness by influencing zinc transporters.	[16,18,19]
Prostate cancer (PCa)	Prostate cancer ranks as the second most common cancer in men globally, with high mortality rates, especially in cases with extracapsular disease. Unlike normal and benign prostate tissue, malignant prostate tissue shows decreased zinc levels, indicating a role for zinc alterations in cancer development. Zinc concentrations drop early in prostate cancer progression, inhibiting citrate oxidation, a key function of prostate cells. This loss of zinc may remove its inhibitory effects on cancer cells, potentially promoting prostate cancer initiation and progression.	[20,21]
Endometrial cancer (EC)	Endometrial cancer is a type of cancer that originates in the lining of the uterus known as the endometrium. Research has examined zinc metabolism in different cancers, including endometrial cancer. Although a direct link between serum zinc levels and endometrial cancer risk or progression hasn't been established, studies suggest zinc may influence pathways related to cancer development and progression. Zinc is thought to possess anti-cancer properties by aiding in DNA repair, regulating cell growth, and impacting immune function. However, further research is required to comprehensively grasp the connection between zinc levels and endometrial cancer.	[22]

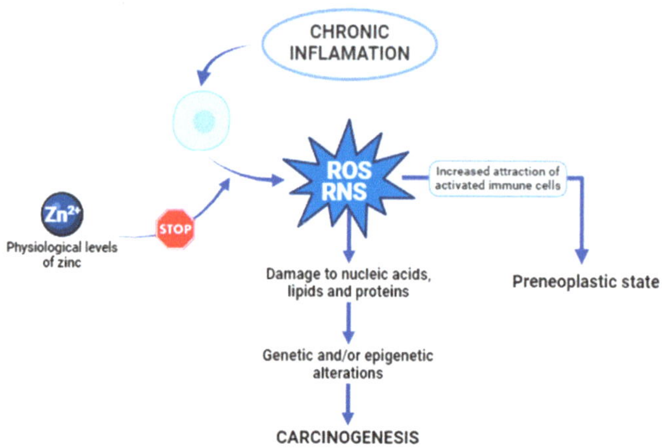

Figure 1. The anticarcinogenic effect of zinc. ROS, reactive oxygen species. RNS, reactive nitrogen species [14,15].

3.2. Copper

3.2.1. Copper's Biological Role

Copper (Cu) plays a crucial role as a vital micronutrient in various fundamental biological processes [23]. The primary source of copper for individuals is typically through their diet, with organ meats and shellfish being among the most copper-rich food options. The recommended daily intake of copper for adults is advised to be within the range of 0.8–2.4 mg/day [24]. Although only small amounts of copper are necessary in our diet, an ample supply of this metal is essential to support the growth and development of the human body [25]. It engages in various biological processes, such as lipid metabolism, energy regulation, and the synthesis of neurotransmitters [26]. In normal conditions, effective homeostatic mechanisms maintain a low concentration of intracellular copper ions [27].

However, an imbalance in cellular metal ion levels, whether in excess or deficiency, can be equally detrimental, as it may lead to an intensified rise in oxidative stress in both scenarios [28]. An abundance of copper ions induces heightened cellular respiration, leading to cytotoxic effects and eventual cell death as the levels progressively surpass a critical threshold [27]. Copper plays a substantial role in altering the function of specific types of superoxide dismutase (SOD) isoforms and other enzymes, including ceruloplasmin (CP). These enzymes are either directly or indirectly engaged in maintaining the balance of redox homeostasis [28]. Ensuring a balance of copper ions in biological systems is crucial to preventing atherosclerosis and cardiovascular diseases associated with it [29]. Disruption in the regulation of copper levels is also believed to play a role in the process of carcinogenesis [30]. The concentrations of copper ions can function as a vital indicator of cancer progression [27]. Irregular accumulation of copper is noticeable in various malignant tumours, and a connection has been identified between increased copper levels in both serum and tissues and the development of multiple types of cancers [31]. Studies on lung cancer, prostate cancer, breast cancer, stomach cancer, and thyroid cancers have identified a notable elevation in serum copper ion levels among tumour patients compared to those without tumours [32]. Additionally, a substantial increase in copper concentration was detected in the serum, bile, and gallbladder tissue of individuals diagnosed with gallbladder carcinoma [33].

3.2.2. Copper in Cancer

A growing body of preclinical studies suggests that copper is crucial for the advancement of metastatic cancer. Copper's involvement may extend to tumour growth, proliferation, epithelial–mesenchymal transition, and the formation of both the tumour microenvironment and the pre-metastatic niche [34]. Copper was thought to exert a pivotal function in signalling pathways associated with receptor tyrosine kinase (RTK). The activated RTK then triggers the phosphorylation of downstream proteins like extracellular regulated protein kinases (ERK) and agammaglobulinemia tyrosine kinase (ATK), ultimately resulting in migration and proliferation of cancer cells [35]. Moreover, copper-induced ATK activation can subsequently accelerate the phosphorylation and redistribution within the cell of forehead box O1a (FoxO1a) and forehead box O4 (FoxO4). This process promotes tumour growth [32]. Furthermore, the amassment of this ion has been found to correlate with cancer angiogenesis [27]. It is showed that copper exhibits precise spatial regulation, moving from perinuclear regions of the cell towards the ends of extending filopodia and traversing the cell membrane into the extracellular space during angiogenic processes [36]. It has been shown that inflammatory cytokines, such as IL-17, drive copper uptake by cells through the induction of the metalloreductase six-transmembrane epithelial antigen of prostate-4 (STEAP4). IL-17-induced intracellular copper elevation leads to the activation of ubiquitin (E3) ligase and X-linked inhibitor of apoptosis protein (XIAP). Tumour metastasis is initiated by the copper uptake facilitated by STEAP4. Moreover colitis-associated colon tumorigenesis is also facilitated by the promotion of STEAP4 [37]. It needs to be added that the absorption of copper through high affinity copper uptake protein-1 (CTR1) triggers allosteric activation of mitogen-activated protein kinase (MEK1), enhancing oncogenic signalling through the mitogen-activated protein kinase (MAP kinase) pathway (Figure 2) [25]. Copper has the capacity to provoke oxidative stress (reactive oxygen species, ROS) by generating highly reactive hydroxyl radicals through the Fenton-like reactions [38]. Enhanced mitochondrial ROS production induced by copper leads to complete autophagy, thereby promoting cancer cell survival. This phenomenon has been confirmed by both in vitro and in vivo studies on cervical cancer [39]. Copper homeostasis is also controlled, among other factors, by the Golgi-localized ATPases transporting Cu, ATP7A, and ATP7B. Expanding tumours actively absorb copper and utilize ATP7A/B to control the presence of this metal for oncogenic enzymes like LOX and LOX-like proteins, enhancing the invasiveness of malignant cells. Additionally, the activity and movement of ATP7A/B enable tumour cells to detoxify certain drugs used in the chemotherapy of various solid tumours [40]. In recent research, a newly discovered type of cell death termed cuproptosis, dependent on copper, has been identified. This mechanism is distinct from all previously known pathways leading to cell death [41]. Cuproptosis takes place when copper binds to lipoylated enzymes in the tricarboxylic acid (TCA) cycle, resulting in the aggregation of proteins, proteotoxic stress, and, ultimately, cancer cell death [42]. Nevertheless, the influence of cuproptosis on malignant tumours remains incompletely comprehended from a clinical standpoint [30]. In investigations concerning the anticancer efficacy of copper (II) complexes and other metallic coordination compounds on breast and lung cancer cells, the copper (II) complex displayed encouraging cytotoxic potency. Its comparable toxicity to cisplatin, the standard chemotherapeutic agent, indicates its potential as an alternative anticancer therapeutic. Moreover, the copper (II) complex exhibited significant cytotoxicity against cancer cells, underscoring its potential utility in anticancer interventions [43]. Recent in vitro and in vivo studies have further elucidated that the anticancer activity of metal compounds, such as copper (II) complexes via intercalative mechanisms, is characterized by their ability to recognize, bind, and induce DNA damage in lung cancer cells treated with complex (2), thereby prompting apoptosis-mediated cell death [43,44]. Furthermore, the combination of the copper (II) complex with cisplatin, another chemotherapeutic agent, manifested synergistic effects, suggesting the promise of this combination as an efficacious anticancer therapeutic strategy [43,45].

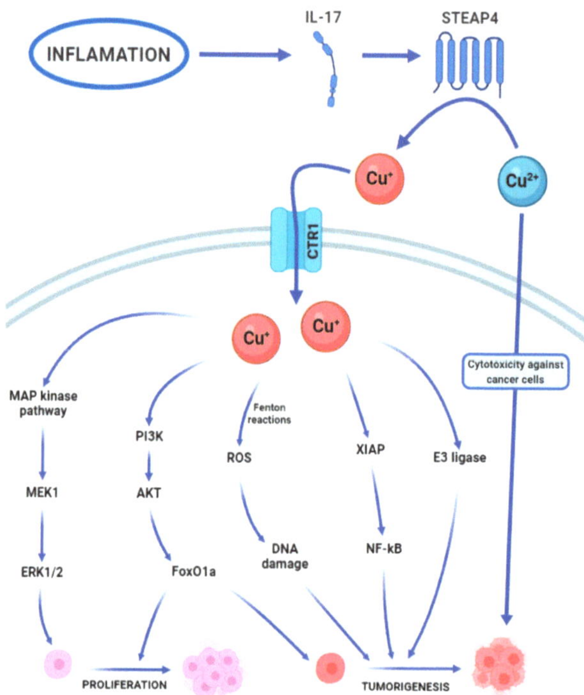

Figure 2. Copper's carcinogenic potential. IL-17, interleukin 17. Cu^{+}, Cu^{2+}, copper ions. CTR1, high-affinity copper uptake protein 1 (copper transporter 1). MAP kinase pathway, mitogen-activated protein kinase pathway. MEK1, mitogen-activated proteinkinase kinase 1. ERK1/2, extracellular signal-regulated kinases 1 and 2. PI3K, phosphoinositide 3-kinase. ATK, agammaglobulinemia tyrosine kinase. FoxO1a, forehead box O1a. ROS, reactive oxygen species. XIAP, X-linked inhibitor of apoptosis. NF-kB, nuclear factor kappa-light-chain-enhancer of activated B cells. E3 ligase, E3 ubiquitin ligase [25,27,32,35–39,43–45].

3.3. Cobalt

Cobalt (Co) is a vital trace element for the human body, existing in both organic and inorganic forms [46]. The main purpose of cobalt in the human body revolves around its involvement in cobalamin (Cbl), also known as vitamin B12 [12]. That is why cobalt plays a vital role in the regulation of red blood cell production and why it is crucial to uphold adequate levels of cobalt in the human body, as a deficiency in this element can result in anaemia [47]. It also plays a significant role in the synthesis of amino acids and certain proteins in nerve cells and contributes to the production of neurotransmitters essential for the proper functioning of the body [46]. Cobalt is most recognized for its essential involvement in alkylcorrinoid cofactors, leveraging the distinctive characteristics of the cobalt-carbon bond to facilitate chemically intricate biotransformation's [48]. It serves as the coenzyme for many crucial enzymes e.g., methylmalonic-CoA mutase or methionine synthase. These enzymes in humans play vital roles in maintaining health [12]. On the other hand, inorganic cobalt in ion form is toxic to the human body, and prolonged retention can lead to increasingly detrimental changes in cells [46]. Being exposed to it can result in conditions such as asthma, hard metal lung disease, contact allergy, and an elevated susceptibility to cancer [49]. Cobalt is commonly found in natural surroundings and can be generated because of human-related activities [46]. The primary route of cobalt absorption is through the respiratory system, although absorption through the skin is also possible [49]. People may encounter exposure to cobalt or its compounds in occupational settings where cobalt is used or manufactured, through cobalt-containing orthopaedic joint replacements, and from environmental sources [50]. Cobalt serves as the primary constituent in metal prostheses used in hip arthroplasty. Research indicates that metal particles, predominantly consisting of cobalt nanoparticles (CoNPs), can induce both systemic and local harmful reactions, attributed to various physical and chemical factors [51]. Biological toxicity was observed in cobalt nanoparticles, manifesting as the inhibition of osteoclast differentiation and proliferation across various concentrations [52]. Moreover, the study showed that nanoparticles of cobalt oxide have an impact on the electromechanical behaviour of heart muscle cells (cardiomyocytes) and the regulation of intracellular calcium. It needs to be added that these nanoparticles trigger the production of reactive oxygen species (ROS), resulting in oxidative stress. This oxidative stress may be linked to DNA damage and could negatively affect the functionality of cardiomyocytes [53]. Exposure of experimental animals to metallic cobalt or cobalt compounds resulted in tumours in rats and/or mice through various exposure routes and in several different tissue locations. Inhalation exposure to metallic cobalt or cobalt sulphate led to tumours in the lungs, pancreas, adrenal glands, and the hematopoietic system [50]. Alveolar/bronchiolar carcinomas in rodents, whether occurring spontaneously or due to chemical exposure, bear resemblance to a specific subtype of lung adenocarcinomas observed in humans. Oxidative stress is a key factor in pulmonary carcinogenesis induced by cobalt metal dust (CMD) in rodents, and these discoveries may have implications for understanding human lung cancers as well [54]. The impact of $CoCl_2$ was also examined on numerous histone modifications at a global level. It was found that in both human lung cancer cells (A549) and human bronchial epithelial cells (Beas-2B), exposure to $CoCl_2$ for 24 h increased the trimethylation of H3K4 and H3K27 through the activation of methyltransferases. Additionally, it elevated the trimethylation of H3K9 and H3K36 by inhibiting the histone demethylation process. It has been shown that cobalt ions disrupt the cellular epigenetic balance. Such changes could potentially result in modified gene expression patterns and contribute to carcinogenesis [55]. Moreover, it has been proven that the hypoxia-mimicking substance cobalt chloride ($CoCl_2$) induces an elevation in the expression of chemokine (C-C motif) ligand-18 (CCL18) [56]. CCL18 serves as an indicator of the M2 macrophage subset, contributing to the immunosuppressive characteristics of the tumour microenvironment and playing a crucial role in cancer immune evasion. As a result, higher levels of CCL18 in both the bloodstream and the tumour are correlated with a poorer prognosis for the patient [57]. Furthermore, it has been confirmed that CCL18 also promotes the migration of endothelial cells and angiogenesis in breast cancer [58]. It has been noted that the available data from studies on humans are insufficient to assess the association between cancer in humans and exposure to cobalt and cobalt compounds. However, there is sufficient evidence of carcinogenicity from experimental animal studies and evidence from studies on the mechanisms of carcinogenesis, indicating that the release of cobalt ions is a crucial event influencing carcinogenicity (Figure 3) [50].

Figure 3. The relationship between cobalt and carcinogenesis. The cobalt atom can interact with DNA, and processes such as oxidative stress and DNA damage are the main mechanisms by which cobalt can induce carcinogenesis. Exposure to cobalt leads to changes in DNA which can result in the development of cancer [54,55].

3.4. Iron

Iron (Fe) stands as the most plentiful metal within the human body, and no autonomous life forms on Earth can thrive without it [59]. It is the predominant metal found in the human brain and a vital trace element that governs various cellular processes [60]. Most of the iron in the human body is found in red blood cells. Despite the high amount of iron in food, many individuals globally experience anaemia. Insufficient iron leads to a hindered synthesis of iron-containing proteins and hampers cell growth [61]. Iron is also necessary for the cellular proliferation as a cofactor of many enzymes [62]. While iron is essential for supporting cell growth and fundamental functions, it can also pose harm and carcinogenic risks [63]. Many medical disorders, like hemochromatosis, prolonged infections with hepatitis B and C, viruses and the presence of endometriosis, are identified as factors associated with an excess of iron, which increases the risk of developing cancer in humans [64]. Moreover, an environment with an excess of iron encourages cellular evolution to become resistant to ferroptosis, constituting a significant factor in the development of cancer [62]. Epidemiological research has shown a link between surplus iron levels and higher rates of cancer occurrence and susceptibility. Meanwhile, laboratory investigations have suggested that iron plays a role in the onset of cancer and fuelling tumour expansion [65]. Iron is essential in the advancement and spread of tumours, primarily because of its critical role in promoting the survival of tumour cells and restructuring the microenvironment within the tumour [66].

An abundance of iron is closely linked to the development of tumours in various human cancer types, operating through diverse mechanisms. These mechanisms encompass impacting signal transduction in cancer cells, influencing DNA replication, repair, and cell cycle advancement, catalysing the creation of mutagenic hydroxyl radicals, and serving as a vital nutrient for the growth of proliferating tumour cells [67]. It was discovered that accumulation of iron facilitated by mitochondria contributes to the development of cancer and the Warburg effect in osteosarcoma cells [68]. The Warburg Effect is characterized by an elevated rate of glucose consumption and the preferential generation of lactate, even when oxygen is present [69]. Simultaneously, inducing a deficiency of iron could emerge as an innovative and effective approach in treating osteosarcoma [68]. Findings also revealed that exposing colorectal cancer (CRC) cells to iron also stimulated the Warburg effect by triggering reactive oxygen species (ROS) and activating nuclear factor erythroid 2-related factor 2 (NRF2). Furthermore, this iron exposure demonstrated an increased resilience of CRC cells to ferroptosis [70]. It was also found that excessive iron accumulation in the liver resulting from F-box and leucine-rich repeat protein-5 (FBXL5) ablation increases the risk of hepatocellular carcinoma (HCC) progression. This condition triggers oxidative stress, tissue damage, inflammation, and compensatory proliferation of hepatocytes [71]. Other studies showed that an abundance of iron saturates the binding capacity of transferrin, leading to the formation of non-transferrin-bound iron (NTBI). This NTBI can initiate free-radical reactions, potentially contributing to oxidant-induced breast carcinogenesis. Additionally, the surplus iron and the disturbance of iron metabolism by local oestrogen in the breast contribute to the production of reactive oxygen species [72]. The data collected from the National Health and Nutrition Examination Survey I and the National Health Evaluation Follow-Up Study showed that increased iron consumption was linked to a heightened risk of colon cancer across the entire colon in both males and females [73], while in other sources we can find that the development of lung cancer is facilitated by iron-dependent cyclin-dependent kinase-1 (CDK-1) activity through the activation of the GP130/STAT3 signalling pathway [74].

Macrophages also play a pivotal role in maintaining iron balance; they capture it through the engulfment of aging red blood cells and serve as a significant reservoir of accessible iron in the body [75]. Considering the various ways in which macrophages have developed mechanisms to acquire, transport, store, and release iron, it can be hypothesized that tumour cells may influence or instruct these macrophages to provide iron, thereby facilitating the growth of the tumour [76]. Elevated concentrations of ferrous (Fe^{2+}) iron have the potential to produce reactive oxygen species through Fenton chemistry reactions. These heightened levels can result in harm to mitochondria and genomic DNA, ultimately fostering the development of cancer [60]. It should be noted that accumulation of iron has been observed in tissues as they age and in diseases associated with the aging process [77]. This induction of senescence in the cancer microenvironment due to therapy is acknowledged as one of the factors that propel tumour advancement [78]. Additionally, an excess of iron might arise unintentionally in specific cancer patients, stemming from the treatment of symptomatic anaemia through inappropriate iron-restoration therapies. This occurs without prior assessment of the body's iron status, and both conditions collaboratively contribute to the exacerbation of the tumour [79]. Moreover, various infections can also increase the risk of cancer development. Reactive oxygen/nitrogen species produced during inflammation not only harm DNA but also impact other large biological molecules like proteins and lipids, leading to impaired functionality. Transferrin, when oxidatively damaged, releases iron ions that can potentially initiate Fenton reactions, producing more reactive oxygen species in the process [80].

To support their growth, cancer cells demonstrate a heightened requirement for iron when compared to normal, non-cancerous cells. This reliance on iron renders cancer cells more susceptible to iron-induced cell death, known as ferroptosis [81]. It is a recently identified type of controlled necrotic cell death and has been shown to be significant in various conditions [82]. Ferroptosis is implicated in cancer development and could potentially serve as a valuable approach for anti-cancer treatment. Various pieces of evidence indicate that ferroptosis is pivotal in inhibiting tumorigenesis [59].

Moreover, the discussion on the role of Fe should include the function of the transferrin receptor and its overexpression in cancer cells, which meets the heightened Fe demand in these cells. Additionally, it should address the efforts to target this receptor for anticancer therapy. Malignant cells frequently overexpress TfR1 due to its pivotal role in cancer cell pathology, and this heightened expression is often linked to poor prognosis across various cancer types. The increased TfR1 levels on malignant cells, combined with its extracellular availability, capacity for internalization, and crucial role in cancer pathology, render this receptor a promising target for antibody-mediated therapy [83].

Furthermore, new research has provided insights into the involvement of iron metabolism in cancer stem cells (CSC). These findings propose that selectively addressing iron metabolism in CSCs could enhance the effectiveness of cancer therapy [84]. The important role in anti-cancer therapy is also played by iron chelators, which have exhibited strong anti-cancer properties in various types of cancers, as evidenced in both laboratory cell culture studies and clinical trials [85].

3.5. Manganese

Manganese (Mn) is a vital metal found abundantly in the environment and is crucial for various essential processes within the human body. Its importance lies in being incorporated into protein structures, acting as a necessary cofactor. Without manganese, crucial functions such as immune response, energy regulation, growth, blood clotting, and the body's ability to manage oxidative stress by-products would be greatly compromised [86,87]. Manganese in its divalent form (Mn(II)), acting as a cofactor for mitochondrial superoxide dismutase (MnSOD), helps eliminate oxygen free radicals like superoxide and hydroxyl radicals. This role is vital for maintaining a balance between oxidative and antioxidative processes, shielding against oxidative stress and its harmful effects. Additionally, studies suggest that at lower concentrations, Mn(II) may protect against the toxic effects of cadmium (Cd) by virtue of its antioxidative properties and its ability to hinder the uptake of this heavy metal into cells [88].

Within cells, free radicals form naturally as part of cellular functions like mitochondrial processes, alongside being generated by external factors such as exposure to ionizing radiation. An imbalance between the creation of free radicals and the body's capacity to counteract them with antioxidants leads to oxidative stress, causing harm to vital cellular structures—a phenomenon termed oxidative damage. This interplay of free radicals, oxidative stress, and oxidative damage is widely acknowledged in various diseases, notably cancer. Beyond their traditional role in cancer development via DNA mutation and genomic instability, free radicals activate pathways that support cell growth, survival, and blood vessel formation, all of which contribute to tumour progression [89]. Manganese superoxide dismutase (MnSOD) significantly influences cancer development owing to its capability to scavenge reactive oxygen species (ROS) [90]—manganese superoxide dismutase serves as the initial defence mechanism against reactive oxygen species ROS by facilitating the conversion of two superoxide molecules into oxygen and hydrogen peroxide (H2O2) [91,92]. This process involves a cyclic exchange of reduction and oxidation reactions at the active metal site [91].

MnSOD has the capacity to directly influence several signalling pathways leading to cell death in cancer, such as apoptosis, proptosis, and autophagy. Elevated levels of MnSOD expression are linked to increased resistance to chemotherapy and radiation therapy across various cancer types. Recently, the focus has shifted towards exploring posttranslational modifications of MnSOD, particularly its acetylation at lysine residue 68, shedding light on its crucial roles in advancing cancer progression. As the significance of the immune system in cancer development garners increased attention, the role of MnSOD in the tumour's immune microenvironment has emerged as a crucial focus. Growing evidence suggests that immune cells infiltrating tumours play a pivotal role in cancer progression, and their activity is closely linked to MnSOD expression across various cancer types. A positive correlation was discovered between MnSOD expression, CXCL8 levels, and neutrophil infiltration, indicating the involvement of the "MnSOD-CXCL8-neutrophil recruitment" pathway in cancer advancement. Additionally, studies have revealed a positive association between MnSOD expression and the infiltration of CD68+ macrophages, potentially indicating unfavourable outcomes in inflammation-induced lung adenocarcinoma. Moreover, heightened MnSOD expression has been noted in aggressive triple-negative breast cancer (TNBC) cases, suggesting it as an adverse prognostic indicator [93]. Assessing the levels of SOD and peroxide-eliminating enzymes is vital for comprehending cellular variations in response to vectors promoting SOD overexpression. Additionally, longstanding evidence indicates that malignant cell lines typically exhibit lower MnSOD levels compared to their normal or non-malignant counterparts [94]. Furthermore, there is ongoing exploration into nanoparticles with potential applications in cancer detection. For instance, in a study conducted by Du et al. (2020), a bi-modal nanopericle-targeted probe for prostate cancer was developed. This probe demonstrated specific accumulation within prostate cancer cells and tumour tissues using optical imaging and MRI in a preclinical model. In vitro and in vivo imaging outcomes suggest that Mn-Msn-Cy7 nanoprobes, which target prostate-specific membrane antigen (PSA), hold promise for detecting prostate cancer [95].

As previously mentioned, at lower concentrations, Mn(II) may protect against the toxic effects of cadmium (Cd) by virtue of its antioxidative properties and its ability to hinder the uptake of this heavy metal into cells [88]. Prior exposure to small amounts of Mn(II) has been shown to increase resistance to cadmium (Cd)−related fatalities and liver damage in both mice and rats. Studies conducted on rat liver mitochondria exposed to cadmium suggest that Mn(II) ions might shield against Cd-triggered lipid peroxidation and the suppression of antioxidant enzymes [96]. Cadmium (Cd) pollution has emerged as a significant worldwide issue, given its extensive presence and severe toxicity, posing a serious risk to both human and animal well-being [97]. According to the International Agency for Research on Cancer (IARC), cadmium falls under the category of Group I carcinogens. Numerous epidemiological studies have consistently identified Cd as a significant risk factor for the development of lung cancer [98]. Nevertheless, the relationship between cadmium exposure and the emergence of tumours in alternate locations like the kidney, breast, and prostate may also hold considerable importance. Moreover, the heightened likelihood of cancer may extend beyond occupations characterized by elevated exposure levels and could also manifest due to environmental factors, such as proximity to locales involved in the processing of heavy metals [99]. Information demonstrates the crucial role of Mn(II) and its protective effect against the Cd−inducted cytotoxic impacts and carcinogenesis (Figure 4).

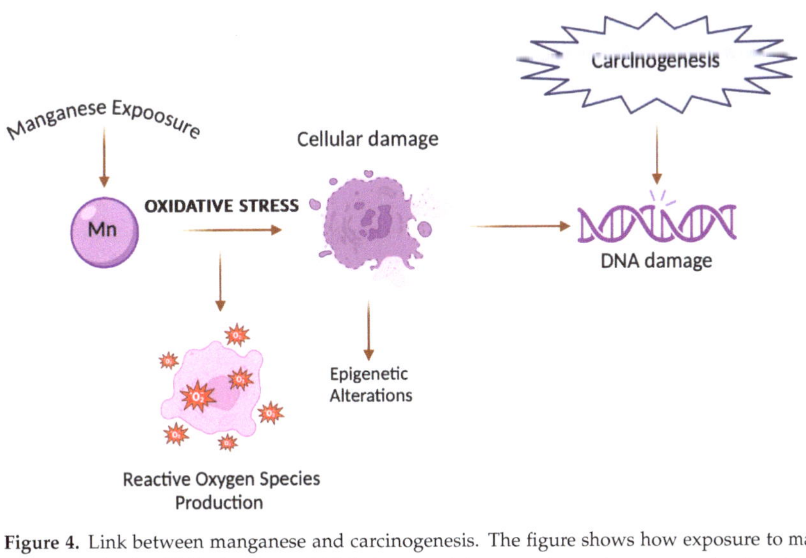

Figure 4. Link between manganese and carcinogenesis. The figure shows how exposure to manganese can lead to DNA damage, which in turn can trigger carcinogenic processes, leading to the development of cancer [92,94].

3.6. Other Heavy Metals

Many other metals share certain common mechanisms of tumour formation, yet each also possesses distinct pathways of its own. Many metal carcinogens such as arsenic, cadmium, beryllium or mercury have been observed to generate reactive oxygen species (ROS) and elevate oxidative stress levels [5]. Additionally, cadmium and arsenic similarly induce many processes that may stimulate carcinogenesis. They disrupt the action of antioxidants such as glutathione (GSH), aggravating the cellular antioxidant capacity [5,100]. It should be noted that these elements competitively interact with or substitute crucial metals like zinc and calcium within proteins, serving as a primary mechanism of cytotoxicity within the cell [5,101]. Moreover, they inhibit cell autophagy, which is critical for tumour suppression. Arsenic reduces SLBP levels and subsequently leads to aberrant polyadenylation of canonical histone mRNA, thus promoting carcinogenesis. Research revealed that Nickel (Ni) also triggered a reduction in SLBP through a similar mechanism [5,100–102]. Ni is also primarily observed to stimulate hypoxia-induced signalling pathways by competing with iron in prolyl hydroxylase [5,103]. Studies have shown that arsenic can induce genotoxicity by disrupting DNA repair and causing chromosomal instability in the cell, which may lead to mutation of tumour suppressor genes such as p53 [104]. It also causes double strand breaks, leading to chromosomal aberrations. On the other hand, the analysis indicated that beryllium metal is improbable to function as a conventional non-threshold mutagen. While effects on DNA repair and cellular transformation were noted, their significance in vivo requires additional investigation. The correlation between beryllium exposure and its potential to cause cancer is still being debated within the scientific community, with ongoing research efforts underway [5,105]. Exposure to mercury is linked to cancer risk, although conflicting data exist. Both cancerous and healthy tissues accumulate mercury differently, potentially contributing to tumour development. Mercury's influence on cell proliferation and its impact on various signalling pathways suggest mechanisms for promoting carcinogenesis. Additionally, the oxidative DNA damage, genotoxicity, and epigenetic effects of mercury may play roles in cancer development [106,107]. There is a possibility that metals may also potentially instigate tumorigenesis through a confluence of their impacts. A thorough grasp of the mechanisms underlying the onset of metal-induced cancer can offer valuable perspectives for therapeutic strategies targeting molecular pathways involved in metal-induced carcinogenesis.

4. Conclusions

Heavy metals play a significant role in cancer formation. Numerous evidence suggests that exposure to heavy metals, may contribute to cancer initiation and progression through various mechanisms. Heavy-metal-induced carcinogenesis itself involves a complex interaction of genetic, epigenetic, and molecular mechanisms. These pathways often intersect and interact, leading to increased carcinogenicity.

It is essential to underscore the critical role of chemical speciation in the biological activity of metals. For instance, Cr (VI) is recognized as a human carcinogen, whereas Cr (III), once thought to be an essential element, likely has no distinct biological function [108]. Metal salts, such as chlorides, used in cell and animal experiments are quickly transformed in biological media into mixtures of complexes with biological ligands, including proteins, which significantly alters their activity [109]. Furthermore, metal speciation in the environment dictates their bioavailability and toxicity to living organisms, including humans [110]. It is also crucial to note that the mechanisms of pro- or anti-cancer activities of soluble metal complexes are likely very different from those of metal nanoparticles [111]. Research on the connections between metal nanoparticles and cancer is rapidly advancing, though this topic is largely beyond the scope of this review.

The carcinogenic potential of heavy metals is dose-dependent, with higher levels of exposure correlating with increased cancer risk. Long-term exposure, even at low concentrations, can cumulatively increase the risk of cancer. Given the widespread presence of heavy metals in the environment and industrial processes, limiting exposure is critical to cancer prevention. This includes implementing stringent regulations, adopting safer industrial practices, and promoting public awareness of potential sources of exposure. Like early detection, due to the latency period between heavy metal exposure and cancer manifestation, early detection strategies are essential to ensure timely intervention and treatment. Surveillance programs targeting high-risk populations can facilitate early diagnosis and improve outcomes.

It is also important to note that metals play a crucial role not only in cancer development but also in the formation of metastases. Metals significantly influence the metastatic process by affecting the function of various proteins and enzymes. Notably, an excess of essential metals such as iron and copper are frequently linked to both carcinogenesis and metastatic disease [112].

It is important to distinguish that essential trace elements (Fe, Cu, Zn, Mn, Co, Mo) naturally contribute to cancer prevention or can promote cancer development when their metabolism is dysregulated, which is the primary focus of this article. Conversely, metal compounds can also cause cancer through environmental or industrial exposure. However, this latter topic has been extensively covered in the literature, unlike the specific area we have chosen to examine.

Although significant progress has been made in understanding the carcinogenicity of heavy metals, knowledge gaps remain, particularly regarding the molecular mechanisms underlying their carcinogenic effects. Further research is needed to elucidate these mechanisms and develop targeted interventions for cancer prevention and treatment.

In summary, solving the problem related to the impact of heavy metals on carcinogenesis processes requires a multi-faceted approach including:

(1) Continuous monitoring of environment with increased levels of harmful substances for humans;
(2) Standard use of protective equipment in accordance with procedures outlined by legal regulations and public health initiatives;
(3) Ongoing monitoring of the health of individuals exposed to the harmful effects of various factors (harmful elements) present in the environment, conducting periodic (standard and additional) examinations;
(4) Early implementation of medical procedures to prevent disease development, limiting the possibility of metastasis;
(5) Establishing a procedural algorithm depending on the diagnosed disease and the impact of the harmful compound on the human body.

The above review of the literature on carcinogenesis provides a detailed overview of the current knowledge on the role of heavy metals in carcinogenesis. Additionally, it highlights areas where research is lacking or where results are inconsistent. Suggests potential directions for future research to address these gaps. Informs you about risk assessments and regulatory guidance on heavy metals.

It is a valuable source of information for researchers, professionals, and students wishing to understand the relationship between heavy metals and cancer.

5. Study Limitations

Although a comprehensive review of carcinogenesis provides valuable insights, it also has several limitations. The vast amount of research available may make it impossible to comprehensively cover all relevant research, so the authors focused on specific aspects. Carcinogenesis involves many interacting factors, which makes it difficult to isolate the specific impact of heavy metals. Genetic and environmental differences among individuals may influence susceptibility to heavy metal-induced carcinogenesis, complicating generalizations.

Author Contributions: Conceptualization, A.G., A.M.-G., M.T. and R.M.; formal analysis, A.M.-G. and R.M.; investigation, A.G., A.M.-G., M.T., M.Ż., J.P. and R.M.; writing—original draft preparation, A.G., A.M.-G., M.T., M.Ż., J.P. and R.M.; writing—review and editing, A.G., A.M.-G., M.T., M.Ż., J.P. and R.M.; visualization, A.M.-G. and M.T.; supervision, A.M.-G. and R.M.; project administration, A.M.-G. and R.M. All authors have read and agreed to the published version of the manuscript.

Funding: This research received no external funding.

Data Availability Statement: Not applicable.

Conflicts of Interest: The authors declare no conflicts of interest.

References

1. Guo, Z.; Zhou, G.; Hu, W. Carcinogenesis induced by space radiation: A systematic review. *Neoplasia* **2022**, *32*, 100828. [CrossRef] [PubMed]
2. Beyersmann, D.; Hartwig, A. Carcinogenic metal compounds: Recent insight into molecular and cellular mechanisms. *Arch. Toxicol.* **2008**, *82*, 493–512. [CrossRef] [PubMed]
3. Kentsis, A. Why do young people get cancer? *Pediatr. Blood Cancer* **2020**, *67*, e28335. [CrossRef] [PubMed]
4. Lewandowska, A.M.; Rudzki, M.; Rudzki, S.; Lewandowski, T.; Laskowska, B. Environmental risk factors for cancer—Review paper. *Ann. Agric. Environ. Med.* **2019**, *26*, 1–7. [CrossRef] [PubMed]
5. Tchounwou, P.B.; Yedjou, C.G.; Patlolla, A.K.; Sutton, D.J. Heavy metal toxicity and the environment. *Exp. Suppl.* **2012**, *101*, 133–164. [CrossRef] [PubMed]
6. Zhu, Y.; Costa, M. Metals, and molecular carcinogenesis. *Carcinogenesis* **2020**, *41*, 1161–1172. [CrossRef] [PubMed]
7. Hussai, A.; Jiang, W.; Wang, X.; Shahid, S.; Saba, N.; Ahmad, M.; Dar, A.; Masood, S.U.; Imran, M.; Mustafa, A. Mechanic impact of zinc deficiency in human development. *Front. Nutr.* **2022**, *9*, 717064. [CrossRef]
8. Guo, H.; Deng, H.; Liu, H.; Jian, Z.; Cui, H.; Fang, J.; Zuo, Z.; Deng, J.; Li, Y.; Wang, X.; et al. Nickel carcinogenesis mechanism: Cell cycle dysregulation. *Environ. Sci. Pollut. Res.* **2021**, *28*, 4893–4901. [CrossRef]
9. Choi, S.; Liu, X.; Pan, Z. Zinc deficiency and cellular oxidative stress: Prognostic implications in cardiovascular diseases. *Acta Pharmacol. Sin.* **2018**, *39*, 1120–1132. [CrossRef]
10. Jomova, K.; Valko, M. Advances in metal-induced oxidative stress and human disease. *Toxicology* **2011**, *283*, 65–87. [CrossRef]
11. Huang, L.; Drake, V.J.; Ho, E. Zinc. *Adv. Nutr.* **2015**, *6*, 224–226. [CrossRef] [PubMed]
12. Jomova, K.; Makova, M.; Alomar, S.Y.; Alwasel, S.H.; Nepovimova, E.; Kuca, K.; Rhodes, C.J.; Valko, M. Essential metals in health and disease. *Chem. Biol. Interact.* **2022**, *367*, 110173. [CrossRef] [PubMed]
13. Caliri, A.W.; Tommasi, S.; Besaratinia, A. Relationship among smoking, oxidative stress, inflammation, macromolecular damage and cancer. *Mutat. Res./Rev. Mutat. Res.* **2021**, *787*, 108365. [CrossRef] [PubMed]
14. Jelic, M.D.; Mandic, A.D.; Maricic, S.M.; Srdjenovic, B.U. Oxidative stress and its role in cancer. *J. Cancer Res. Ther.* **2021**, *17*, 22–28. [CrossRef] [PubMed]
15. Olechnowicz, J.; Tinkov, A.; Skalny, A.; Suliburska, J. Zinc status is associated with inflammation, oxidative stress, lipid, and glucose metabolism. *J. Physiol. Sci.* **2018**, *68*, 19–31. [CrossRef] [PubMed]
16. Zannetti, A. Breast cancer: From pathophysiology to novel therapeutic approaches 2.0. *Int. J. Mol. Sci.* **2023**, *24*, 2542. [CrossRef] [PubMed]
17. Yang, Z.; Pu, M.; Dong, X.; Ji, F.; Veeraraghavan, V.P.; Yang, H. Piperine loaded zinc oxide nanocomposite inhibits the PI3K/AKT/mTOR signalling pathway via attenuating the development of gastric carcinoma: In vitro and in vivo studies. *Arab. J. Chem.* **2020**, *13*, 5501–5516. [CrossRef]
18. Qu, Z.; Liu, Q.; Kong, X.; Wang, X.; Wang, Z.; Wang, J.; Fang, Y. A Systematic study on zinc-related metabolism in breast cancer. *Nutrients* **2023**, *15*, 1703. [CrossRef] [PubMed]
19. Dean-Colomb, W.; Esteva, F.J. Her-2 positive breast cancer: Herceptin and beyond. *Eur. J. Cancer* **2008**, *44*, 2806–2812. [CrossRef]
20. Li, D.; Stovall, D.B.; Wang, W.; Sui, G. Advances of zinc signalling studies in prostate cancer. *Int. J. Mol. Sci.* **2021**, *21*, 667. [CrossRef]
21. Wang, J.; Zhao, H.; Xu, Z.; Cheng, X. Zinc dysregulation in cancers and its potential as a therapeutic target. *Cancer Biol. Med.* **2020**, *17*, 612–625. [CrossRef] [PubMed]
22. Atakul, T.; Altinkaya, S.O.; Abas, B.I.; Yenisey, C. Serum copper and zinc levels in patients with endometrial cancer. *Biol. Trace Elem. Res.* **2020**, *195*, 46–54. [CrossRef] [PubMed]
23. Li, Y. Copper homeostasis: Emerging target for cancer treatment. *IUBMB Life* **2020**, *72*, 1900–1908. [CrossRef] [PubMed]
24. Bost, M.; Houdart, S.; Oberli, M.; Kalonji, E.; Huneau, J.F.; Margaritis, I. Dietary copper, and human health: Current evidence and unresolved issues. *J. Trace Elem. Med. Biol.* **2016**, *35*, 107–115. [CrossRef] [PubMed]
25. Shanbhag, V.C.; Gudekar, N.; Jasmer, K.; Papageorgiou, C.; Singh, K.; Petris, M.J. Copper metabolism as a unique vulnerability in cancer. *Biochim. Biophys. Acta Mol. Cell Res.* **2021**, *1868*, 118893. [CrossRef] [PubMed]
26. Zhou, C.; Yang, J.; Liu, T.; Jia, R.; Yang, L.; Sun, P.; Zhao, W. Copper metabolism and hepatocellular carcinoma: Current insights. *Front. Oncol.* **2023**, *13*, 1186659. [CrossRef] [PubMed]
27. Wang, M.; Zheng, L.; Ma, S.; Lin, R.; Li, J.; Yang, S. Cuproptosis: Emerging biomarkers and potential therapeutics in cancers. *Front. Oncol.* **2023**, *13*, 1288504. [CrossRef] [PubMed]
28. Martín Giménez, V.M.; Bergam, I.; Reiter, R.J.; Manucha, W. Metal ion homeostasis with emphasis on zinc and copper: Potential crucial link to explain the non-classical antioxidative properties of vitamin D and melatonin. *Life Sci.* **2021**, *281*, 119770. [CrossRef] [PubMed]
29. Denoyer, D.; Clatworthy, S.A.S.; Cater, M.A. Copper complexes in cancer therapy. *Met. Ions Life Sci.* **2018**, *18*, 469–506. [CrossRef]
30. Bian, C.; Zheng, Z.; Su, J.; Chang, S.; Yu, H.; Bao, J.; Xin, Y.; Jiang, X. Copper homeostasis and cuproptosis in tumour pathogenesis and therapeutic strategies. *Front. Pharmacol.* **2023**, *14*, 1271613. [CrossRef]
31. Zhao, Q.; Qi, T. The implications and prospect of cuproptosis-related genes and copper transporters in cancer progression. *Front. Oncol.* **2023**, *13*, 1117164. [CrossRef] [PubMed]
32. Xie, J.; Yang, Y.; Gao, Y.; He, J. Cuproptosis: Mechanisms and links with cancers. *Mol. Cancer* **2023**, *22*, 46. [CrossRef] [PubMed]

33. Basu, S.; Singh, M.K.; Singh, T.B.; Bhartiya, S.K.; Singh, S.P.; Shukla, V.K. Heavy and trace metals in carcinoma of the gallbladder. *World J. Surg.* **2013**, *37*, 2641–2646. [CrossRef] [PubMed]
34. Lopez, J.; Ramchandani, D.; Vahdat, L. Copper depletion as a therapeutic strategy in cancer. *Met. Ions Life Sci.* **2019**, *19*, 19. [CrossRef]
35. He, F.; Chang, C.; Liu, B.; Li, Z.; Li, H.; Cai, N.; Wang, H.H. Copper (II) ions activate ligand-independent receptor tyrosine kinase (RTK) signalling pathway. *Biomed. Res. Int.* **2019**, *2019*, 4158415. [CrossRef] [PubMed]
36. Wang, X.; Zhou, M.; Liu, Y.; Si, Z. Cope with copper: From copper linked mechanisms to copper-based clinical cancer therapies. *Cancer Lett.* **2023**, *561*, 216157. [CrossRef] [PubMed]
37. Liao, Y.; Zhao, J.; Bulek, K.; Tang, F.; Chen, X.; Cai, G.; Jia, S.; Fox, P.L.; Huang, E.; Pizarro, T.T.; et al. Inflammation mobilizes copper metabolism to promote colon tumorigenesis via an IL-17-STEAP4-XIAP axis. *Nat. Commun.* **2020**, *11*, 900. [CrossRef] [PubMed]
38. Pham, A.N.; Xing, G.; Miller, C.J.; Waite, T.D. Fenton-like copper redox chemistry revisited: Hydrogen peroxide and superoxide mediation of copper-catalysed oxidant production. *J. Catal.* **2013**, *301*, 54–64. [CrossRef]
39. Guan, D.; Zhao, L.; Shi, X.; Ma, X.; Chen, Z. Copper in cancer: From pathogenesis to therapy. *Biomed. Pharmacother.* **2023**, *163*, 114791. [CrossRef]
40. Petruzzelli, R.; Polishchuk, R.S. Activity, and trafficking of copper-transporting ATPases in tumour development and defence against Platinum-Based Drugs. *Cells* **2019**, *8*, 1080. [CrossRef]
41. Chen, L.; Min, J.; Wang, F. Copper homeostasis and cuproptosis in health and disease. *Signal Transduct. Target Ther.* **2022**, *7*, 378. [CrossRef] [PubMed]
42. Tsvetkov, P.; Coy, S.; Petrova, B.; Dreishpoon, M.; Verma, A.; Abdusamad, M.; Rossen, J.; Joesch-Cohen, L.; Humeidi, R.; Spangler, R.D.; et al. Copper induces cell death by targeting lipoylated TCA cycle proteins. *Science* **2022**, *375*, 1254–1261. [CrossRef] [PubMed]
43. Mariani, D.; Ghasemishahrestani, Z.; Freitas, W.; Pezzuto, P.; Costa-da-Silva, A.C.; Tanuri, A.; Kanashiro, M.M.; Fernandes, C.; Horn, A., Jr.; Pereira, M.D. Antitumoral synergism between a copper (II) complex and cisplatin improves in vitro and in vivo anticancer activity against melanoma, lung and breast cancer cells. *Biochim. Biophys. Acta Gen. Subj.* **2021**, *1865*, 129963. [CrossRef] [PubMed]
44. Maciel, L.L.F.; de Freitas, W.R.; Bull, E.S.; Fernandes, C.; Horn, A., Jr.; de Aquino Almeida, J.C.; Kanashiro, M.M. In vitro and in vivo anti-proliferative activity and ultrastructure investigations of a copper (II) complex toward human lung cancer cell NCI-H460. *J. Inorg. Biochem.* **2020**, *210*, 111166. [CrossRef] [PubMed]
45. Romani, A.M.P. Cisplatin in cancer treatment. *Biochem. Pharmacol.* **2022**, *206*, 115323. [CrossRef] [PubMed]
46. Czarnek, K.; Terpiłowska, S.; Siwicki, A.K. Selected aspects of the action of cobalt ions in the human body. *Cent. Eur. J. Immunol.* **2015**, *40*, 236–242. [CrossRef] [PubMed]
47. Ćwiertnia, A.; Kozłowski, M.; Cymbaluk-Płoska, A. The role of iron and cobalt in gynaecological diseases. *Cells* **2022**, *12*, 117. [CrossRef] [PubMed]
48. Marques, H.M. The inorganic chemistry of the cobalt corrinoids—An update. *J. Inorg. Biochem.* **2023**, *242*, 112154. [CrossRef] [PubMed]
49. Wahlqvist, F.; Bryngelsson, I.L.; Westberg, H.; Vihlborg, P.; Andersson, L. Dermal and inhalable cobalt exposure-uptake of cobalt for workers at Swedish hard metal plants. *PLoS ONE* **2020**, *15*, e0237100. [CrossRef]
50. National Toxicology Program. *Report on Carcinogens Monograph on Cobalt and Cobalt Compounds That Release Cobalt Ions In Vivo: RoC Monograph 06*; National Toxicology Program: Research Triangle Park, NC, USA, 2016. [PubMed]
51. Zhang, W.; Wang, C.; Zhu, W.; Liu, F.; Liu, Y. Ferrostatin-1 alleviates cytotoxicity of cobalt nanoparticles by inhibiting ferroptosis. *Bioengineered* **2022**, *13*, 6163–6172. [CrossRef]
52. Liu, Y.K.; Ye, J.; Han, Q.L.; Tao, R.; Liu, F.; Wang, W. Toxicity, and bioactivity of cobalt nanoparticles on the monocytes. *Orthop. Surg.* **2015**, *7*, 168–173. [CrossRef] [PubMed]
53. Savi, M.; Bocchi, L.; Cacciani, F.; Vilella, R.; Buschini, A.; Perotti, A.; Galati, S.; Montalbano, S.; Pinelli, S.; Frati, C.; et al. Cobalt oxide nanoparticles induce oxidative stress and alter electromechanical function in rat ventricular myocytes. *Part Fibre Toxicol.* **2021**, *18*, 1. [CrossRef] [PubMed]
54. Ton, T.T.; Kovi, R.C.; Peddada, T.N.; Chhabria, R.M.; Shockley, K.R.; Flagler, N.D.; Gerrish, K.E.; Herbert, R.A.; Behl, M.; Hoenerhoff, M.J.; et al. Cobalt-induced oxidative stress contributes to alveolar/bronchiolar carcinogenesis in B6C3F1/N mice. *Arch. Toxicol.* **2021**, *95*, 3171–3190. [CrossRef] [PubMed]
55. Li, Q.; Ke, Q.; Costa, M. Alterations of histone modifications by cobalt compounds. *Carcinogenesis* **2009**, *30*, 1243–1251. [CrossRef] [PubMed]
56. Grochans, S.; Korbecki, J.; Simińska, D.; Żwierełło, W.; Rzeszotek, S.; Kolasa, A.; Kojder, K.; Tarnowski, M.; Chlubek, D.; Baranowska-Bosiacka, I. CCL18 expression is higher in a glioblastoma multiforme tumor than in the peritumoral area and causes the migration of tumour cells sensitized by hypoxia. *Int. J. Mol. Sci.* **2022**, *23*, 8536. [CrossRef] [PubMed]
57. Korbecki, J.; Olbromski, M.; Dzięgiel, P. CCL18 in the progression of cancer. *Int. J. Mol. Sci.* **2020**, *21*, 7955. [CrossRef] [PubMed]
58. Lin, L.; Chen, Y.S.; Yao, Y.D.; Chen, J.Q.; Chen, J.N.; Huang, S.Y.; Zeng, Y.J.; Yao, H.R.; Zeng, S.H.; Fu, Y.S.; et al. CCL18 from tumor-associated macrophages promotes angiogenesis in breast cancer. *Oncotarget* **2015**, *6*, 34758–34773. [CrossRef] [PubMed]

59. Pandrangi, S.L.; Chittineedi, P.; Chikati, R.; Lingareddy, J.R.; Nagoor, M.; Ponnada, S.K. Role of dietary iron revisited: In metabolism, ferroptosis and pathophysiology of cancer. *Am. J. Cancer Res.* **2022**, *12*, 974–985. [PubMed]
60. Halcrow, P.W.; Lynch, M.L.; Geiger, J.D.; Ohm, J.E. Role of endolysosome function in iron metabolism and brain carcinogenesis. *Semin. Cancer Biol.* **2021**, *76*, 74–85. [CrossRef]
61. Jomova, K.; Valko, M. Importance of iron chelation in free radical-induced oxidative stress and human disease. *Curr. Pharm. Des.* **2011**, *17*, 3460–3473. [CrossRef]
62. Toyokuni, S.; Kong, Y.; Zheng, H.; Mi, D.; Katabuchi, M.; Motooka, Y.; Ito, F. Double-edged sword role of iron-loaded ferritin in extracellular vesicles. *J. Cancer Prev.* **2021**, *26*, 244–249. [CrossRef] [PubMed]
63. Wang, Y.; Yu, L.; Ding, J.; Chen, Y. Iron metabolism in cancer. *Int. J. Mol. Sci.* **2018**, *20*, 95. [CrossRef] [PubMed]
64. Toyokuni, S.; Kong, Y.; Cheng, Z.; Sato, K.; Hayashi, S.; Ito, F.; Jiang, L.; Yanatori, I.; Okazaki, Y.; Akatsuka, S. Carcinogenesis as side effects of iron and oxygen utilization: From the unveiled truth toward ultimate bioengineering. *Cancers* **2020**, *12*, 3320. [CrossRef] [PubMed]
65. Torti, S.V.; Manz, D.H.; Paul, B.T.; Blanchette-Farra, N.; Torti, F.M. Iron and cancer. *Annu. Rev. Nutr.* **2018**, *38*, 97–125. [CrossRef] [PubMed]
66. Jung, M.; Mertens, C.; Tomat, E.; Brüne, B. Iron as a central player and promising target in cancer progression. *Int. J. Mol. Sci.* **2019**, *20*, 273. [CrossRef] [PubMed]
67. Ying, J.F.; Lu, Z.B.; Fu, L.Q.; Tong, Y.; Wang, Z.; Li, W.F.; Mou, X.Z. The role of iron homeostasis and iron-mediated ROS in cancer. *Am. J. Cancer Res.* **2021**, *11*, 1895–1912. [PubMed Central]
68. Ni, S.; Kuang, Y.; Yuan, Y.; Yu, B. Mitochondrion-mediated iron accumulation promotes carcinogenesis and Warburg effect through reactive oxygen species in osteosarcoma. *Cancer Cell Int.* **2020**, *20*, 399. [CrossRef] [PubMed]
69. Liberti, M.V.; Locasale, J.W. The Warburg effect: How does it benefit cancer cells? *Trends Biochem. Sci.* **2016**, *41*, 211–218. [CrossRef] [PubMed]
70. Yuan, Y.; Ni, S.; Zhuge, A.; Li, B.; Li, L. Iron regulates the Warburg effect and ferroptosis in colorectal cancer. *Front. Oncol.* **2021**, *11*, 614778. [CrossRef]
71. Muto, Y.; Moroishi, T.; Ichihara, K.; Nishiyama, M.; Shimizu, H.; Eguchi, H.; Moriya, K.; Koike, K.; Mimori, K.; Mori, M.; et al. Disruption of FBXL5-mediated cellular iron homeostasis promotes liver carcinogenesis. *J. Exp. Med.* **2019**, *216*, 950–965. [CrossRef]
72. Islam, S.; Hoque, N.; Nasrin, N.; Hossain, M.; Rizwan, F.; Biswas, K.; Asaduzzaman, M.; Rahman, S.; Hoskin, D.W.; Sultana, S.; et al. Iron overload and breast cancer: Iron chelation as a potential therapeutic approach. *Life* **2022**, *12*, 963. [CrossRef] [PubMed]
73. Ploug, M.; Kroijer, R.; Qvist, N.; Lindahl, C.H.; Knudsen, T. Iron deficiency in colorectal cancer patients: A cohort study on prevalence and associations. *Color. Dis.* **2020**, *23*, 853–859. [CrossRef] [PubMed]
74. Kuang, Y.; Guo, W.; Ling, J.; Xu, D.; Liao, Y.; Zhao, H.; Du, X.; Wang, H.; Xu, M.; Song, H.; et al. Iron-dependent CDK1 activity promotes lung carcinogenesis via activation of the GP130/STAT3 signalling pathway. *Cell Death Dis.* **2019**, *10*, 297. [CrossRef] [PubMed]
75. Jung, M.; Mertens, C.; Brüne, B. Macrophage iron homeostasis and polarization in the context of cancer. *Immunobiology* **2015**, *220*, 295–304. [CrossRef] [PubMed]
76. Jung, M.; Mertens, C.; Bauer, R.; Rehwald, C.; Brüne, B. Lipocalin-2 and iron trafficking in the tumour microenvironment. *Pharmacol Res.* **2017**, *120*, 146–156. [CrossRef] [PubMed]
77. Crescenzi, E.; Leonardi, A.; Pacifico, F. Iron metabolism in cancer and senescence: A cellular perspective. *Biology* **2023**, *12*, 989. [CrossRef]
78. Özdemir, A.; Şimay Demir, Y.D.; Yeşilyurt, Z.E.; Ark, M. Senescent cells and SASP in cancer microenvironment: New approaches in cancer therapy. *Adv. Protein Chem. Struct. Biol.* **2023**, *133*, 115–158. [CrossRef] [PubMed]
79. Basak, T.; Kanwar, R.K. Iron imbalance in cancer: Intersection of deficiency and overload. *Cancer Med.* **2022**, *11*, 3837–3853. [CrossRef] [PubMed]
80. Murata, M. Inflammation, and cancer. *Environ. Health Prev. Med.* **2018**, *23*, 50. [CrossRef]
81. Hassannia, B.; Vandenabeele, P.; Vanden Berghe, T. Targeting ferroptosis to iron out cancer. *Cancer Cell* **2019**, *35*, 830–849. [CrossRef]
82. Liang, D.; Minikes, A.M.; Jiang, X. Ferroptosis at the intersection of lipid metabolism and cellular signalling. *Mol. Cell* **2022**, *82*, 2215–2227. [CrossRef] [PubMed]
83. Candelaria, P.V.; Leoh, L.S.; Penichet, M.L.; Daniels-Wells, T.R. Antibodies targeting the transferrin receptor 1 (TfR1) as direct anti-cancer agents. *Front. Immunol.* **2021**, *12*, 607692. [CrossRef] [PubMed]
84. El Hout, M.; Dos Santos, L.; Hamaï, A.; Mehrpour, M. A promising new approach to cancer therapy: Targeting iron metabolism in cancer stem cells. *Semin. Cancer Biol.* **2018**, *53*, 125–138. [CrossRef] [PubMed]
85. Mertens, C.; Akam, E.A.; Rehwald, C.; Brüne, B.; Tomat, E.; Jung, M. Intracellular iron chelation modulates the macrophage iron phenotype with consequences on tumour progression. *PLoS ONE* **2016**, *11*, e0166164. [CrossRef] [PubMed]
86. Evans, G.R.; Masullo, L.N. Manganese toxicity. In *StatPearls*; StatPearls Publishing: Treasure Island, FL, USA, 2023. [PubMed]
87. Aschner, M.; Erikson, K. Manganese. *Adv. Nutr.* **2017**, *8*, 520–521. [CrossRef] [PubMed]

88. Brzóska, M.M.; Gałażyn-Sidorczuk, M.; Kozłowska, M.; Smereczański, N.M. The body status of manganese and activity of this element-dependent mitochondrial superoxide dismutase in a rat model of human exposure to cadmium and Co-administration of *Aronia melanocarpa* L. Extract. *Nutr.* **2022**, *14*, 4773. [CrossRef] [PubMed]
89. Kim, A. Modulation of MnSOD in cancer: Epidemiological and experimental evidence. *Toxicol. Res.* **2010**, *26*, 83–93. [CrossRef] [PubMed]
90. Holley, A.K.; Dhar, S.K.; St Clair, D.K. Curbing cancer's sweet tooth: Is there a role for MnSOD in regulation of the Warburg effect? *Mitochondrion.* **2013**, *13*, 170–188. [CrossRef] [PubMed]
91. Bonetta Valentino, R. The structure-function relationships and physiological roles of MnSOD mutants. *Biosci. Rep.* **2022**, *42*, BSR20220202. [CrossRef]
92. Funke, S.; Risch, A.; Nieters, A.; Hoffmeister, M.; Stegmaier, C.; Seiler, C.M.; Brenner, H.; Chang-Claude, J. Genetic polymorphisms in genes related to oxidative stress (GSTP1, GSTM1, GSTT1, CAT, MnSOD, MPO, eNOS) and survival of rectal cancer patients after radiotherapy. *J. Cancer Epidemiol.* **2009**, *2009*, 302047. [CrossRef]
93. Liu, M.; Sun, X.; Chen, B.; Dai, R.; Xi, Z.; Xu, H. Insights into manganese superoxide dismutase and human diseases. *Int. J. Mol. Sci.* **2022**, *23*, 15893. [CrossRef]
94. Weydert, C.J.; Waugh, T.A.; Ritchie, J.M.; Iyer, K.S.; Smith, J.L.; Li, L.; Spitz, D.R.; Oberley, L.W. Overexpression of manganese or copper-zinc superoxide dismutase inhibits breast cancer growth. *Free Radic. Biol. Med.* **2006**, *41*, 226–237. [CrossRef]
95. Du, D.; Fu, H.J.; Ren, W.; Li, X.L.; Guo, L.H. PSA targeted dual-modality manganese oxide-mesoporus silica nanoparticles for prostate cancer imaging. *Biomed. Pharmacother.* **2020**, *121*, 109614. [CrossRef] [PubMed]
96. Eybl, V.; Kotyzová, D. Protective effect of manganese in cadmium-induced hepatic oxidative damage, changes in cadmium distribution and trace elements level in mice. *Interdiscip. Toxicol.* **2010**, *3*, 68–72. [CrossRef]
97. Zhang, W.; Li, H.; Tan, X.; Li, Z.; Zhong, C.; Xiao, W.; Xiong, Y.; Zhang, W.; Yang, L.; Wu, G. Fe-Mn plaque formation mechanism underlying the inhibition of cadmium absorption by rice under oxidation conditions. *Environ. Eng. Sci.* **2021**, *38*, 676–684. [CrossRef]
98. Lener, M.R.; Reszka, E.; Marciniak, W.; Lesicka, M.; Baszuk, P.; Jabłońska, E.; Białkowska, K.; Muszyńska, M.; Pietrzak, S.; Derkacz, R.; et al. Blood cadmium levels as a marker for early lung cancer detection. *J. Trace Elem. Med. Biol.* **2021**, *64*, 126682. [CrossRef]
99. Ebrahimi, M.; Khalili, N.; Razi, S.; Keshavarz-Fatki, M.; Khalili, N.; Rezaei, N. Effects of lead and cadmium on the immune system and cancer progression. *J. Environ. Health Sci. Eng.* **2020**, *18*, 335–343. [CrossRef] [PubMed]
100. Zwolak, I. The role of selenium in arsenic and cadmium toxicity: An updated review of scientific literature. *Biol. Trace Elem. Res.* **2020**, *193*, 44–63. [CrossRef] [PubMed]
101. Massányi, P.; Massányi, M.; Madeddu, R.; Stawarz, R.; Lukáč, N. Effects of cadmium, lead, and mercury on the structure and function of reproductive organs. *Toxics* **2020**, *8*, 94. [CrossRef]
102. Saintilnord, W.N.; Fondufe-Mittendorf, Y. Arsenic-induced epigenetic changes in cancer development. *Semin. Cancer Biol.* **2021**, *76*, 195–205. [CrossRef]
103. Son, Y.O. Molecular mechanisms of nickel-induced carcinogenesis. *Endocr. Metab. Immune Disord. Drug Targets* **2020**, *20*, 1015–1023. [CrossRef] [PubMed]
104. Ozturk, M.; Metin, M.; Altay, V.; Bhat, R.A.; Ejaz, M.; Gul, A.; Unal, B.T.; Hasanuzzaman, M.; Nibir, L.; Nahar, K.; et al. Arsenic and human health: Genotoxicity, epigenomic effects, and cancer signalling. *Biol. Trace Elem. Res.* **2022**, *200*, 988–1001. [CrossRef]
105. Strupp, C. Beryllium metal II. A review of the available toxicity data. *Ann. Occup. Hyg.* **2011**, *55*, 43–56. [CrossRef]
106. Skalny, A.V.; Aschner, M.; Sekacheva, M.I.; Santamaria, A.; Barbosa, F.; Ferrer, B.; Aaseth, J.; Paoliello, M.M.; Rocha, J.B.; Tinkov, A.A. Mercury and cancer: Where are we now after two decades of research? *Food Chem. Toxicol.* **2022**, *164*, 113001. [CrossRef]
107. Sánchez-Alarcón, J.; Milić, M.; Bustamante-Montes, L.P.; Isaac-Olivé, K.; Valencia-Quintana, R.; Ramírez-Durán, N. Genotoxicity of mercury and its derivatives demonstrated in vitro and in vivo in human populations studies. Systematic review. *Toxics* **2021**, *9*, 326. [CrossRef]
108. Vincent, J.B. New evidence against chromium as an essential trace element. *J. Nutr.* **2017**, *147*, 2212–2219. [CrossRef]
109. Levina, A.; Crans, D.C.; Lay, P.A. Speciation of metal drugs, supplements and toxins in media and bodily fluids controls in vitro activities. *Coord. Chem. Rev.* **2017**, *352*, 473–498. [CrossRef]
110. Pontoni, L.; La Vecchia, C.; Boguta, P.; Sirakov, M.; D'Aniello, E.; Fabbricino, M.; Locascio, A. Natural organic matter controls metal speciation and toxicity for marine organisms: A review. *Environ. Chem. Lett.* **2022**, *20*, 797–812. [CrossRef]
111. Medici, S.; Peana, M.; Pelucelli, A.; Zoroddu, M.A. An updated overview on metal nanoparticles toxicity. *Semin. Cancer Biol.* **2021**, *76*, 17–26. [CrossRef]
112. Fouani, L.; Menezes, S.V.; Paulson, M.; Richardson, D.R.; Kovacevic, Z. Metals and metastasis: Exploiting the role of metals in cancer metastasis to develop novel anti-metastatic agents. *Pharmacol. Res.* **2017**, *115*, 275–287. [CrossRef]

International Journal of
Molecular Sciences

MDPI

Review

Enhancing Colorectal Cancer Immunotherapy: The Pivotal Role of Ferroptosis in Modulating the Tumor Microenvironment

Yanqing Li [1] and Xiaofei Cheng [2,*]

[1] Department of Pathology, The Second Affiliated Hospital, Zhejiang University School of Medicine, Hangzhou 310009, China; yanqingli@zju.edu.cn
[2] Department of Colorectal Surgery, The First Affiliated Hospital, Zhejiang University School of Medicine, Hangzhou 310003, China
* Correspondence: xfcheng@zju.edu.cn

Abstract: Colorectal cancer (CRC) represents a significant challenge in oncology, with increasing incidence and mortality rates worldwide, particularly among younger adults. Despite advancements in treatment modalities, the urgent need for more effective therapies persists. Immunotherapy has emerged as a beacon of hope, offering the potential for improved outcomes and quality of life. This review delves into the critical interplay between ferroptosis, an iron-dependent form of regulated cell death, and immunotherapy within the CRC context. Ferroptosis's influence extends beyond tumor cell fate, reshaping the tumor microenvironment (TME) to enhance immunotherapy's efficacy. Investigations into Ferroptosis-related Genes (OFRGs) reveal their pivotal role in modulating immune cell infiltration and TME composition, closely correlating with tumor responsiveness to immunotherapy. The integration of ferroptosis inducers with immunotherapeutic strategies, particularly through novel approaches like ferrotherapy and targeted co-delivery systems, showcases promising avenues for augmenting treatment efficacy. Furthermore, the expression patterns of OFRGs offer novel prognostic tools, potentially guiding personalized and precision therapy in CRC. This review underscores the emerging paradigm of leveraging ferroptosis to bolster immunotherapy's impact, highlighting the need for further research to translate these insights into clinical advancements. Through a deeper understanding of the ferroptosis-immunotherapy nexus, new therapeutic strategies can be developed, promising enhanced efficacy and broader applicability in CRC treatment, ultimately improving patient outcomes and quality of life in the face of this formidable disease.

Keywords: ferroptosis; immunotherapy; colorectal cancer (CRC); tumor microenvironment (TME); ferroptosis-related genes (OFRGs)

Citation: Li, Y.; Cheng, X. Enhancing Colorectal Cancer Immunotherapy: The Pivotal Role of Ferroptosis in Modulating the Tumor Microenvironment. *Int. J. Mol. Sci.* **2024**, *25*, 9141. https://doi.org/10.3390/ijms25179141

Academic Editor: Peter J. K. Kuppen

Received: 29 July 2024
Revised: 19 August 2024
Accepted: 20 August 2024
Published: 23 August 2024

1. Introduction

Colorectal cancer (CRC) ranks among the top three in both incidence and mortality rates globally [1]. Most CRCs originate from polyps, beginning with abnormalities in glandular crypts and progressing into polyps and then cancer. This process can take 5 to 15 years, with villous adenomas exhibiting a high malignancy rate of up to 40% [2]. In recent years, there has been a significant increase in the incidence and mortality rates of CRC among young adults aged 18 to 49 years across nearly all regions worldwide, signaling an alarming trend that underscores the urgent need for effective treatment modalities [3].

Sex differences in the progression of CRC have become an increasingly important area of research. Studies show that men and women exhibit significant differences in CRC incidence, progression, and response to treatment. Generally, men have a higher incidence of CRC, which may be linked to lifestyle-related risk factors such as smoking, alcohol consumption, and diet [4]. In contrast, women may benefit from the protective effects of hormones like estrogen, which may delay CRC onset. However, this protective effect diminishes after menopause, leading to a convergence in CRC risk between men and women.

At the genetic level, there are sex-based differences in the frequency and types of mutations in CRC-related genes, such as *APC* and *TP53*, which can influence tumor behavior and treatment response [5]. Estrogen, for example, is thought to exert anti-tumor effects by interacting with intracellular signaling pathways that inhibit CRC development [6]. Additionally, estrogen may modulate the immune microenvironment, potentially enhancing the effectiveness of immunotherapy [7]. Sex differences also extend to the immune response, with women generally exhibiting a stronger immune reaction compared to men [8]. This heightened immune response could lead to better outcomes in immunotherapy for women but also increases the risk of immune-related side effects.

Immunotherapy has shown promising prospects in cancer treatment, with the potential to enhance treatment outcomes, prolong survival, and improve patients' quality of life. The aspiration for long-term survival with the disease or complete tumor eradication is becoming an increasingly tangible goal, thanks to immunotherapy. The significance of immunotherapy in cancer treatment is evident in several aspects. Firstly, it is highly specific and efficient, mobilizing the body's immune cells to recognize and combat cancer, thereby indirectly eliminating and controlling the disease with minimal side effects, ensuring safety and efficacy [9]. Secondly, immunotherapy encompasses a diverse range of approaches, such as molecular targeted therapy, immune checkpoint inhibitors (ICIs), adoptive cell therapy, cytokine therapy, and cancer vaccines, among others [10]. Unlike traditional treatments such as surgery, radiotherapy, and chemotherapy, immunotherapy does not directly kill cancer cells. Instead, it activates the body's immune cells, enhancing the immune system's ability to combat the disease. With advances in tumor immunology research and continuous technological advancements, immunotherapy is poised to achieve new breakthroughs in tumor eradication, aspiring to become a mainstream method in cancer treatment [11].

When discussing CRC and immunotherapy, understanding the concept of ferroptosis becomes particularly relevant. Ferroptosis, an iron-dependent form of regulated cell death, has garnered significant interest in recent years due to its potential role in tumor immunity. There exists a complex interplay between ferroptosis and tumor immunity. On the one hand, ferroptosis can influence the immunogenicity of tumor cells, thereby affecting the recognition and attack by immune cells [12]. On the other hand, ferroptosis also impacts the activity of immune cells, as iron plays a crucial role in their activation, proliferation, and functionality [13]. Moreover, the relationship between ferroptosis and ICIs, a class of drugs used to enhance the immune system's attack on tumors, has been elucidated. Research suggests that ferroptosis may be associated with the efficacy of ICIs. Ferroptosis can affect the expression of certain molecules on tumor cells, thereby influencing the expression of immune checkpoints and subsequently impacting the effectiveness of immunotherapy [14]. Furthermore, the metabolic pathways of immune cells are crucial for their functionality. Ferroptosis can influence the metabolic pathways of immune cells, affecting their activity, proliferation, and functionality within the TME [15]. Consequently, ferroptosis may serve as a critical factor in regulating the role of immune cells in the TME.

In conclusion, the concept of ferroptosis holds promise in reshaping our understanding of tumor immunity and immunotherapy. Its intricate interactions with tumor cells and immune cells underscore its potential as a novel therapeutic target for enhancing the efficacy of immunotherapy in CRC and other malignancies.

2. Ferroptosis: Understanding Its Mechanism and Impact

Introduced by Brent R. Stockwell from Columbia University in 2012, ferroptosis is recognized as a unique, regulated cellular demise mechanism driven by iron dependency, setting it apart from apoptosis, necrosis, and autophagy [16]. It pivots around the essential roles of glutathione depletion and the inactivation of glutathione peroxidase 4 (*GPX4*), hindering the cellular capacity to neutralize lipid peroxides. This deficiency fosters an environment ripe for iron accumulation, escalating lipid oxidation, and the formation of reactive oxygen species (ROS), culminating in cell death [17]. Distinguished by iron buildup, lipid peroxidation, and enhanced ROS levels, ferroptotic cells also display genetic shifts in iron homeostasis and lipid peroxidation regulation. Microscopically, these cells are marked by diminished mitochondrial volume, with notable alterations in membrane and cristae structures, while nuclear morphology remains largely unaffected [18].

Ferroptosis intersects with several metabolic pathways, including those involving amino acids, iron, and polyunsaturated fatty acids, along with the synthesis of glutathione, phospholipids, NADPH, and coenzyme Q10. Its implication in various mammalian degenerative conditions and diseases highlights its pathological significance [19].

Central to triggering ferroptosis is the rampant accumulation of lipid peroxides, propagated through two primary routes: enzymatic lipid peroxidation and the iron-catalyzed Fenton reaction. The enzymatic pathway transforms polyunsaturated fatty acids (*PUFAs*) into reactive lipid peroxides, with arachidonic (AA) and adrenic acids as key PUFA contributors [20]. This process is meticulously orchestrated by enzymes such as acyl-CoA synthetase long-chain family member 4 (*ACSL4*), lysophosphatidylcholine acyltransferase 3 (*LPCAT3*), and lipoxygenases (*LOXs*), which sequentially activate and oxidize AA to form lipid peroxides, thus playing pivotal roles in ferroptosis induction [21]. Conversely, the Fenton reaction, facilitated by free iron ions, produces lipid peroxides from peroxyl radical interactions with lipids. This mechanism is exacerbated by fluctuations in cellular iron levels, intensifying lipid peroxide formation and propelling the cell toward ferroptosis [22]. To counteract lipid peroxide accumulation, cells employ *GPX4*, which utilizes glutathione (*GSH*), synthesized from imported cysteine, to revert lipid peroxides to their benign state. Given *GPX4*'s exclusivity in this reduction, its role is deemed indispensable in the ferroptotic pathway, making it a prime target for therapeutic interventions [23].

Beyond the *GSH* pathway, the coenzyme Q10 (*CoQ10*)-dependent reduction pathway also mitigates ferroptosis. Here, the ferroptosis suppressor protein FSP1 plays a crucial role, leveraging NADH to restore *CoQ10* to its antioxidative ubiquinol form, thus reducing lipid peroxides [24]. Other noteworthy mechanisms include the *GCH1* pathway, producing the antioxidant BH4 and facilitating *CoQ10* synthesis, and *DHODH's* contribution, paralleling FSP1's function in the mitochondria to dampen lipid peroxide production, emphasizing the complexity and multifaceted nature of cellular defenses against ferroptosis [25]. Figure 1 illustrates the primary signaling pathways regulating ferroptosis.

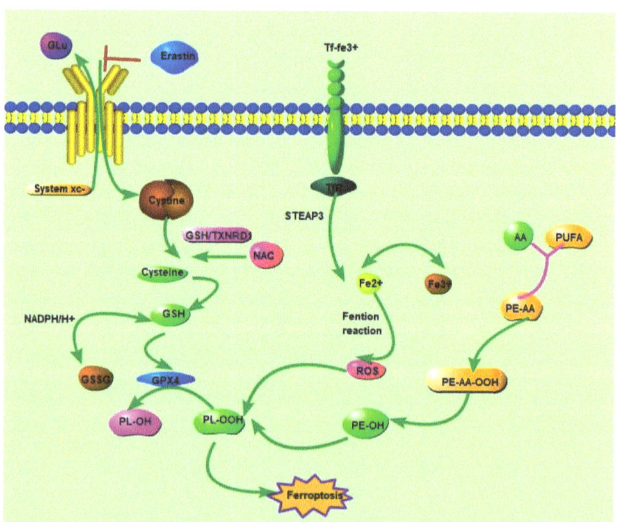

Figure 1. This figure shows key pathways and molecules involved in ferroptosis in tumor cells. System xc- imports cystine in exchange for glutamate, maintaining glutathione (*GSH*) levels, which act as antioxidants. *GPX4* uses *GSH* to prevent lipid peroxidation. Erastin inhibits system xc-, reducing

cystine uptake and *GSH* and increasing lipid peroxidation. The transferrin receptor (TfR) uptakes iron, which, through the Fenton reaction, produces ROS that promotes lipid peroxidation. Polyunsaturated fatty acids (*PUFAs*) and arachidonic acid (AA) contribute to ferroptosis through lipid peroxidation. Antioxidants like N-acetylcysteine (*NAC*) and *GSH/TXNRD* reduce oxidative stress, inhibiting ferroptosis. This balance between lipid peroxidation and antioxidant defenses determines cell fate through ferroptosis. Abbreviations: *GSH*: Glutathione; *GPX4*: Glutathione peroxidase 4; GSR: Glutathione reductase; TXNRD1: Thioredoxin reductase 1; NAC: N-Acetylcysteine; *CoQ10*: Coenzyme Q10; FSP1: Ferroptosis suppressor protein 1; GCH1: GTP cyclohydrolase 1; BH4: Tetrahydrobiopterin; PUFA: Polyunsaturated fatty acid; PE-PUFA-OOH: Phosphatidylethanolamine-linked PUFA hydroperoxides; α-TOH: Alpha-tocopherol (Vitamin E); IPP: Isopentenyl pyrophosphate; CoA: Coenzyme A; BCNU: Carmustine; RSL3: RAS-selective lethal 3.

3. Effect of Tumor Microenvironment (TME) on Ferroptosis

Hypoxia, a hallmark of the TME, induces changes in iron metabolism that impact ferroptosis regulation. Under hypoxic conditions, iron regulatory protein 2 (*IRP2*) is upregulated post-translationally, leading to increased expression of iron transporters (TFRC, *SLC11A2*) and decreased expression of iron storage proteins (FTH) [26]. Carbonic anhydrase IX (CAIX), a key regulator of tumor hypoxia, controls intracellular pH and prevents cancer cells from undergoing ferroptosis by modulating the cystine/glutamate antiporter xCT [27,28]. Hypoxia-induced factors such as *HIF-1α/lncRNA-PMAN* and the *CBSLR/CBS* signal axis further protect cancer cells from ferroptosis by regulating key genes involved in iron metabolism [29]. Additionally, hypoxia reduces the expression of NCOA4 in macrophages, leading to increased ferritin levels and decreased susceptibility to ferroptosis [22,30]. Recently, dual hypoxia-sensitive polymeric nanocarriers have been designed to sensitize hypoxic tumor cells to ferroptosis by depleting NADPH, *GSH*, and Trx, thereby inhibiting *GPX4* activity and enhancing the efficacy of ferroptosis-inducing agents, specifically in hypoxic tumors [31,32].

Lactate accumulation in the TME, resulting from cancer cell glycolysis, has been found to inhibit ferroptosis. High levels of lactic acid upregulate the expression of hydroxycarboxylic acid receptor 1 (*HCAR1*) and monocarboxylate transporter 1 *(MCT1)*, thereby reducing lipid peroxidation and inhibiting ferroptosis in hepatoma cells [33]. Targeting the HCAR1/MCT1 axis enhances ferroptosis sensitivity, suggesting a potential therapeutic strategy for overcoming lactate-mediated ferroptosis resistance in cancer treatment [33].

Furthermore, cancer-related inflammation within the TME can either promote or inhibit tumor ferroptosis. Proinflammatory factors such as *IL-6* and *HMGB1* have been implicated in regulating ferroptosis in various cancer types through different signaling pathways, including the *JAK2/STAT3* pathway and the *RAS-JNK/p38* pathway [34]. Additionally, non-coding RNAs like miR-539 and circRNAs like *CircABCB10* and *Circ-IL-4* receptors have been shown to modulate ferroptosis by targeting key regulators such as xCT and *GPX4*, thereby influencing cancer progression [35].

Recent studies have shed light on the intricate relationship between ferroptosis and immune effector cells within the TME, underscoring its implications for cancer immunotherapy. CD8+ T cells, pivotal players in antitumor immunity, secrete cytokines such as *IL-2*, *IL-12*, and *IFNγ* within the TME, enhancing their ability to target and eliminate tumor cells. Interestingly, *IFNγ* produced by activated CD8+ T cells inhibits the expression of *SLC3A2* and Solute Carrier Family 7 Member 11 *(SLC7A11)*, components of the cystine/glutamate antiporter system xc-, promoting lipid peroxidation and ferroptosis in tumor cells [36,37]. Additionally, *IFNγ*, in conjunction with arachidonic acid, induces tumor cell ferroptosis through *ACSL4*, highlighting the potential of targeting tumor ferroptosis metabolism to enhance cancer immunotherapy [38,39].

Moreover, Oxidized Low-Density Lipoprotein (*oxLDL*) in the TME induces ferroptosis and p38 phosphorylation in CD8+ T cells via CD36-dependent mechanisms. Activation of p38 leads to CD8+ T cell death, suppression of *IFNγ* and TNFα production, and depletion of CD8+ T cells, impairing antitumor immunity. Notably, combination therapy involving anti-PD-1 antibodies and CD36 deletion in CD8+ T cells exhibits superior antitumor effects, suggesting that targeting CD36 and ferroptosis could enhance T cell-based immunotherapy efficacy [40–42]. NK cells, crucial for tumor surveillance and immunotherapy, exert cytotoxic effects on tumor cells through perforin, granzyme, and *IFN*γ release. Furthermore, activating mitochondrial apoptosis in cancer cells enhances NK cell-mediated killing, and the synergistic effect of BH3 mimetics and ferroptosis induction enhances cancer cell death. Clinical-grade iron oxide nanoparticles induce ferroptosis in prostate cancer cells, activating NK cells and augmenting their cytotoxic function. Combining ferroptosis induction with NK cell therapy results in tumor regression, indicating that ferroptosis enhances NK cell activity [42,43].

However, while ferroptosis releases cytokines and damage-associated molecular patterns (DAMPs), it does not activate antitumor immune responses. Moreover, ferroptosis inhibits the cross-presentation of soluble antigens to dendritic cells (DCs), impairs DC maturation, and inhibits DC-mediated phagocytosis of tumor cells. Consequently, cancer cell ferroptosis may not represent an immunogenic form of cell death, highlighting the complex interplay between ferroptosis and immune responses within the TME [44,45].

To fully understand the complex regulatory networks of ferroptosis in the TME, it is essential to integrate interdisciplinary approaches, including computational biology, bioinformatics, and systems biology. These fields offer powerful tools and methodologies to dissect the intricate interactions and pathways involved in ferroptosis, providing deeper insights into its role in cancer therapy.

Computational biology utilizes mathematical models and computational techniques to simulate biological processes. In the context of ferroptosis, computational models can predict the dynamics of lipid peroxidation, iron metabolism, and antioxidant defenses under various conditions. By simulating these processes, researchers can identify key regulatory nodes and potential therapeutic targets. For instance, computational models can help predict how different ferroptosis inducers interact with cellular pathways, aiding in the design of more effective and specific treatments [46].

Bioinformatics involves the application of computational tools to analyze biological data. High-throughput omics technologies, such as genomics, transcriptomics, proteomics, and metabolomics, generate vast amounts of data that can be analyzed to uncover the molecular mechanisms underlying ferroptosis. Bioinformatics approaches can identify gene expression patterns, protein-protein interactions, and metabolic alterations associated with ferroptosis [47]. For example, integrating transcriptomic data from CRC patients with ferroptosis-related gene expression profiles can reveal biomarkers for patient stratification and treatment response prediction [48,49].

Systems biology adopts a holistic approach to understanding biological systems by integrating data from multiple sources to build comprehensive models. In ferroptosis research, systems biology can elucidate the complex interactions between different cellular components and pathways within the TME. This approach can identify emergent properties and network behaviors that are not apparent from studying individual components in isolation [50]. Systems biology models can also simulate how interventions, such as ferroptosis inducers or inhibitors, impact the overall system, providing insights into potential therapeutic strategies and their systemic effects [51].

Combining computational biology, bioinformatics, and systems biology can lead to a more comprehensive understanding of ferroptosis in the TME. For instance, computational models can be informed by bioinformatics analyses of omics data, while systems biology can integrate these models to simulate complex interactions within the TME [52]. This interdisciplinary approach can identify novel regulatory mechanisms, potential biomarkers, and therapeutic targets, ultimately enhancing the efficacy of ferroptosis-based therapies in CRC [53].

In CRC research, predictive modeling can predict patient-specific responses to ferroptosis-based therapies, allowing for personalized treatment plans. Bioinformatics analyses can identify biomarkers associated with ferroptosis sensitivity and resistance, facilitating patient stratification and monitoring [50]. Systems biology can reveal key regulatory nodes and pathways that can be targeted to enhance ferroptosis and improve therapeutic outcomes. By leveraging these interdisciplinary approaches, researchers can gain a deeper understanding of the regulatory networks of ferroptosis in the TME, paving the way for more effective and targeted cancer therapies.

In conclusion, the relationship between ferroptosis and the TME is intricate and multifaceted, impacting not only the survival and death of tumor cells but also the function of immune cells, thereby influencing tumor development and treatment response. Table 1 summarizes the regulatory interactions between ferroptosis and the TME, highlighting the mechanisms and effects. Understanding these interactions is crucial for developing novel cancer therapeutic strategies.

Table 1. Regulatory Interplay between Ferroptosis and the Tumor Microenvironment: Mechanisms and Impacts.

TME Factor/ Regulator	Mechanism of Impact	Effect on Ferroptosis	Reference Numbers
Hypoxia	Induction of IRP2 post-translational upregulation - Regulation of iron transporter and storage protein expression	Promotes ferroptosis	[26]
Carbonic Anhydrase IX (CAIX)	Modulation of the cystine/glutamate antiporter xCT	Prevents ferroptosis	[27,28]
HIF-1α/lncRNA-PMAN	Regulation of key genes involved in iron metabolism	Protects cancer cells from ferroptosis	[29]
CBSLR/CBS Signal Axis	Same as above	Protects cancer cells from ferroptosis	[29]
NCOA4 Expression Reduction	Reduced expression in macrophages, leading to increased ferritin levels and decreased ferroptosis susceptibility	Decreases ferroptosis susceptibility	[22,30]
Polymeric Nanocarriers	Depleting *NADPH*, *GSH*, and Trx to sensitize hypoxic tumor cells to ferroptosis	Sensitizes hypoxic tumor cells to ferroptosis	[31,32]
Lactate Accumulation	Upregulation of HCAR1 and MCT1 reducing lipid peroxidation	Inhibits ferroptosis	[33]
Inflammatory Factors	Regulation of ferroptosis through various cancer types and signaling pathways, such as *JAK2/STAT3* and *RAS-JNK/p38*	May promote or inhibit ferroptosis	[34]
Non-coding RNAs	Modulation of ferroptosis by targeting key regulators like xCT and *GPX4*	Influences cancer progression and ferroptosis	[35]
CD8 T Cells	Secretion of cytokines like *IFNγ* inhibiting the expression of *SLC3A2* and *SLC7A11*	Promotes ferroptosis in tumor cells	[36,37]
oxLDL and CD36	Inducing ferroptosis and p38 phosphorylation in CD8+ T cells via CD36-dependent mechanisms	Affects T cell function and antitumor immunity	[40–42]
NK Cells	Cytotoxic effects on tumor cells through perforin, granzyme, and *IFNγ* release	Enhances NK cell activity and promotes ferroptosis	[42,43]
Ferroptosis and Immune Response	Release of cytokines and DAMPs without activating antitumor immune responses, inhibiting dendritic cell functions	May not represent an immunogenic cell death form	[44,45]

4. The Role of Ferroptosis in Immunotherapy for Cancer

The role of ferroptosis, a non-apoptotic form of cell death induced by iron-dependent lipid peroxidation, is increasingly recognized in cancer immunotherapy. Not only does ferroptosis directly cause tumor cell death, but it also enhances the efficacy of immunotherapy by modulating the activity of immune cells within the TME.

In the treatment of melanoma, activation of ferroptosis sensitizes tumor cells to chemotherapy and enhances immunotherapy efficacy by regulating the activity of immune cells within the TME. For instance, ferroptosis-induced tumor cell death promotes the polarization of M1-like macrophages while inhibiting the functions of M2-like macrophages that promote tumor growth and metastasis [54,55]. In pancreatic cancer therapy, ferroptosis induction selectively kills tumor cells and activates the immune system, bolstering antitumor immune responses [56–58]. Studies have shown that ferroptosis inducers such as RSL3 and FIN56 increase tumor cell sensitivity to radiotherapy, thereby improving treatment efficacy when used in combination [59,60]. Furthermore, ferroptosis promotes the release of DAMPs from tumor cells, activating dendritic cells and T cells to enhance tumor-specific immune responses [61,62]. In breast cancer therapy, ferroptosis induction also shows promise. When combined with ICIs, it effectively enhances tumor cell ferroptosis and improves treatment outcomes. For example, CD8+ T cells suppress the expression of *SLC7A11* and *SLC3A2* in tumor cells via IFNγ release, increasing tumor cell sensitivity to ferroptosis [37,63]. Moreover, ferroptosis inducers such as sulfasalazine and elastin enhance breast cancer cell sensitivity to chemotherapy drugs [60].

In summary, the role of ferroptosis in cancer immunotherapy is multifaceted. It not only directly induces tumor cell death but also enhances immunotherapy efficacy by modulating immune responses within the TME. Induction and regulation of ferroptosis hold promise for improving treatment outcomes in various types of cancer, including melanoma, pancreatic cancer, breast cancer, and CRC.

5. The Challenges of Immunotherapy in CRC

Immunotherapy holds immense promise in the treatment of CRC, yet it faces several significant challenges. Firstly, its applicability is limited to a subset of patients, primarily those with microsatellite instability-high (MSI-H) or deficient mismatch repair (dMMR) status. This restriction means that a large proportion of CRC patients, particularly those with microsatellite stable (MSS) or proficient mismatch repair (pMMR) status, currently do not benefit from immunotherapy [64]. Expanding the eligible patient population for immunotherapy remains a major hurdle [65,66]. Secondly, accurately predicting treatment efficacy poses a challenge. MSI-H/dMMR status is currently the best predictor of immunotherapy response. However, even within this subgroup, only about 40% of patients respond effectively, leaving a significant portion without benefit. Improving the accuracy of predicting which patients will benefit from immunotherapy is another critical challenge [65,67,68]. Thirdly, treatment strategies require optimization. While combination immunotherapy with *CTLA-4* inhibitors has shown good efficacy in MSI-H/dMMR patients, it also comes with increased toxicity. Balancing efficacy and reducing side effects through optimized treatment strategies is a current imperative [65,69,70]. Fourthly, transforming "cold tumors" like MSS CRC into "hot tumors" that respond well to immunotherapy presents a substantial challenge. Efforts to achieve this transformation have not yet succeeded. Effectively converting "cold tumors" into "hot tumors" represents another significant challenge [71–74]. Lastly, there is a need for the development of novel immunotherapy strategies and drugs. While some immunotherapy agents like PD-1 inhibitors and CAR-T therapy are currently used in CRC treatment, they have limitations [75,76]. Developing new and more effective immunotherapy strategies and drugs to meet the needs of a broader patient population is an ongoing and future imperative.

Addressing these challenges is crucial for advancing the field of immunotherapy in CRC treatment. Interestingly, emerging research suggests that ferroptosis may offer a potential avenue for addressing some of these challenges.

6. The Role of Ferroptosis in Immunotherapy for CRC

In CRC, ferroptosis not only influences the fate of tumor cells but also exerts profound effects on the efficacy of immunotherapy by altering the composition and function of the TME. Studies investigating the expression patterns of OFRGs in CRC have revealed the crucial role of ferroptosis in regulating immune cell infiltration and the formation of the TME. These studies have found a close correlation between the expression levels of OFRGs and the sensitivity of tumors to immunotherapy. In immune-desert tumors characterized by low immune cell infiltration, induction of ferroptosis can increase immune cell infiltration, particularly CD8+ T cells, thereby enhancing tumor immunogenicity [77]. Conversely, in immune-inflamed and immune-excluded tumors, ferroptosis may influence the efficacy of immunotherapy by modulating the number and function of immune suppressor cells such as regulatory T cells (Tregs) and myeloid-derived suppressor cells (MDSCs) [78,79]. Thus, ferroptosis not only acts directly on tumor cells but also indirectly enhances the efficacy of immunotherapy by modulating immune cell activity and the state of the TME [80,81]. Furthermore, the roles of molecules such as Coactivator Associated Arginine Methyltransferase 1 (*CARM1*)and heat shock protein family B (small) member 1 *(HSPB1)* in regulating ferroptosis further emphasize the importance of ferroptosis in the tumor immune microenvironment. CARM1 promotes tumor progression by methylating *ACSL4* while inhibiting CARM1 can increase tumor cell sensitivity to ferroptosis, thereby promoting immunogenic cell death of tumor cells [82]. On the other hand, HSPB1 negatively regulates ferroptosis by reducing iron-mediated ROS production [83]. These findings suggest that targeting these molecules can modulate tumor cell sensitivity to ferroptosis, thereby enhancing the effectiveness of immunotherapy [84,85].

The combination of ferroptosis inducers and immunotherapy represents a promising approach to cancer treatment, particularly in CRC therapy. Ferrotherapy is an emerging cancer treatment modality that induces ferroptosis by catalyzing the decomposition of hydrogen peroxide (H_2O_2) into highly toxic hydroxyl radicals (·OH) using iron ions [86]. This therapy not only directly kills tumor cells but also activates the immune system by inducing immunogenic cell death (ICD), thereby promoting tumor-specific immune responses [87]. ICD, characterized by the release of tumor-specific antigens, activates dendritic cells and T cells, eliciting an immune response against the tumor. Moreover, the co-delivery of specific drugs, such as dihydroartemisinin (DHA) and pyropheophorbide-iron (Pyro-Fe), can further enhance the efficacy of ferroptosis. This co-delivery strategy not only increases the concentration of drugs at the tumor site but also enhances ferroptosis induction by generating more ROS [88]. This enhanced ferroptotic response can more effectively activate T cells, especially CD8+ T cells, which are crucial for antitumor immunity [89]. Figure 2 illustrates the interconnected components involved in enhancing CRC immunotherapy through ferroptosis modulation.

The expression patterns of ferroptosis-related genes provide new prognostic tools for the treatment of CRC, holding significant implications for personalized medicine and precision therapy. Studies investigating OFRG expression patterns have revealed different molecular clusters associated with clinical outcomes and biological pathways [78]. These molecular clusters help distinguish between "cold" and "hot" tumors in CRC. "Cold" tumors, characterized by low immune cell infiltration and poor response to immunotherapy, can be differentiated from "hot" tumors, which exhibit high immune cell infiltration and better response to immunotherapy [90]. This differentiation allows clinicians to select treatment regimens more accurately, providing patients with more effective treatments. Additionally, these studies suggest that the expression levels of ferroptosis-related genes can serve as biomarkers for predicting patient responses to PD-1-based immunotherapy [77,80]. For instance, high expression of ferroptosis-related genes may indicate a better response to PD-1 inhibitors, thereby facilitating more effective immunotherapy. The development of such predictive tools not only improves the success rate of treatment but also reduces unnecessary side effects and treatment costs. Some studies further emphasize the role of apolipoprotein L3 (*APOL3*) in ferroptosis and immunotherapy. APOL3 enhances the anti-tumor immune capacity of CD8+ T cells by promoting lactate dehydrogenase A (LDHA)-mediated ferroptosis [91]. This action increases the sensitivity of tumor cells to ferroptosis and promotes tumor immunogenic death by increasing interferon-gamma (IFNγ) production and reducing lactate concentration, enhancing tumor immunogenic death. This finding provides new predictive factors for the efficacy of immunotherapy and potential targets for the development of new treatment strategies.

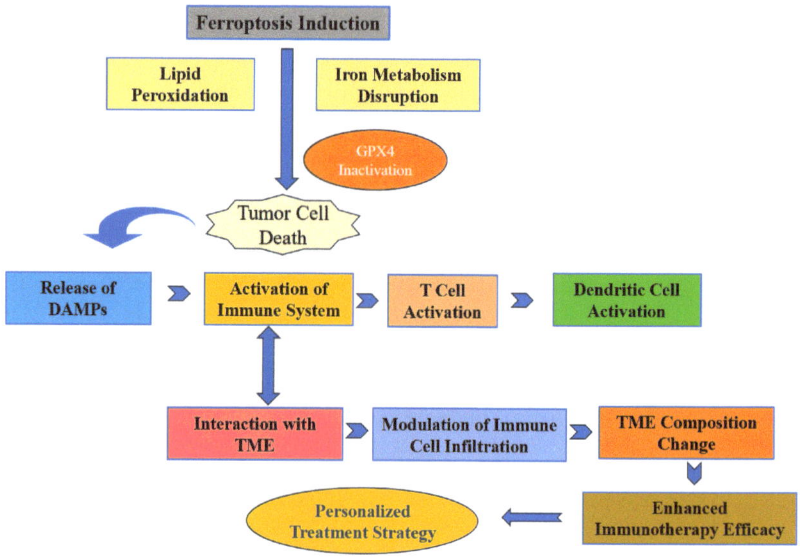

Figure 2. The figure depicts the sequence of events from ferroptosis induction to its impact on tumor microenvironment (TME) and immunotherapy. Ferroptosis triggers lipid peroxidation and *GPX4* inactivation, leading to tumor cell death and DAMPs release. This activates the immune system, enhancing T cell and dendritic cell functions, and modulates TME for improved immunotherapy efficacy, paving the way for personalized treatment strategies.

Emerging treatment strategies incorporating the concept of ferroptosis into CRC immunotherapy offer new avenues for improving treatment efficacy and overcoming drug resistance [92]. The method discussed in studies involving the combination of ferroptosis inducers and immunotherapy demonstrates the potential of drug combinations and nanotechnology applications to enhance antitumor immune responses [93–95].

Of particular note is the strategy mentioned in a study, which involves the co-delivery of dihydroartemisinin (DHA) and pyropheophorbide-iron (Pyro-Fe). This co-delivery approach not only enhances the anti-cancer efficacy of DHA in vivo but also sensitizes non-immunogenic CRC tumors to PD-L1 checkpoint blockade immunotherapy by increasing tumor immunogenicity. The key to this strategy lies in the intervention of ferroptosis inducers, which alter the TME, promote the release and presentation of tumor antigens, activate dendritic cells and T cells, and trigger specific immune responses against the tumor [96]. Furthermore, the application of nanotechnology offers new possibilities for targeted delivery of ferroptosis inducers. Nanoparticles can be designed with specific targeting properties to deliver drugs more accurately to tumor cells while reducing damage to normal cells. This targeting not only enhances treatment efficacy but also reduces treatment side effects. For example, ZnP@DHA/Pyro-Fe core-shell nanoparticles mentioned in the literature can be prepared using nanotechnology to release DHA and Pyro-Fe at the tumor site, effectively inducing ferroptosis [97].

Recent advancements in understanding the molecular mechanisms of ferroptosis in CRC have shed light on its potential as a therapeutic target. The cellular processes of iron metabolism and lipid peroxidation are central to ferroptosis induction, with agents like elastin and sulfasalazine effectively triggering this form of cell death in CRC cells. This occurs through the inhibition of the xCT system, leading to significant changes within the cell, including reduced cystine uptake, glutathione depletion, and increased lipid peroxidation, offering a novel approach to cancer treatment [98,99].

One of the most promising aspects of ferroptosis research is its potential to overcome drug resistance in CRC. By specifically targeting ferroptosis regulators such as *GPX4* and *ACSL4*, alongside pathways involved in iron metabolism, researchers have been able to increase the sensitivity of CRC cells to both chemotherapy and immunotherapy [100]. This approach not only enhances the efficacy of these treatments but also reverses established drug resistance, marking a significant advancement in CRC treatment strategies. Furthermore, the interplay between ferroptosis and immunotherapy has opened new avenues for enhancing antitumor immunity [101]. The process of ferroptosis in tumor cells leads to the release of DAMPs, which can activate the immune system and promote antitumor responses. Particularly, the induction of ferroptosis has been shown to bolster the activity of CD8+ T cells within the TME, thereby amplifying the effectiveness of immunotherapeutic approaches [79].

Ferroptosis is an emerging form of cell death that has garnered significant attention for its potential in cancer treatment. Here is a summary of clinical studies involving ferroptosis-related molecules and drugs in CRC (https://clinicaltrials.gov). Sulfasalazine targets SLC7A11 and is being investigated in patients with metastatic CRC under clinical trial ID NCT06134388. The study aims to explore the effectiveness of sulfasalazine in this patient group. CNSI-Fe (II), which targets iron ions, is the focus of clinical trial NCT06048367. This study aims to evaluate the efficacy of carbon nanoparticle-loaded iron (CNSI-Fe (II)) in treating advanced solid tumors, including CRC. Neratinib, another drug targeting iron ions, is being examined in combination with Trastuzumab or Cetuximab for patients with KRAS/NRAS/BRAF/PIK3CA wild-type metastatic CRC. This phase II trial, identified as NCT03457896, seeks to determine the effectiveness of these combinations. Lapatinib, which also targets iron ions, is being evaluated in several studies for advanced or metastatic CRC. Clinical trials such as NCT00536809 (Phase I/II) [102], NCT00574171 (Phase II), NCT01184482 (Phase II), NCT04831528 (Phase II), and NCT03418558 (Phase II) are investigating Lapatinib in combination with various drugs. Some studies suggest that these combinations show efficacy in specific cancer types, but further research is needed to confirm their effectiveness in CRC.

Sorafenib, targeting *SLC7A11*, is involved in multiple clinical trials, including NCT00780169 (Phase I), NCT00869570 (Phase I) [103], NCT00989469 (Phase II) [104], NCT01715441 (Phase II) [105], NCT00826540 (Phase II) [106], NCT01290926 (Phase II) [107], NCT00134069 (Phase I/II), NCT00865709 (Phase II), NCT00703638 (Phase I), and NCT01376453 (Phase II) [108]. These studies investigate the efficacy of Sorafenib combined with different chemotherapy regimens for metastatic or locally advanced CRC. Some trials indicate that Sorafenib combination therapies are active and have acceptable toxicity, particularly in KRAS-mutant CRC. Simvastatin, targeting HMGCR, is the subject of clinical trial NCT01238094. This study assesses the efficacy of Simvastatin combined with XELIRI/FOLFIRI regimens in patients with metastatic CRC. Detailed information about these clinical trials can be found in Table 2.

Table 2. Clinical Studies on Ferroptosis-related Molecules and Drugs in CRC.

Drug/Molecule	Target	Clinical Trial ID	Phase	Combination/Specific Aim	Reference
Sulfasalazine	*SLC7A11*	NCT06134388	Phase I	Metastatic CRC	-
CNSI-Fe (II)	Iron ions	NCT06048367	Phase I	Advanced solid tumors including CRC	-
Neratinib	Iron ions	NCT03457896	Phase II	Combination with Trastuzumab or Cetuximab	-
Lapatinib	Iron ions	NCT00536809 NCT00574171 NCT01184482 NCT04831528 NCT03418558	Phase I/II	Combination with various drugs	[102]

Table 2. *Cont.*

Drug/Molecule	Target	Clinical Trial ID	Phase	Combination/Specific Aim	Reference
Sorafenib	*SLC7A11*	NCT00780169 NCT00869570 NCT00989469 NCT01715441 NCT00826540 NCT01290926 NCT00134069 NCT00865709 NCT00703638 NCT01376453	Phase I/II	Combination with chemotherapy	[103–108]
Simvastatin	HMGCR	NCT01238094	Phase I	Combination with XELIRI/FOLFIRI	-

Note: "-" indicates that the clinical research is still ongoing, and there are no related publications available at this time.

Overall, Sulfasalazine and Sorafenib target SLC7A11, inhibiting glutamine transport and reducing the synthesis of the antioxidant *GSH*, thereby enhancing cellular sensitivity to oxidative stress and promoting ferroptosis [109,110]. CNSI-Fe (II) increases intracellular iron ion concentration, promoting ROS generation and lipid peroxidation, directly inducing ferroptosis [111]. Neratinib and Lapatinib studied primarily in combination with other therapies, show some preliminary efficacy in metastatic CRC, but further research is necessary to confirm their mechanisms and effectiveness. Simvastatin, though targeting HMGCR, might have potential efficacy when combined with chemotherapy in CRC. Research on ferroptosis-related therapeutic strategies in CRC is still in its early stages. Initial findings are promising, but extensive clinical studies are needed to validate these treatments' efficacy and safety further [98].

7. Global Research Trends

From 2019 to 2024, there has been a significant increase in the number of publications on ferroptosis, particularly in the context of cancer research. The number of citations has also risen, indicating the high impact and relevance of these studies. Notable contributions have been made by researchers from the United States, China, and European countries, reflecting a broad international interest [112]. A substantial portion of the research focuses on understanding the biochemical pathways and molecular mechanisms underlying ferroptosis. This includes the roles of lipid peroxidation, iron metabolism, and antioxidant defenses in regulating cell death [20,49]. Another major theme is the exploration of ferroptosis as a therapeutic strategy. Studies investigate the potential of ferroptosis inducers to enhance the efficacy of existing cancer treatments, including chemotherapy, radiotherapy, and immunotherapy [113].

Research has also concentrated on identifying biomarkers for ferroptosis sensitivity and resistance. These biomarkers can help predict patient responses to ferroptosis-based therapies and tailor personalized treatment plans [101]. There is ongoing work in the development of novel ferroptosis inducers and inhibitors. This includes both synthetic compounds and natural products, with a focus on improving specificity and reducing side effects [114]. The field has seen robust international collaboration, with multi-institutional studies becoming more common. Collaborative networks often include partnerships between academic institutions, research hospitals, and pharmaceutical companies. This multidisciplinary approach is essential for translating basic research findings into clinical applications [115].

The future of research on ferroptosis in CRC immunotherapy is poised to address several critical areas. Further studies are needed to optimize the combination of ferroptosis inducers with other therapeutic modalities, such as ICIs. Translational research efforts are expected to focus on conducting clinical trials to evaluate the safety and efficacy of ferroptosis-based therapies in CRC patients. The development of advanced diagnostic tools to monitor ferroptosis in real time and assess treatment responses is anticipated. In-depth studies to understand and overcome resistance mechanisms to ferroptosis will be crucial for enhancing therapeutic outcomes.

By analyzing global research trends, it is evident that the study of ferroptosis in CRC immunotherapy is a rapidly evolving field with significant potential for improving cancer treatment. Continued international collaboration and multidisciplinary research will be key drivers of innovation and clinical translation.

8. Challenges and Limitations

One of the primary challenges in leveraging ferroptosis for CRC immunotherapy is the specificity and targeting of ferroptosis inducers. While ferroptosis has been shown to effectively induce cell death in tumor cells, ensuring that these inducers selectively target cancer cells without affecting normal cells is crucial. The off-target effects can lead to unwanted toxicity and damage to healthy tissues, which limits the clinical application of ferroptosis inducers [100]. The TME plays a significant role in modulating ferroptosis. Hypoxia, a common feature of the TME, can inhibit ferroptosis through various mechanisms, including the upregulation of hypoxia-inducible factors and iron metabolism regulators like IRP2 and CAIX. These factors protect cancer cells from ferroptosis, making it difficult to induce ferroptosis effectively in a hypoxic environment [113,116]. Cancer cells can develop resistance to ferroptosis through several pathways. For instance, the upregulation of antioxidant systems like the cystine/glutamate antiporter (system Xc-) and *GPX4* can counteract lipid peroxidation and prevent ferroptotic cell death [117]. Additionally, genetic alterations in ferroptosis-related genes and compensatory metabolic pathways can also contribute to resistance, posing a challenge to sustained therapeutic efficacy [118].

While ferroptosis can influence the immune response, it does not always result in an immunogenic form of cell death. Ferroptosis can inhibit dendritic cell maturation and antigen presentation, which are critical for initiating robust anti-tumor immune responses. Furthermore, the release of DAMPs during ferroptosis might not be sufficient to activate a strong anti-tumor immune response, limiting its effectiveness as an adjunct to immunotherapy [45]. Determining the optimal dosing and timing for ferroptosis inducers in combination with immunotherapy is another significant challenge. The therapeutic window must be carefully balanced to maximize tumor cell death while minimizing toxicity and adverse effects. Additionally, the timing of administration relative to immunotherapeutic agents needs to be optimized to achieve synergistic effects [56]. Translating preclinical findings into clinical practice requires robust biomarkers to identify patients who are likely to benefit from ferroptosis-based therapies. Currently, there is a lack of reliable biomarkers to predict ferroptosis sensitivity in CRC patients. The heterogeneity of tumors and the variability in ferroptosis-related gene expression further complicate patient stratification and the design of personalized treatment regimens [119]. Combining ferroptosis inducers with existing immunotherapies (e.g., checkpoint inhibitors) holds promise but also presents challenges. The interactions between different therapeutic modalities need to be thoroughly understood to avoid antagonistic effects. Additionally, the development of effective co-delivery systems for ferroptosis inducers and immunotherapeutic agents is crucial to enhance treatment efficacy and reduce systemic toxicity [120]. Finally, the development and approval of new ferroptosis-based therapies face significant regulatory hurdles. Ensuring the safety, efficacy, and quality of these therapies through rigorous clinical trials is essential. Long-term studies are required to assess potential late-onset toxicities and the overall impact on patient survival and quality of life [121]. Addressing these challenges requires a multidisciplinary approach, integrating insights from molecular biology, immunology, pharmacology, and clinical sciences. Continued research and innovation in this field are essential to overcome these limitations and harness the full potential of ferroptosis in CRC immunotherapy.

9. Future Prospect

The clinical application prospects of ferroptosis-related molecules as potential therapeutic targets represent a burgeoning area of research, particularly in the treatment of CRC [122]. Ferroptosis, a form of programmed cell death induced by lipid peroxidation, is associated with various pathophysiological processes and disease states, including cancer. In tumor therapy, modulation of key molecules related to ferroptosis can enhance the sensitivity of tumor cells to ferroptosis, thereby serving as a novel therapeutic strategy to improve the efficacy of immunotherapy.

Several key ferroptosis-related molecules and their potential clinical applications include *SLC3A2* and *SLC7A11*, which encode proteins comprising the system Xc- involved in the exchange of intracellular cystine and glutamate. Cystine is a crucial antioxidant that maintains the cellular reducing environment, while glutamate participates in cellular energy metabolism. Inhibiting *SLC3A2* and *SLC7A11* can decrease intracellular cystine levels in tumor cells, increasing their sensitivity to ferroptosis induction [29,36,123]. Therefore, drugs targeting these molecules may help induce ferroptosis in tumor cells and enhance the efficacy of immunotherapy.

Another significant molecule is *GPX4*, an antioxidant enzyme essential for maintaining cell membrane integrity and preventing lipid peroxidation. Overexpression of *GPX4* can protect tumor cells from the effects of ferroptosis. Therefore, inhibiting *GPX4* activity or expression may increase the sensitivity of tumor cells to ferroptosis, making them more susceptible to clearance by the immune system [124].

Ferroptosis inducers such as elastin and sorafenib can directly induce ferroptosis in tumor cells by interfering with intracellular iron metabolism and lipid peroxidation pathways, leading to tumor cell death [125]. In clinical treatment, these ferroptosis inducers can be used in combination with ICIs to enhance anti-tumor immune responses. Combination therapy strategies involving ferroptosis inducers and other treatment modalities such as radiotherapy, chemotherapy, or immunotherapy can enhance treatment efficacy. For example, certain chemotherapy drugs can increase tumor cell sensitivity to ferroptosis while inducing tumor cell death, thereby enhancing subsequent immunotherapy efficacy [126]. Furthermore, the development of biomarkers based on ferroptosis-related molecules can predict tumor responses to treatment. For instance, the expression levels of *SLC3A2* and *SLC7A11* may correlate with tumor sensitivity to ferroptosis inducers, serving as predictive factors for treatment outcomes.

In conclusion, the clinical application prospects of ferroptosis-related molecules lie in the development of novel treatment strategies. By modulating the expression or activity of these molecules, tumor cell sensitivity to ferroptosis can be enhanced, thereby improving the efficacy of immunotherapy. Future research needs to further explore the roles of these molecules in different types of tumors and evaluate their safety and effectiveness as therapeutic targets.

10. Summary

In conclusion, the relationship between ferroptosis and the immune system presents a promising frontier in colorectal cancer (CRC) therapy. By integrating ferroptosis inducers with current immunotherapies, we can enhance anti-tumor efficacy and overcome resistance. Key future directions include developing ferroptosis-based diagnostic tools to detect early tumor changes in the microenvironment, using ferroptosis inducers alongside immunotherapies to target resistant cancer cells, and identifying reliable biomarkers to predict ferroptosis sensitivity, optimizing treatment windows, and understanding therapeutic interactions. Conducting extensive research and clinical trials is crucial to translating these strategies into effective treatments.

Ferroptosis not only kills tumor cells but also enhances the immune system's ability to recognize and attack them. By understanding the expression profiles of ferroptosis-related genes, we can tailor treatments more precisely, distinguishing between 'cold' and 'hot' tumors to optimize therapy. This approach aligns with the principles of precision medicine, allowing clinicians to navigate treatment plans with greater accuracy and efficacy.

However, challenges remain, including understanding the molecular interactions between ferroptosis and immune responses and integrating these insights into clinical practice. The journey from bench to bedside requires a nuanced understanding of the underlying biological mechanisms and careful consideration of the therapeutic window to balance efficacy with safety. Continued research and collaboration among the scientific and medical communities are essential to fully realize the potential of ferroptosis in CRC therapy. By leveraging the tumor's own vulnerabilities and enhancing the immune response, we edge closer to a new era in cancer treatment where the demise of tumor cells is orchestrated by a synergistic approach combining ferroptosis and immunotherapy.

Author Contributions: Conceptualization, Y.L. and X.C.; methodology, Y.L.; software, Y.L.; validation, Y.L. and X.C.; formal analysis, Y.L.; investigation, Y.L.; resources, Y.L.; data curation, Y.L.; writing—original draft preparation, Y.L.; writing—review and editing, X.C.; visualization, Y.L.; supervision, X.C.; project administration, X.C.; funding acquisition, Y.L. All authors have read and agreed to the published version of the manuscript.

Funding: This research received no external funding.

Conflicts of Interest: The author declares no conflicts of interest.

References

1. Siegel, R.L.; Miller, K.D.; Wagle, N.S.; Jemal, A. Cancer statistics, 2023. *CA Cancer J. Clin.* **2023**, *73*, 17–48. [CrossRef] [PubMed]
2. Galandiuk, S.; Fazio, V.W.; Jagelman, D.G.; Lavery, I.C.; Weakley, F.A.; Petras, R.E.; Badhwar, K.; McGonagle, B.; Eastin, K.; Sutton, T. Villous and tubulovillous adenomas of the colon and rectum: A retrospective review, 1964–1985. *Am. J. Surg.* **1987**, *153*, 41–47. [CrossRef] [PubMed]
3. Sinicrope, F.A. Increasing Incidence of Early-Onset Colorectal Cancer. *N. Engl. J. Med.* **2022**, *386*, 1547–1558. [CrossRef] [PubMed]
4. Aleksandrova, K.; Pischon, T.; Jenab, M.; Bueno-De-Mesquita, H.B.; Fedirko, V.; Norat, T.; Romaguera, D.; Knüppel, S.; Boutron-Ruault, M.-C.; Dossus, L.; et al. Combined impact of healthy lifestyle factors on colorectal cancer: A large European cohort study. *BMC Med.* **2014**, *12*, 1–68. [CrossRef] [PubMed]
5. Ali, R.H.; Marafie, M.J.; Bitar, M.S.; Al-Dousari, F.; Ismael, S.; Bin Haider, H.; Al-Ali, W.; Jacob, S.P.; Al-Mulla, F. Gender-Associated Genomic Differences in Colorectal Cancer: Clinical Insight from Feminization of Male Cancer Cells. *Int. J. Mol. Sci.* **2014**, *15*, 17344–17365. [CrossRef]
6. Caiazza, F.; Ryan, E.J.; Doherty, G.; Winter, D.C.; Sheahan, K. Estrogen Receptors and Their Implications in Colorectal Carcinogenesis. *Front. Oncol.* **2015**, *5*, 19. [CrossRef]
7. Benslimane, Y.; Amalfi, K.; Lapin, S.; Perrino, S.; Brodt, P. Estrogen receptor blockade potentiates immunotherapy for liver metastases by altering the liver immunosuppressive microenvironment. *Cancer Res. Commun.* **2024**, *4*, 1963–1977. [CrossRef]
8. Klein, S.L.; Flanagan, K.L. Sex differences in immune responses. *Nat. Rev. Immunol.* **2016**, *16*, 626–638. [CrossRef]
9. Dhar, R.; Kumar, A.; Karmakar, S. Checkmate with checkpoint inhibitors: New paradigm in immunotherapy. *Asian J. Med. Sci.* **2023**, *14*, 1–2. [CrossRef]
10. Poeta, V.M.; Massara, M.; Capucetti, A.; Bonecchi, R. Chemokines and Chemokine Receptors: New Targets for Cancer Immunotherapy. *Front. Immunol.* **2019**, *10*, 379. [CrossRef]
11. Alard, E.; Butnariu, A.-B.; Grillo, M.; Kirkham, C.; Zinovkin, D.A.; Newnham, L.; Macciochi, J.; Pranjol, Z.I. Advances in Anti-Cancer Immunotherapy: Car-T Cell, Checkpoint Inhibitors, Dendritic Cell Vaccines, and Oncolytic Viruses, and Emerging Cellular and Molecular Targets. *Cancers* **2020**, *12*, 1826. [CrossRef] [PubMed]
12. Chung, C.; Lin, C.; Chen, C.; Hsueh, C.; Chang, Y.; Wang, C.; Chu, P.; Tai, S.; Yang, M. Ferroptosis Signature Shapes the Immune Profiles to Enhance the Response to Immune Checkpoint Inhibitors in Head and Neck Cancer. *Adv. Sci.* **2023**, *10*, 2204514. [CrossRef] [PubMed]
13. Deng, J.; Zhou, M.; Liao, T.; Kuang, W.; Xia, H.; Yin, Z.; Tan, Q.; Li, Y.; Song, S.; Zhou, E.; et al. Targeting Cancer Cell Ferroptosis to Reverse Immune Checkpoint Inhibitor Therapy Resistance. *Front. Cell Dev. Biol.* **2022**, *10*, 818453. [CrossRef]
14. Fan, F.; Liu, P.; Bao, R.; Chen, J.; Zhou, M.; Mo, Z.; Ma, Y.; Liu, H.; Zhou, Y.; Cai, X.; et al. A Dual PI3K/HDAC Inhibitor Induces Immunogenic Ferroptosis to Potentiate Cancer Immune Checkpoint Therapy. *Cancer Res.* **2021**, *81*, 6233–6245. [CrossRef]
15. Chen, W.M.; Deng, J.M.; Zhou, Y. The construction of a novel ferroptosis-related lncRNA model to predict prognosis in colorectal cancer patients. *Medicine* **2023**, *102*, e33114. [CrossRef] [PubMed]
16. Dixon, S.J.; Lemberg, K.M.; Lamprecht, M.R.; Skouta, R.; Zaitsev, E.M.; Gleason, C.E.; Patel, D.N.; Bauer, A.J.; Cantley, A.M.; Yang, W.S.; et al. Ferroptosis: An Iron-Dependent Form of Nonapoptotic Cell Death. *Cell* **2012**, *149*, 1060–1072. [CrossRef]
17. Yang, W.S.; Stockwell, B.R. Ferroptosis: Death by Lipid Peroxidation. *Trends Cell Biol.* **2016**, *26*, 165–176. [CrossRef]

18. Latunde-Dada, G.O. Ferroptosis: Role of lipid peroxidation, iron and ferritinophagy. *Biochim. Biophys. Acta (BBA)-Gen. Subj.* **2017**, *1861*, 1893–1900. [CrossRef]
19. Seibt, T.M.; Proneth, B.; Conrad, M. Role of GPX4 in ferroptosis and its pharmacological implication. *Free Radic. Biol. Med.* **2018**, *133*, 144–152. [CrossRef]
20. Lee, J.-Y.; Nam, M.; Son, H.Y.; Hyun, K.; Jang, S.Y.; Kim, J.W.; Kim, M.W.; Jung, Y.; Jang, E.; Yoon, S.-J.; et al. Polyunsaturated fatty acid biosynthesis pathway determines ferroptosis sensitivity in gastric cancer. *Proc. Natl. Acad. Sci. USA* **2020**, *117*, 32433–32442. [CrossRef]
21. Sha, W.; Hu, F.; Xi, Y.; Chu, Y.; Bu, S. Mechanism of Ferroptosis and Its Role in Type 2 Diabetes Mellitus. *J. Diabetes Res.* **2021**, *2021*, 9999612. [CrossRef] [PubMed]
22. Liu, J.; Kang, R.; Tang, D. Signaling pathways and defense mechanisms of ferroptosis. *FEBS J.* **2021**, *289*, 7038–7050. [CrossRef] [PubMed]
23. Ursini, F.; Maiorino, M. Lipid peroxidation and ferroptosis: The role of GSH and GPx. *Free Radic. Biol. Med.* **2020**, *152*, 175–185. [CrossRef]
24. Doll, S.; Freitas, F.P.; Shah, R.; Aldrovandi, M.; da Silva, M.C.; Ingold, I.; Grocin, A.G.; da Silva, T.N.X.; Panzilius, E.; Scheel, C.H.; et al. FSP1 is a glutathione-independent ferroptosis suppressor. *Nature* **2019**, *575*, 693–698. [CrossRef]
25. Bersuker, K.; Hendricks, J.M.; Li, Z.; Magtanong, L.; Ford, B.; Tang, P.H.; Roberts, M.A.; Tong, B.; Maimone, T.J.; Zoncu, R.; et al. The CoQ oxidoreductase FSP1 acts parallel to GPX4 to inhibit ferroptosis. *Nature* **2019**, *575*, 688–692. [CrossRef]
26. Zimmer, M.; Lamb, J.; Ebert, B.L.; Lynch, M.; Neil, C.; Schmidt, E.; Golub, T.R.; Iliopoulos, O. The Connectivity Map Links Iron Regulatory Protein-1–Mediated Inhibition of Hypoxia-Inducible Factor-2a Translation to the Anti-inflammatory 15-deoxy-Δ12,14-Prostaglandin J2. *Cancer Res.* **2010**, *70*, 3071–3079. [CrossRef]
27. Lou, Y.; McDonald, P.C.; Oloumi, A.; Chia, S.; Ostlund, C.; Ahmadi, A.; Kyle, A.; Keller, U.A.D.; Leung, S.; Huntsman, D.; et al. Targeting Tumor Hypoxia: Suppression of Breast Tumor Growth and Metastasis by Novel Carbonic Anhydrase IX Inhibitors. *Cancer Res.* **2011**, *71*, 3364–3376. [CrossRef]
28. Venkateswaran, G.; McDonald, P.C.; Chafe, S.C.; Brown, W.S.; Gerbec, Z.J.; Awrey, S.J.; Parker, S.J.; Dedhar, S. A Carbonic Anhydrase IX/SLC1A5 Axis Regulates Glutamine Metabolism Dependent Ferroptosis in Hypoxic Tumor Cells. *Mol. Cancer Ther.* **2023**, *22*, 1228–1242. [CrossRef]
29. Koppula, P.; Zhuang, L.; Gan, B. Cystine transporter SLC7A11/xCT in cancer: Ferroptosis, nutrient dependency, and cancer therapy. *Protein Cell* **2020**, *12*, 599–620. [CrossRef]
30. Fuhrmann, D.C.; Mondorf, A.; Beifuß, J.; Jung, M.; Brüne, B. Hypoxia inhibits ferritinophagy, increases mitochondrial ferritin, and protects from ferroptosis. *Redox Biol.* **2020**, *36*, 101670. [CrossRef]
31. Li, K.; Xu, K.; He, Y.; Yang, Y.; Tan, M.; Mao, Y.; Zou, Y.; Feng, Q.; Luo, Z.; Cai, K. Oxygen Self-Generating Nanoreactor Mediated Ferroptosis Activation and Immunotherapy in Triple-Negative Breast Cancer. *ACS Nano* **2023**, *17*, 4667–4687. [CrossRef] [PubMed]
32. Fu, J.; Li, T.; Yang, Y.; Jiang, L.; Wang, W.; Fu, L.; Zhu, Y.; Hao, Y. Activatable nanomedicine for overcoming hypoxia-induced resistance to chemotherapy and inhibiting tumor growth by inducing collaborative apoptosis and ferroptosis in solid tumors. *Biomaterials* **2020**, *268*, 120537. [CrossRef] [PubMed]
33. Zhao, Y.; Li, M.; Yao, X.; Fei, Y.; Lin, Z.; Li, Z.; Cai, K.; Zhao, Y.; Luo, Z. HCAR1/MCT1 Regulates Tumor Ferroptosis through the Lactate-Mediated AMPK-SCD1 Activity and Its Therapeutic Implications. *Cell Rep.* **2020**, *33*, 108487. [CrossRef] [PubMed]
34. Ye, F.; Chai, W.; Xie, M.; Yang, M.; Yu, Y.; Cao, L.; Yang, L. HMGB1 regulates erastin-induced ferroptosis via RAS-JNK/p38 signaling in HL-60/NRASQ61L cells. *Am. J. Cancer Res.* **2019**, *9*, 730–739.
35. Xu, Q.; Zhou, L.; Yang, G.; Meng, F.; Wan, Y.; Wang, L.; Zhang, L. CircIL4R facilitates the tumorigenesis and inhibits ferroptosis in hepatocellular carcinoma by regulating the miR-541-3p/GPX4 axis. *Cell Biol. Int.* **2020**, *44*, 2344–2356. [CrossRef]
36. Wang, W.; Green, M.; Choi, J.E.; Gijón, M.; Kennedy, P.D.; Johnson, J.K.; Liao, P.; Lang, X.; Kryczek, I.; Sell, A.; et al. CD8+ T cells regulate tumour ferroptosis during cancer immunotherapy. *Nature* **2019**, *569*, 270–274. [CrossRef]
37. Wang, W.; Green, M.; Choi, J.E.; Gijón, M.; Kennedy, P.; Liao, P.; Lang, X.; Kryczek, I.; Sell, A.; Johnson, J.; et al. CD8+ T cells regulate tumor ferroptosis by targeting the system xc^- during cancer immunotherapy. *J. Immunol.* **2019**, *202*, 137.11. [CrossRef]
38. Lei, G.; Zhang, Y.; Koppula, P.; Liu, X.; Zhang, J.; Lin, S.H.; Ajani, J.A.; Xiao, Q.; Liao, Z.; Wang, H.; et al. The role of ferroptosis in ionizing radiation-induced cell death and tumor suppression. *Cell Res.* **2020**, *30*, 146–162. [CrossRef]
39. Liu, Y.; Niu, R.; Deng, R.; Song, S.; Wang, Y.; Zhang, H. Multi-enzyme Co-expressed Dual-Atom Nanozymes Induce Cascade Immunogenic Ferroptosis via Activating Interferon-γ and Targeting Arachidonic Acid Metabolism. *J. Am. Chem. Soc.* **2023**, *145*, 8965–8978. [CrossRef]
40. Min, K.-J.; Um, H.J.; Cho, K.-H.; Kwon, T.K. Curcumin inhibits oxLDL-induced CD36 expression and foam cell formation through the inhibition of p38 MAPK phosphorylation. *Food Chem. Toxicol.* **2013**, *58*, 77–85. [CrossRef]
41. Ma, X.; Xiao, L.; Liu, L.; Ye, L.; Su, P.; Bi, E.; Wang, Q.; Yang, M.; Qian, J.; Yi, Q. CD36-mediated ferroptosis dampens intratumoral CD8+ T cell effector function and impairs their antitumor ability. *Cell Metab.* **2021**, *33*, 1001–1012.e5. [CrossRef]
42. Xu, S.; Chaudhary, O.; Rodríguez-Morales, P.; Sun, X.; Chen, D.; Zappasodi, R.; Xu, Z.; Pinto, A.F.; Williams, A.; Schulze, I.; et al. Uptake of oxidized lipids by the scavenger receptor CD36 promotes lipid peroxidation and dysfunction in CD8+ T cells in tumors. *Immunity* **2021**, *54*, 1561–1577.e7. [CrossRef] [PubMed]
43. Huang, Y.; Lin, J.; Xiong, Y.; Chen, J.; Du, X.; Liu, Q.; Liu, T. Superparamagnetic Iron Oxide Nanoparticles Induce Ferroptosis of Human Ovarian Cancer Stem Cells by Weakening Cellular Autophagy. *J. Biomed. Nanotechnol.* **2020**, *16*, 1612–1622. [CrossRef]

44. Kepp, O.; Kroemer, G. Is ferroptosis immunogenic? The devil is in the details! *Oncoimmunology* **2022**, *11*, 2127273. [CrossRef] [PubMed]
45. Han, L.; Bai, L.; Qu, C.; Dai, E.; Liu, J.; Kang, R.; Zhou, D.; Tang, D.; Zhao, Y. PPARG-mediated ferroptosis in dendritic cells limits antitumor immunity. *Biochem. Biophys. Res. Commun.* **2021**, *576*, 33–39. [CrossRef]
46. Chen, X.; Li, J.; Kang, R.; Klionsky, D.J.; Tang, D. Ferroptosis: Machinery and regulation. *Autophagy* **2020**, *17*, 2054–2081. [CrossRef]
47. He, J.; Ding, H.; Li, H.; Pan, Z.; Chen, Q. Intra-Tumoral Expression of SLC7A11 Is Associated with Immune Microenvironment, Drug Resistance, and Prognosis in Cancers: A Pan-Cancer Analysis. *Front. Genet.* **2021**, *12*, 770857. [CrossRef] [PubMed]
48. Tang, B.; Zhu, J.; Li, J.; Fan, K.; Gao, Y.; Cheng, S.; Kong, C.; Zheng, L.; Wu, F.; Weng, Q.; et al. The ferroptosis and iron-metabolism signature robustly predicts clinical diagnosis, prognosis and immune microenvironment for hepatocellular carcinoma. *Cell Commun. Signal.* **2020**, *18*, 174. [CrossRef]
49. Jiang, X.; Stockwell, B.R.; Conrad, M. Ferroptosis: Mechanisms, biology and role in disease. *Nat. Rev. Mol. Cell Biol.* **2021**, *22*, 266–282. [CrossRef]
50. Dahal, S.; Yurkovich, J.T.; Xu, H.; Palsson, B.O.; Yang, L. Synthesizing Systems Biology Knowledge from Omics Using Genome-Scale Models. *Proteomics* **2020**, *20*, e1900282. [CrossRef]
51. Buetti-Dinh, A.; Herold, M.; Christel, S.; El Hajjami, M.; Delogu, F.; Ilie, O.; Bellenberg, S.; Wilmes, P.; Poetsch, A.; Sand, W.; et al. Reverse engineering directed gene regulatory networks from transcriptomics and proteomics data of biomining bacterial communities with approximate Bayesian computation and steady-state signalling simulations. *BMC Bioinform.* **2020**, *21*, 23. [CrossRef] [PubMed]
52. Misra, B.B.; Langefeld, C.D.; Olivier, M.; Cox, L.A. Integrated Omics: Tools, Advances, and Future Approaches. *J. Mol. Endocrinol.* **2019**, *62*, R21–R45. [CrossRef] [PubMed]
53. Henao, J.D.; Lauber, M.; Azevedo, M.; Grekova, A.; Theis, F.; List, M.; Ogris, C.; Schubert, B. Multi-omics regulatory network inference in the presence of missing data. *Briefings Bioinform.* **2023**, *24*, bbad309. [CrossRef]
54. Wen, Q.; Liu, J.; Kang, R.; Zhou, B.; Tang, D. The release and activity of HMGB1 in ferroptosis. *Biochem. Biophys. Res. Commun.* **2019**, *510*, 278–283. [CrossRef] [PubMed]
55. Yi, C.; Wu, S.; Duan, Q.; Liu, L.; Li, L.; Luo, Y.; Wang, A. Ferroptosis-dependent breast cancer cell-derived exosomes inhibit migration and invasion of breast cancer cells by suppressing M2 macrophage polarization. *PeerJ* **2023**, *11*, e15060. [CrossRef] [PubMed]
56. Efimova, I.; Catanzaro, E.; Van der Meeren, L.; Turubanova, V.D.; Hammad, H.; Mishchenko, T.A.; Vedunova, M.V.; Fimognari, C.; Bachert, C.; Coppieters, F.; et al. Vaccination with early ferroptotic cancer cells induces efficient antitumor immunity. *J. Immunother. Cancer* **2020**, *8*, e001369. [CrossRef]
57. Gu, X.; Liu, Y.; Dai, X.; Yang, Y.-G.; Zhang, X. Deciphering the potential roles of ferroptosis in regulating tumor immunity and tumor immunotherapy. *Front. Immunol.* **2023**, *14*, 1137107. [CrossRef]
58. Li, J.; Liu, J.; Zhou, Z.; Wu, R.; Chen, X.; Yu, C.; Stockwell, B.; Kroemer, G.; Kang, R.; Tang, D. Tumor-specific GPX4 degradation enhances ferroptosis-initiated antitumor immune response in mouse models of pancreatic cancer. *Sci. Transl. Med.* **2023**, *15*, eadg3049. [CrossRef]
59. Ghoochani, A.; Hsu, E.-C.; Aslan, M.; Rice, M.A.; Nguyen, H.M.; Brooks, J.D.; Corey, E.; Paulmurugan, R.; Stoyanova, T. Ferroptosis Inducers Are a Novel Therapeutic Approach for Advanced Prostate Cancer. *Cancer Res.* **2021**, *81*, 1583–1594. [CrossRef]
60. Zhang, X.; Guo, Y.; Li, H.; Han, L. FIN56, a novel ferroptosis inducer, triggers lysosomal membrane permeabilization in a TFEB-dependent manner in glioblastoma. *J. Cancer* **2021**, *12*, 6610–6619. [CrossRef]
61. Liu, Y.; Huang, P.; Li, Z.; Xu, C.; Wang, H.; Jia, B.; Gong, A.; Xu, M. Vitamin C Sensitizes Pancreatic Cancer Cells to Erastin-Induced Ferroptosis by Activating the AMPK/Nrf2/HMOX1 Pathway. *Oxid. Med. Cell. Longev.* **2022**, *2022*, 5361241. [CrossRef] [PubMed]
62. Zhao, Y.; Lian, J.; Lan, Z.; Zou, K.; Wang, W.; Yu, G. Ferroptosis promotes anti-tumor immune response by inducing immunogenic exposure in HNSCC. *Oral Dis.* **2021**, *29*, 933–941. [CrossRef]
63. Lee, N.; Carlisle, A.E.; Peppers, A.; Park, S.J.; Doshi, M.B.; Spears, M.E.; Kim, D. xCT-Driven Expression of GPX4 Determines Sensitivity of Breast Cancer Cells to Ferroptosis Inducers. *Antioxidants* **2021**, *10*, 317. [CrossRef]
64. Lizardo, D.Y.; Kuang, C.; Hao, S.; Yu, J.; Huang, Y.; Zhang, L. Immunotherapy efficacy on mismatch repair-deficient colorectal cancer: From bench to bedside. *Biochim. Biophys. Acta (BBA)-Rev. Cancer* **2020**, *1874*, 188447. [CrossRef]
65. Ganesh, K.; Stadler, Z.K.; Cercek, A.; Mendelsohn, R.B.; Shia, J.; Segal, N.H.; Diaz, L.A., Jr. Immunotherapy in colorectal cancer: Rationale, challenges and potential. *Nat. Rev. Gastroenterol. Hepatol.* **2019**, *16*, 361–375. [CrossRef]
66. Ding, K.; Mou, P.; Wang, Z.; Liu, S.; Liu, J.; Lu, H.; Yu, G. The next bastion to be conquered in immunotherapy: Microsatellite stable colorectal cancer. *Front. Immunol.* **2023**, *14*, 1298524. [CrossRef] [PubMed]
67. Mulet-Margalef, N.; Linares, J.; Badia-Ramentol, J.; Jimeno, M.; Monte, C.S.; Mozo, J.L.M.; Calon, A. Challenges and Therapeutic Opportunities in the dMMR/MSI-H Colorectal Cancer Landscape. *Cancers* **2023**, *15*, 1022. [CrossRef]
68. Fan, A.; Wang, B.; Wang, X.; Nie, Y.; Fan, D.; Zhao, X.; Lu, Y. Immunotherapy in colorectal cancer: Current achievements and future perspective. *Int. J. Biol. Sci.* **2021**, *17*, 3837–3849. [CrossRef] [PubMed]
69. Ciardiello, D.; Vitiello, P.P.; Cardone, C.; Martini, G.; Troiani, T.; Martinelli, E.; Ciardiello, F. Immunotherapy of colorectal cancer: Challenges for therapeutic efficacy. *Cancer Treat. Rev.* **2019**, *76*, 22–32. [CrossRef] [PubMed]

70. Zhang, J.; Yang, L.; Kong, F.; Wu, D.; Hu, B.; Yang, J.; He, J.; Liu, L. Case report: A combined immunotherapy strategy as a promising therapy for MSI-H colorectal carcinomas with multiple HPD risk factors. *Front. Med.* **2023**, *10*, 1051034. [CrossRef]
71. Li, D.-D.; Tang, Y.-L.; Wang, X. Challenges and exploration for immunotherapies targeting cold colorectal cancer. *World J. Gastrointest. Oncol.* **2023**, *15*, 55–68. [CrossRef] [PubMed]
72. Thibaudin, M.; Ghiringhelli, F. Chemoimmunotherapy triggers immune responses targeting microsatellite stable colorectal cancer. *Oncoimmunology* **2023**, *12*, 2257098. [CrossRef]
73. Ros, J.; Balconi, F.; Baraibar, I.; Gonzalez, N.S.; Salva, F.; Tabernero, J.; Elez, E. Advances in immune checkpoint inhibitor combination strategies for microsatellite stable colorectal cancer. *Front. Oncol.* **2023**, *13*, 1112276. [CrossRef] [PubMed]
74. Yu, G.; Wang, W.; He, X.; Xu, J.; Xu, R.; Wan, T.; Wu, Y. Synergistic Therapeutic Effects of Low Dose Decitabine and NY-ESO-1 Specific TCR-T Cells for the Colorectal Cancer With Microsatellite Stability. *Front. Oncol.* **2022**, *12*, 895103. [CrossRef] [PubMed]
75. Makaremi, S.; Asadzadeh, Z.; Hemmat, N.; Baghbanzadeh, A.; Sgambato, A.; Ghorbaninezhad, F.; Safarpour, H.; Argentiero, A.; Brunetti, O.; Bernardini, R.; et al. Immune Checkpoint Inhibitors in Colorectal Cancer: Challenges and Future Prospects. *Biomedicines* **2021**, *9*, 1075. [CrossRef]
76. Shi, G.; Yang, Q.; Zhang, Y.; Jiang, Q.; Lin, Y.; Yang, S.; Wang, H.; Cheng, L.; Zhang, X.; Li, Y.; et al. Modulating the Tumor Microenvironment via Oncolytic Viruses and CSF-1R Inhibition Synergistically Enhances Anti-PD-1 Immunotherapy. *Mol. Ther.* **2019**, *27*, 244–260. [CrossRef]
77. Luo, W.; Dai, W.; Li, Q.; Mo, S.; Han, L.; Xiao, X.; Gu, R.; Xiang, W.; Ye, L.; Wang, R.; et al. Ferroptosis-associated molecular classification characterized by distinct tumor microenvironment profiles in colorectal cancer. *Int. J. Biol. Sci.* **2022**, *18*, 1773–1794. [CrossRef]
78. Yang, C.; Huang, S.; Cao, F.; Zheng, Y. Role of ferroptosis-related genes in prognostic prediction and tumor immune microenvironment in colorectal carcinoma. *PeerJ* **2021**, *9*, e11745. [CrossRef]
79. Lv, Y.; Tang, W.T.; Xu, Y.Q.; Chang, W.J.; Zhang, Z.Y.; Lin, Q.; Ji, M.L.; Feng, Q.Y.; He, G.D.; Xu, J.M. Apolipoprotein L3 enhances CD8+ T cell antitumor immunity of colorectal cancer by promoting LDHA-mediated ferroptosis. *Int. J. Biol. Sci.* **2023**, *19*, 1284–1298. [CrossRef]
80. Shao, Y.; Jia, H.; Huang, L.; Li, S.; Wang, C.; Aikemu, B.; Yang, G.; Hong, H.; Yang, X.; Zhang, S.; et al. An Original Ferroptosis-Related Gene Signature Effectively Predicts the Prognosis and Clinical Status for Colorectal Cancer Patients. *Front. Oncol.* **2021**, *11*, 711776. [CrossRef]
81. Chen, W.; Chen, Y.; Liu, L.; Wu, Y.; Fu, P.; Cao, Y.; Xiong, J.; Tu, Y.; Li, Z.; Liu, Y.; et al. Comprehensive Analysis of Immune Infiltrates of Ferroptosis-Related Long Noncoding RNA and Prediction of Colon Cancer Patient Prognoses. *J. Immunol. Res.* **2022**, *2022*, 9480628. [CrossRef]
82. Feng, S.; Rao, Z.; Zhang, J.; She, X.; Chen, Y.; Wan, K.; Li, H.; Zhao, C.; Feng, Y.; Wang, G.; et al. Inhibition of CARM1-Mediated Methylation of ACSL4 Promotes Ferroptosis in Colorectal Cancer. *Adv. Sci.* **2023**, *10*, e2303484. [CrossRef] [PubMed]
83. Long, S.; Peng, F.; Song, B.; Wang, L.; Chen, J.; Shang, B. Heat Shock Protein Beta 1 is a Prognostic Biomarker and Correlated with Immune Infiltrates in Hepatocellular Carcinoma. *Int. J. Gen. Med.* **2021**, *14*, 5483–5492. [CrossRef]
84. Yang, Y.; Zhu, T.; Wang, X.; Xiong, F.; Hu, Z.; Qiao, X.; Yuan, X.; Wang, D. ACSL3 and ACSL4, Distinct Roles in Ferroptosis and Cancers. *Cancers* **2022**, *14*, 5896. [CrossRef] [PubMed]
85. Tang, B.; Xu, W.; Wang, Y.; Zhu, J.; Wang, H.; Tu, J.; Weng, Q.; Kong, C.; Yang, Y.; Qiu, R.; et al. Identification of critical ferroptosis regulators in lung adenocarcinoma that RRM2 facilitates tumor immune infiltration by inhibiting ferroptotic death. *Clin. Immunol.* **2021**, *232*, 108872. [CrossRef] [PubMed]
86. Li, Q.; Su, R.; Bao, X.; Cao, K.; Du, Y.; Wang, N.; Wang, J.; Xing, F.; Yan, F.; Huang, K.; et al. Glycyrrhetinic acid nanoparticles combined with ferrotherapy for improved cancer immunotherapy. *Acta Biomater.* **2022**, *144*, 109–120. [CrossRef] [PubMed]
87. Cai, H.; Ren, Y.; Chen, S.; Wang, Y.; Chu, L. Ferroptosis and tumor immunotherapy: A promising combination therapy for tumors. *Front. Oncol.* **2023**, *13*, 1119369. [CrossRef]
88. Chen, Q.; Ma, X.; Xie, L.; Chen, W.; Xu, Z.; Song, E.; Zhu, X.; Song, Y. Iron-based nanoparticles for MR imaging-guided ferroptosis in combination with photodynamic therapy to enhance cancer treatment. *Nanoscale* **2021**, *13*, 4855–4870. [CrossRef]
89. Wang, W.; Ling, Y.; Zhong, Y.; Li, Z.; Tan, C.; Mao, Z. Ferroptosis-Enhanced Cancer Immunity by a Ferrocene-Appended Iridium(III) Diphosphine Complex. *Angew. Chem. Int. Ed.* **2021**, *61*, e202115247. [CrossRef]
90. Wang, X.; Xu, Y.; Dai, L.; Yu, Z.; Wang, M.; Chan, S.; Sun, R.; Han, Q.; Chen, J.; Zuo, X.; et al. A novel oxidative stress- and ferroptosis-related gene prognostic signature for distinguishing cold and hot tumors in colorectal cancer. *Front. Immunol.* **2022**, *13*, 1043738. [CrossRef]
91. Lv, Y.; Zheng, P.; Mao, Y.; Xu, Y.; Chang, W.; Lin, Q.; Ji, M.; Ye, L.; Tang, W.; Xu, J. Intratumor APOL3 delineates a distinctive immunogenic ferroptosis subset with prognosis prediction in colorectal cancer. *Cancer Sci.* **2023**, *115*, 257–269. [CrossRef] [PubMed]
92. Cheng, X.; Zhao, F.; Ke, B.; Chen, D.; Liu, F. Harnessing Ferroptosis to Overcome Drug Resistance in Colorectal Cancer: Promising Therapeutic Approaches. *Cancers* **2023**, *15*, 5209. [CrossRef] [PubMed]
93. Yun, T.; Liu, Z.; Wang, J.; Wang, R.; Zhu, L.; Zhu, Z.; Wang, X. Microenvironment immune response induced by tumor ferroptosis—The application of nanomedicine. *Front. Oncol.* **2022**, *12*, 1019654. [CrossRef] [PubMed]
94. Wu, Y.; Yu, C.; Luo, M.; Cen, C.; Qiu, J.; Zhang, S.; Hu, K. Ferroptosis in Cancer Treatment: Another Way to Rome. *Front. Oncol.* **2020**, *10*, 571127. [CrossRef]

95. Li, Y.; Wei, C.; Yan, J.; Li, F.; Chen, B.; Sun, Y.; Luo, K.; He, B.; Liang, Y. The application of nanoparticles based on ferroptosis in cancer therapy. *J. Mater. Chem. B* **2023**, *12*, 413–435. [CrossRef]
96. Duan, X.; Chan, C.; Guo, N.; Han, W.; Weichselbaum, R.R.; Lin, W. Photodynamic Therapy Mediated by Nontoxic Core–Shell Nanoparticles Synergizes with Immune Checkpoint Blockade To Elicit Antitumor Immunity and Antimetastatic Effect on Breast Cancer. *J. Am. Chem. Soc.* **2016**, *138*, 16686–16695. [CrossRef]
97. Wang, D.; Zhou, J.; Chen, R.; Shi, R.; Xia, G.; Zhou, S.; Liu, Z.; Zhang, N.; Wang, H.; Guo, Z.; et al. Magnetically guided delivery of DHA and Fe ions for enhanced cancer therapy based on pH-responsive degradation of DHA-loaded Fe_3O_4 @C@MIL-100(Fe) nanoparticles. *Biomaterials* **2016**, *107*, 88–101. [CrossRef]
98. Wang, K.; Zhang, Z.; Tsai, H.-I.; Liu, Y.; Gao, J.; Wang, M.; Song, L.; Cao, X.; Xu, Z.; Chen, H.; et al. Branched-chain amino acid aminotransferase 2 regulates ferroptotic cell death in cancer cells. *Cell Death Differ.* **2020**, *28*, 1222–1236. [CrossRef]
99. Hu, Q.; Wei, W.; Wu, D.; Huang, F.; Li, M.; Li, W.; Yin, J.; Peng, Y.; Lu, Y.; Zhao, Q.; et al. Blockade of GCH1/BH4 Axis Activates Ferritinophagy to Mitigate the Resistance of Colorectal Cancer to Erastin-Induced Ferroptosis. *Front. Cell Dev. Biol.* **2022**, *10*, 810327. [CrossRef]
100. Liu, S.; Zhao, X.; Shui, S.; Wang, B.; Cui, Y.; Dong, S.; Yuwen, T.; Liu, G. PDTAC: Targeted Photodegradation of GPX4 Triggers Ferroptosis and Potent Antitumor Immunity. *J. Med. Chem.* **2022**, *65*, 12176–12187. [CrossRef]
101. Feng, Y.; Dai, Y. APOL3-LDHA axis related immunity activation and cancer ferroptosis induction. *Int. J. Biol. Sci.* **2023**, *19*, 1401–1402. [CrossRef]
102. Dennie, T.W.; Fleming, R.A.; Bowen, C.J.; Dar, M.M.; Alberti, D.; Oliver, K.; Loconte, N.; Mulkerin, D.; Holen, K.D. A Phase I Study of Capecitabine, Oxaliplatin, and Lapatinib in Metastatic or Advanced Solid Tumors. *Clin. Colorectal Cancer* **2011**, *10*, 57–62. [CrossRef]
103. von Moos, R.; Koeberle, D.; Schacher, S.; Hayoz, S.; Winterhalder, R.C.; Roth, A.; Bodoky, G.; Samaras, P.; Berger, M.D.; Rauch, D.; et al. Neoadjuvant radiotherapy combined with capecitabine and sorafenib in patients with advanced *KRAS*-mutated rectal cancer: A phase I/II trial (SAKK 41/08). *Eur. J. Cancer* **2018**, *89*, 82–89. [CrossRef] [PubMed]
104. Samalin, E.; Bouché, O.; Thézenas, S.; Francois, E.; Adenis, A.; Bennouna, J.; Taieb, J.; Desseigne, F.; Seitz, J.F.; Conroy, T.; et al. Sorafenib and irinotecan (NEXIRI) as second- or later-line treatment for patients with metastatic colorectal cancer and KRAS-mutated tumours: A multicentre Phase I/II trial. *Br. J. Cancer* **2014**, *110*, 1148–1154. [CrossRef] [PubMed]
105. Samalin, E.; de la Fouchardière, C.; Thézenas, S.; Boige, V.; Senellart, H.; Guimbaud, R.; Taïeb, J.; François, E.; Galais, M.-P.; Lièvre, A.; et al. Sorafenib Plus Irinotecan Combination in Patients With RAS-mutated Metastatic Colorectal Cancer Refractory To Standard Combined Chemotherapies: A Multicenter, Randomized Phase 2 Trial (NEXIRI-2/PRODIGE 27). *Clin. Colorectal Cancer* **2020**, *19*, 301–310.e1. [CrossRef] [PubMed]
106. Xie, H.; Lafky, J.M.; Morlan, B.W.; Stella, P.J.; Dakhil, S.R.; Gross, G.G.; Loui, W.S.; Hubbard, J.M.; Alberts, S.R.; Grothey, A. Dual VEGF inhibition with sorafenib and bevacizumab as salvage therapy in metastatic colorectal cancer: Results of the phase II North Central Cancer Treatment Group study N054C (Alliance). *Ther. Adv. Med. Oncol.* **2020**, *12*, 1758835920910913. [CrossRef] [PubMed]
107. Hendlisz, A.; Deleporte, A.; Delaunoit, T.; Maréchal, R.; Peeters, M.; Holbrechts, S.; Eynde, M.V.D.; Houbiers, G.; Filleul, B.; Van Laethem, J.-L.; et al. The Prognostic Significance of Metabolic Response Heterogeneity in Metastatic Colorectal Cancer. *PLoS ONE* **2015**, *10*, e0138341. [CrossRef]
108. Kim, R.; Prithviraj, G.K.; Shridhar, R.; Hoffe, S.E.; Jiang, K.; Zhao, X.; Chen, D.-T.; Almhanna, K.; Strosberg, J.; Campos, T.; et al. Phase I study of pre-operative continuous 5-FU and sorafenib with external radiation therapy in locally advanced rectal adenocarcinoma. *Radiother. Oncol.* **2016**, *118*, 382–386. [CrossRef]
109. Zheng, J.; Sato, M.; Mishima, E.; Sato, H.; Proneth, B.; Conrad, M. Sorafenib fails to trigger ferroptosis across a wide range of cancer cell lines. *Cell Death Dis.* **2021**, *12*, 698. [CrossRef]
110. Leung, W.-H.; Shih, J.-W.; Chen, J.-S.; Mokgautsi, N.; Wei, P.-L.; Huang, Y.-J. Preclinical Identification of Sulfasalazine's Therapeutic Potential for Suppressing Colorectal Cancer Stemness and Metastasis through Targeting KRAS/MMP7/CD44 Signaling. *Biomedicines* **2022**, *10*, 377. [CrossRef]
111. Yang, J.; Zhou, Y.; Xie, S.; Wang, J.; Li, Z.; Chen, L.; Mao, M.; Chen, C.; Huang, A.; Chen, Y.; et al. Metformin induces Ferroptosis by inhibiting UFMylation of SLC7A11 in breast cancer. *J. Exp. Clin. Cancer Res.* **2021**, *40*, 206. [CrossRef] [PubMed]
112. Cheng, K.; Guo, Q.; Shen, Z.; Yang, W.; Zhou, Y.; Sun, Z.; Yao, X.; Wu, H. Frontiers of ferroptosis research: An analysis from the top 100 most influential articles in the field. *Front. Oncol.* **2022**, *12*, 948389. [CrossRef]
113. Xu, H.; Ye, D.; Ren, M.; Zhang, H.; Bi, F. Ferroptosis in the tumor microenvironment: Perspectives for immunotherapy. *Trends Mol. Med.* **2021**, *27*, 856–867. [CrossRef]
114. Zhu, L.; Meng, D.; Wang, X.; Chen, X. Ferroptosis-Driven Nanotherapeutics to Reverse Drug Resistance in Tumor Microenvironment. *ACS Appl. Bio Mater.* **2022**, *5*, 2481–2506. [CrossRef] [PubMed]
115. Wang, H.; Lin, D.; Yu, Q.; Li, Z.; Lenahan, C.; Dong, Y.; Wei, Q.; Shao, A. A Promising Future of Ferroptosis in Tumor Therapy. *Front. Cell Dev. Biol.* **2021**, *9*, 629150. [CrossRef] [PubMed]
116. Chafe, S.C.; McDonald, P.C.; Saberi, S.; Nemirovsky, O.; Venkateswaran, G.; Burugu, S.; Gao, D.; Delaidelli, A.; Kyle, A.H.; Baker, J.H.E.; et al. Targeting Hypoxia-Induced Carbonic Anhydrase IX Enhances Immune-Checkpoint Blockade Locally and Systemically. *Cancer Immunol. Res.* **2019**, *7*, 1064–1078. [CrossRef]

117. Hou, J.; Wang, B.; Li, J.; Liu, W. Ferroptosis and its role in gastric and colorectal cancers. *Korean J. Physiol. Pharmacol.* **2024**, *28*, 183–196. [CrossRef]
118. Eaton, J.K.; Furst, L.; Ruberto, R.A.; Moosmayer, D.; Hilpmann, A.; Ryan, M.J.; Zimmermann, K.; Cai, L.L.; Niehues, M.; Badock, V.; et al. Selective covalent targeting of GPX4 using masked nitrile-oxide electrophiles. *Nat. Chem. Biol.* **2020**, *16*, 497–506. [CrossRef]
119. Yang, Y.-B.; Zhou, J.-X.; Qiu, S.-H.; He, J.-S.; Pan, J.-H.; Pan, Y.-L. Identification of a Novel Ferroptosis-Related Gene Prediction Model for Clinical Prognosis and Immunotherapy of Colorectal Cancer. *Dis. Mark.* **2021**, *2021*, 4846683. [CrossRef]
120. Yu, B.; Choi, B.; Li, W.; Kim, D.-H. Magnetic field boosted ferroptosis-like cell death and responsive MRI using hybrid vesicles for cancer immunotherapy. *Nat. Commun.* **2020**, *11*, 3637. [CrossRef]
121. Lei, H.; Li, Q.; Pei, Z.; Liu, L.; Yang, N.; Cheng, L. Nonferrous Ferroptosis Inducer Manganese Molybdate Nanoparticles to Enhance Tumor Immunotherapy. *Small* **2023**, *19*, e2303438. [CrossRef] [PubMed]
122. Sun, L.-L.; Linghu, D.-L.; Hung, M.-C. Ferroptosis: A promising target for cancer immunotherapy. *Am. J. Cancer Res.* **2021**, *11*, 5856–5863. [PubMed]
123. Lang, X.; Green, M.D.; Wang, W.; Yu, J.; Choi, J.E.; Jiang, L.; Liao, P.; Zhou, J.; Zhang, Q.; Dow, A.; et al. Radiotherapy and Immunotherapy Promote Tumoral Lipid Oxidation and Ferroptosis via Synergistic Repression of SLC7A11. *Cancer Discov.* **2019**, *9*, 1673–1685. [CrossRef] [PubMed]
124. Zeng, C.; Tang, H.; Chen, H.; Li, M.; Xiong, D. Ferroptosis: A new approach for immunotherapy. *Cell Death Discov.* **2020**, *6*, 122. [CrossRef]
125. Lachaier, E.; Louandre, C.; Godin, C.; Saidak, Z.; Baert, M.; Diouf, M.; Chauffert, B.; Galmiche, A. Sorafenib induces ferroptosis in human cancer cell lines originating from different solid tumors. *Anticancer Res.* **2014**, *34*, 6417–6422. [PubMed]
126. Sato, M.; Kusumi, R.; Hamashima, S.; Kobayashi, S.; Sasaki, S.; Komiyama, Y.; Izumikawa, T.; Conrad, M.; Bannai, S.; Sato, H. The ferroptosis inducer erastin irreversibly inhibits system xc^- and synergizes with cisplatin to increase cisplatin's cytotoxicity in cancer cells. *Sci. Rep.* **2018**, *8*, 968. [CrossRef]

International Journal of *Molecular Sciences*

MDPI

Article

Transferrin-Conjugated Nanostructured Lipid Carriers for Targeting Artemisone to Melanoma Cells

Njoud Altuwaijri [1,*,†] and Eman Atef [2,*]

1 Pharmaceutical Sciences Department, MCPHS University, 179 Longwood Ave, Boston, MA 02115, USA
2 Pharmacy College, West Coast University, 590 N Vermont Ave, Los Angeles, CA 90005, USA
* Correspondence: naltuwaijri@ksu.edu.sa (N.A.); eatef@westcoastuniversity.edu (E.A.)
† Present address: Pharmaceutics Department, College of Pharmacy, King Saud University, Prince Turki Ibn Abdulaziz Al Awwal Rd, Riyadh 12371, Saudi Arabia.

Abstract: We report a successful formulation of Artemisone (ATM) in transferrin (Tf)-conjugated nanostructured lipid carriers (NLCs), achieving nearly a five-times increase in cell toxicity. The escalating cost of new drug discoveries led to the repurposing of approved drugs for new indications. This study incorporated Artemisone, an antimalarial drug, into a nanostructured lipid carrier (NLC) and tested for possible anticancer effects. The aim was to develop NLCs, and transferrin-conjugated NLCs (NLC-Tf) encapsulating Artemisone to enhance its delivery and anticancer activity. NLC formulations were prepared using high-pressure homogenization followed by ultrasonication and were characterized by particle size, zeta potential, and PDI. The conjugation of (Tf) to (NLC) was confirmed using IR, and the anticancer activity was tested using MTS assay. All formulations were in the nanometer size range (140–167 nm) with different zeta potential values. IR spectroscopy confirmed the successful conjugation of transferrin to NLC. Upon testing the formulations on melanoma cell lines using MTS assay, there was a significant decrease in viability and an increase in the encapsulated ATM-Tf toxicity compared to positive control ATM. The NLCs presented a promising potential carrier for delivering ATM to melanoma cells, and further conjugation with Tf significantly improved the ATM cytotoxicity.

Keywords: lipid nanoparticles (LNPs); transferrin; targeted drug delivery; repurposing; nanostructured lipid carriers (NLCs); cancer

check for updates

Citation: Altuwaijri, N.; Atef, E. Transferrin-Conjugated Nanostructured Lipid Carriers for Targeting Artemisone to Melanoma Cells. *Int. J. Mol. Sci.* **2024**, *25*, 9119. https://doi.org/10.3390/ijms25169119

Academic Editor: Kang-Nan Wang

Received: 7 July 2024
Revised: 27 July 2024
Accepted: 20 August 2024
Published: 22 August 2024

1. Introduction

With the increased cost of new drug investigations, some pharmaceutical companies are considering drug repurposing or repositioning. This involves investigating new uses and applications for already approved drugs, thereby shortening development timeframes and reducing total development costs. This was evident in the early stages of COVID-19 when numerous institutions were rigorously investigating the efficacy of already approved drugs against the virus [1–3].

Artemisinins, a class of antimalarial drugs, exhibit potential for repurposing as anticancer medications. The groups' major challenge of use is poor solubility and permeability (BSC IV). Scientists developed three generations of artemisinins to overcome these challenges, including semisynthetic and fully synthetic compounds [4].

Several studies have investigated the mechanism of action of Artemisone (ATM) as a promising anticancer drug for use in various combinations [5,6]. Our focus is to improve the delivery and, thus, the efficacy of Artemisone by incorporating it in a nanostructured lipid carrier conjugated to a targeting ligand. One of the suggested anticancer mechanisms of action of artemisinin is the reaction between the iron in cancer cells and the endoperoxide moiety of artemisinins [5,7–10], cleavage of the endoperoxide bridge releases ROS, leading to a series of processes that kill cancer cells [11].

Artemisone aqueous solubility at pH 7.4 is 89 μg/mL, and its log P at pH 7.4 is 2.49. The low lipophilicity of ATM has contributed to its low cytotoxicity and neurotoxicity in different in vivo and in vitro assays [4,12]. The organic solvent-free first-generation solid lipid nanoparticles (SLNs) were a nanoparticle breakthrough, yet their high crystallinity resulted in a significant drug-loading challenge. The latter challenge was addressed by incorporating liquid lipids in nanostructured lipid carriers (NLCs), resulting in solid matrix imperfections; in other words, a less organized lipid–solid matrix with a higher capacity to incorporate significant amounts of the drug and increased physical stability during storage [13–17]. In addition to these benefits, the lipids and surfactants used in NLCs are FDA-approved. Another benefit is that the NLC preparation is simple and easily scaled up [18,19].

Nanoparticles are frequently used to facilitate the delivery of hydrophobic drugs, improve their pharmacokinetic properties, and subsequently enhance their effect [20,21]. In addition, a major formulation advantage of nanoparticles is their ability to target the anticancer drug to the tumor by the addition of targeting ligands such as transferrin. These nanocarrier features eventually change the biodistribution of the anticancer agent, favoring its accumulation at the tumor site [22–24].

Transferrin receptors are surface homodimers responsible for iron uptake and cell growth regulation. Fe (III) is solubilized by binding to Tf. It forms holo-transferrin (holo-Tf), and then is transported into the cells by TfR [25]. The binding of the holo-Tf to the TfR forms an internalized complex into the cell through endocytosis. Next, the iron is released from Tf due to the low pH, the Tf–TfR complex is recycled back to the surface, and, finally, Tf is released [26,27].

The codelivery of Tf and artemisinins in one system results in the internalization of Tf, iron, and artemisinins into the cell by endosomes in one entity. Once iron is released inside the cell, artemisinin activation starts, producing ROS, resulting in mitochondrial damage, caspase activation, and cell death [28–30].

Scientists have conducted studies using Tf as a targeting ligand with artemisinins and approved their higher efficacy over nontargeted formulations against different cancers [30,31]. In summary, transferrin emerges as a promising targeting ligand when conjugated with NLCs for encapsulating Artemisone against melanoma, aligning with the mechanism of action typical of artemisinins. The novelty of our research lies in utilizing the cancer cells' iron uptake mechanism by transferrin receptors and the endoperoxide moiety of the drug to enhance its uptake and toxicity. Utilizing Tf-conjugated NLCs shows potential as an effective carrier for delivering Artemisone to melanoma cells, enhancing cytotoxicity through synergistic features. This study aimed to develop NLCs and transferrin-conjugated NLCs encapsulating Artemisone to facilitate its delivery and improve its anticancer activity against melanoma.

2. Results

2.1. Preparation and Characterization of NLC Formulation

The characterization of different tested NLC formulations is presented in Tables 1 and 2. Formulation A yielded the smallest particle size with an acceptable polydispersity index and zeta potential. As a result, it was chosen to be used throughout the study. The parameters necessary to evaluate NLC formulations are particle size, zeta potential, polydispersity index, and encapsulation efficiency, and they are presented in Table 3.

Table 1. Characterization of NLC formulations with different surfactants (Tween 80 or Kolliphor).

Formulation	Tween 80 (mg)	Kolliphor (mg)	Particle Size (nm)	Polydispersity Index
* A	200	0	140	0.2
B	50	50	315	0.4
C	25	25	420	0.4

Table 1. *Cont.*

Formulation	Tween 80 (mg)	Kolliphor (mg)	Particle Size (nm)	Polydispersity Index
D	25	0	500	0.5
E	0	25 or 50	**	
F	0	100	>1 µm	**
G	0	200	>1 µm	**

** Very large particle size or polydispersity index that the particle size analyzer cannot measure. * Optimum formulation used in the study. All formulations contained the same amounts of lipids, surfactant, and water (compritol® 200 mg, labrafil® 200 mg, lecithin 50 mg, water 10 mL).

Table 2. Characterization of different NLC formulations prepared using different ratios of solid and liquid lipids.

Formulation	Compritol®:Labrafil®	Particle Size (nm)	Polydispersity Index	Zeta Potential (mV)
1	70:30	168	0.2	+24.9
2	60:40	160	0.2	+20.6
3	50:50	140	0.2	+17.8
4	40:60	150	0.2	+22.3
5	30:70	155	0.2	+20.7

+ A constant weight of lecithin, tween 80, and water were used in the formulations: (lecithin 50 mg, tween 80 200 mg, water 10 mL).

Table 3. Characterization of NLC formulations and encapsulation efficiencies of blank, encapsulated, and conjugated NLCs.

Formulation	Particle Size (nm)	Zeta Potential (mV)	Polydispersity Index (PDI)	Encapsulation Efficiency (EE%)
NLC	140 ± 4	+15 ± 0.4	0.2 ± 0.0	—
NLC-ATM	167 ± 15	+15 ± 2	0.2 ± 0.0	95 ± 5%
NLC-TF	167 ± 10	* −19 ± 7	0.2 ± 0.1	—
NLC-TF-ATM	154 ± 5	−18 ± 2	0.2 ± 0.0	60 ± 9%

* The change in the zeta potential charge could be due to the nanoparticle's coating with Tf, which changed the zeta potential toward a negative value.

2.2. Coupling Efficiency

No variations in coupling efficiency were observed upon varying the ratios of lipid to Tf (95:5, 90:10, 80:20). The ratio of lipid to Tf used in the study was 90:10, which was reported in several studies [32,33]. The percent of Tf coupled to NLC was 30% for NLC-TF and 26% for NLC-TF-ATM. No significant difference was found between the percent of Tf coupled to NLC-TF and NLC-TF-ATM. The statistical differences were tested using a *t*-test.

2.3. IR Spectroscopic Analysis

The IR spectra of the formulations NLC, NLC-TF, and transferrin are presented in Figure 1 and the IR identification key peaks are present in Table 4.

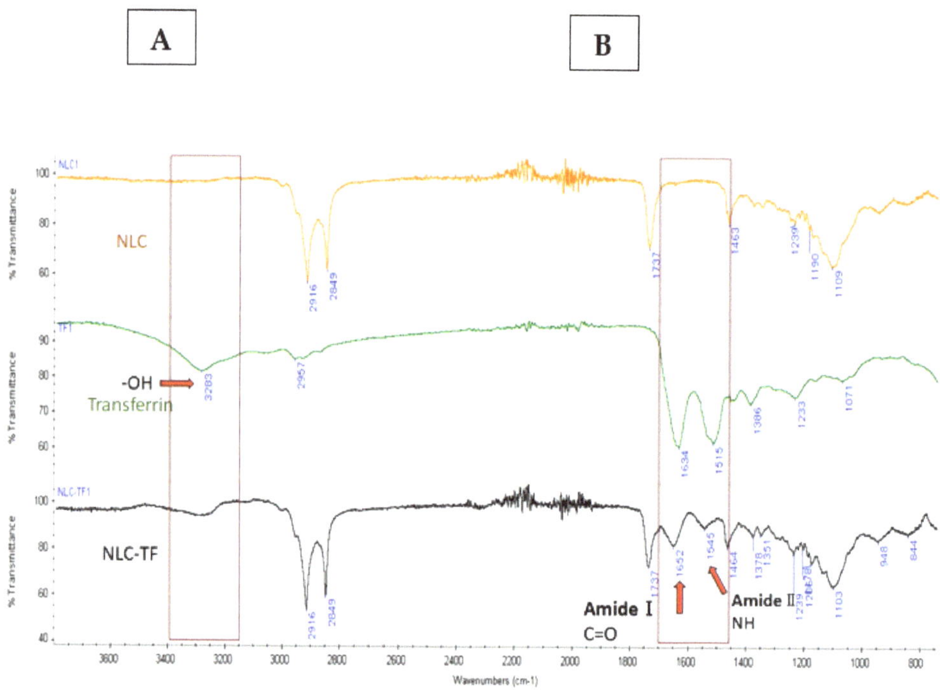

Figure 1. The IR spectra of NLC, TF, and NLC-TF. Panels (**A**,**B**) define the peak range where changes due to conjugation are observed. The amide bond formation in the NLC-TF spectrum confirmed the conjugation of Tf to the NLC. The presence of amide I bond at 1652 cm^{-1} (C=O stretching) and amide II at 1545 cm^{-1} (CN stretching and NH bending), as well as the reduction in the carboxylic peak at 3283 cm^{-1}, confirms a conjugation reaction (n = 3).

Table 4. The identification of the IR key peaks.

Material	Peak Number	Peak Assignment	Monitored Peak Changes to Confirm Conjugation
NLC	2915 cm^{-1} 2849 cm^{-1}	C-H stretching vibration	
	1736 cm^{-1}	carbonyl band -C=O	
	1463 cm^{-1}	C=C stretching vibration	
Transferrin	1634 cm^{-1}	Amide I bond C=O stretching	
	1515 cm^{-1}	Amide II CN stretching and NH bending	
	3282 cm^{-1}	-OH of the carboxylic acid	
NLC-TF	1652 cm^{-1}	Amide I bond C=O stretching	Appearance
	1545 cm^{-1}	Amide II CN stretching and NH bending	Appearance
	3283 cm^{-1}	-OH of the carboxylic acid	Reduction

2.4. In Vitro Cytotoxicity Study

The MTS assay results of ATM, NLC-ATM, and NLC-TF-ATM on SK-MEL-19 after 24 and 48 h are shown in Figure 2.

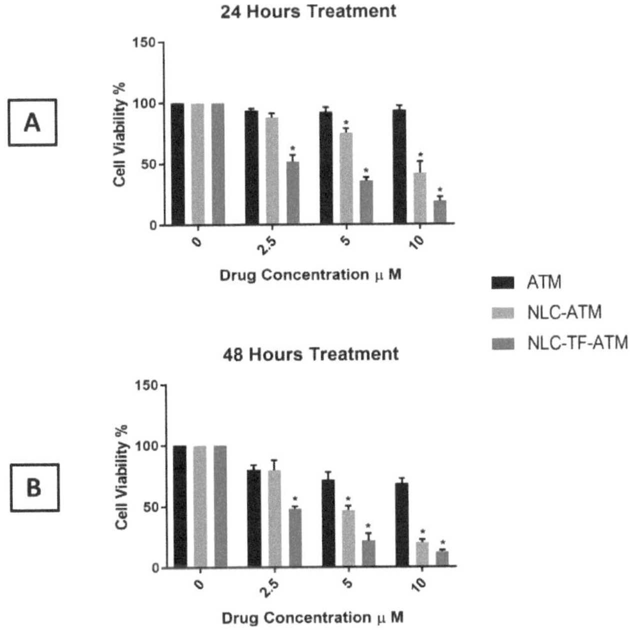

Figure 2. Cell viability of SK-MEL-19 cells after treatment with Artemisone (ATM), NLC-ATM, and NLC-TF-ATM (**A**) after 24 h of treatments; (**B**) after 48 h of treatment. Data are plotted as mean ± SD (n = 3) (*) $p < 0.001$.

The results indicate that the cell viability of SK-MEL-19 decreased significantly after treatment with NLC-ATM and NLC-TF-ATM compared to free ATM using 5 and 10 μM after 24 and 48 h. For the 2.5 μM concentration, the cell viability after treatment with NLC-TF-ATM was significantly reduced compared to NLC-ATM and ATM treatment after 24 and 48 h.

An increase in toxicity by almost five times was observed with the drug upon encapsulation and conjugation with Tf compared to the drug alone at both 24 h and 48 h. The significance of the results was determined using one-way ANOVA followed by Tukey's test.

2.5. Stability Studies

The NLC and NLC-ATM formulations maintained their stability under both 4 °C and room temperature for 30 days, while the NLC-TF and NLC-TF-ATM formulations were only stable when stored at 4 °C. There was a significant increase in the particle size of the NLC-TF and NLC-TF-ATM formulations stored at room temperature after seven days. A reduction in the zeta potential and an increase in the polydispersity index was also observed. Storing the formulations at 4 °C proved to be suitable for maintaining the stability of all formulations.

3. Discussion

We have demonstrated that ATM encapsulation into the NLCs increased cell internalization and improved its anticancer effect by increasing its accumulation and cellular uptake using Tf-conjugated NLCs. ATM has two challenging properties: low solubility and short half-life [12]. The NLCs were used to encapsulate the drug and deliver it to melanoma cells to improve ATM's efficacy and physicochemical properties.

In addition, the Tf conjugation increased cell uptake and, hence, drug toxicity, which could be due to the uptake mechanism of the NLC-TF by TfR on cancer. The abundance of the TfR receptors on the cancer cell's surface and the high iron uptake contributed to the selective cytotoxicity of artemisinins toward cancer cells and not normal cells [28], in addition to other mechanisms that have not yet been investigated.

The homogenization followed by ultrasonication used in this study is the most commonly used preparation method for NLCs [34]. It yields a small particle size formulation with a narrow distribution and is prepared in aqueous media. The particle size of the NLC and NLC-TF formulations in this study was in the nanometer size range below 200 nm, which is recommended in anticancer treatments to utilize the EPR effect [35,36]. After conjugating the NLCs with Tf, the particle's size did not change significantly.

The zeta potential of the NLCs and NLC-ATMs was around +15 mV; however, after conjugation with Tf, it changed to −19 mV. These zeta potential values are expected to result in moderate stability because for the formulations to be stable, a zeta potential > ±30 mV is recommended by some references [37,38].

Our studies confirmed the stability of the formulations for up to 30 days at 4 °C even with low zeta potential. When considering the stability of nanoparticle formulations, the particle size, in addition to the zeta potential, plays an important role [39]. Several published studies have demonstrated stable formulations of nanostructured lipid carriers (NLCs) despite having a low zeta potential. These findings underline the complexity of factors influencing the stability of NLCs [32,40–42].

Highly cationic nanoparticles, with positive potential, are well known to interact strongly with cell membranes, resulting in higher uptake but more toxicity [39]. Our NLC formulations had positive zeta potential, while the NLC-TF possessed a negative charge. The effect of the charge on the interaction with cell membranes and toxicity would need further in vivo investigation.

Another critical parameter for characterizing NLCs is the polydispersity index (PDI), analyzed by dynamic light scattering technique, which indicates the particles' homogeneity and monodispersity. In our study, the PDI values of different formulations were around 0.2 ± 0.1, a value that is commonly acceptable for lipid-based nanocarriers. A PDI below 0.25 is generally regarded as evidence of monodispersity and homogeneity in lipid nanocarrier formulations, as reported by several references [43–45]. Particle size, PDI, and zeta potential are crucial parameters that collectively influence the stability, size distribution, and surface charge of the nanostructure. Controlling these parameters plays an important role in the pharmacokinetics of the nanocarriers and, eventually, their clinical applications [43,46].

A relatively high drug loading of 90%, a known advantage of NLCs, was achieved before conjugation. However, upon conjugation, the encapsulation efficiency dropped to 60%. Similar phenomena were reported by Khajavinia et al. upon conjugating Tf to the NLC-encapsulated etoposide [47]. According to several studies, the drug leaking during the conjugation process may be the cause of the decrease in drug loading following conjugation [32,47].

The FT-IR is used to confirm the chemical reaction between the amine of the SA in NLCs and the carboxylic acid in Tf to form an amide bond. The coupling reaction was conducted using EDC as a coupling agent and NHS as an additive to stabilize the reaction in PBS. EDC activates the carboxyl group by forming an amine-reactive O-acylisourea intermediate. The intermediate is unstable in an aqueous solution and needs to be reacted quickly with the amino group to form an amide bond and release iso-urea byproducts. If the intermediate fails to react with the amine, a hydrolysis reaction will occur, and the carboxyl group will be regenerated. NHS is added to the coupling reaction to stabilize the intermediate and allow the reaction between the carboxylic acid and the amino group (Figure 3). For the amide reaction to succeed, the amination reaction should be faster than the hydrolysis of the activated carboxylic acid intermediate [48–50].

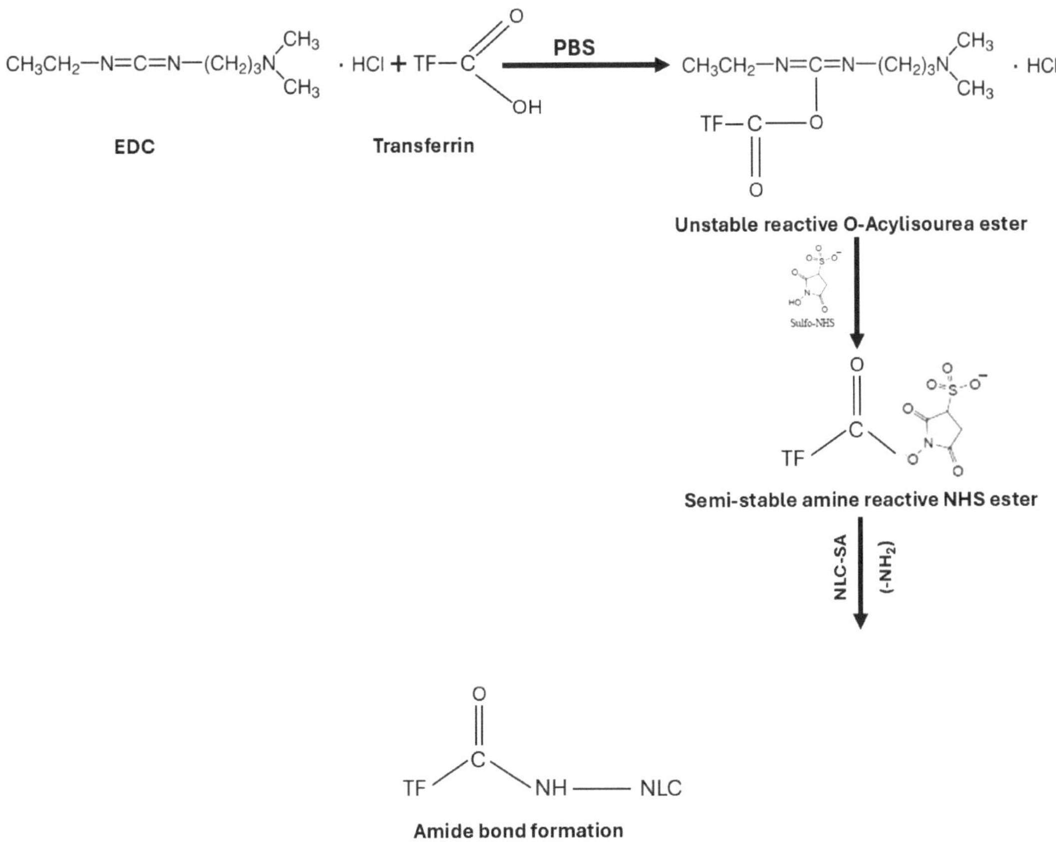

Figure 3. Schematic representation of transferrin conjugation with NLCs using EDC/NHS coupling reaction.

The spectra of NLC-TF in Figure 1 show the two amide peaks at 1652 and 1545 cm^{-1}, which confirm the conjugation of Tf to the SA in the NLC formulations. In reference to peaks 2916 and 2849 cm^{-1}, it is apparent that most NLC peaks represent the peak composition of NLC-TF. As a result, the 1652 and 1545 cm^{-1} are new peaks, which are expected due to the small weight percentage of TF compared to NLC (Table 4). The presence of amide peaks on different nanoparticle formulations was reported in different studies after conjugation with Tf [47,51].

In the cell viability assay experiments, it was observed that encapsulating the ATM into NLC and NLC-Tf significantly increased its cytotoxic effect at lower concentrations. Using anticancer drugs at lower concentrations would reduce some side effects as well as reduce the toxicity on normal cells. We conclude from these experiments that encapsulating the drug into a nanoparticle would increase its internalization into the cells, which could be due to its size and lipid nature. In addition, conjugating the NLCs with Tf caused higher toxicity to the cells, and that could be due to the uptake mechanism of the NLC-TFs by TfR on cancer cells. The hypothesis suggested by Sahoo and Labhasetwar could explain the increase in the cytotoxicity of conjugated NLCs [52]. They stated that the TF-conjugated nanoparticles could have different intracellular sorting pathways after their uptake by TfR, compared to the unconjugated nanoparticles. That specific pathway would increase the accumulation of the conjugated nanoparticles inside the cells and increase therapeutic efficacy. Different studies confirmed the increase in the cytotoxicity after conjugating their nanoparticle formulations with Tf. Leto et al. demonstrated an increase in the cytotoxicity of artemisinin encapsulated into a liposome coupled to Tf by the terminal PEG. There was a reduction in the IC50 from 127 ± 8.5 μM for the free drug to 69 ± 23 μM for the TF-conjugated liposome [30]. In conclusion, different studies on different nanoparticle formulations conjugated with Tf have demonstrated an increase in the anticancer activity of the drugs encapsulated [40,49–52].

4. Materials and Methods

4.1. Preparation of Nanostructured Lipid Carriers (NLCs) Formulations

The nanostructured lipid carriers were prepared by homogenization followed by ultrasonication. Different ratios of solid lipids, liquid lipids, and surfactants were evaluated before finalizing the optimum formulation. The lipid phase composed of Compritol®, Labrafil® (Gattefosse, Paramus, NJ, USA), and lecithin (Letco Medical, Decatur, AL, USA), with or without ATM (Alsachim, Graffenstaden, France), was heated to melt at 80 °C; simultaneously, the aqueous phase containing Tween-80 (Letco Medical, Decatur, AL, USA) was heated to the same temperature. Next, the aqueous phase was dispersed in the lipid phase and mixed using a homogenizer (Ultra-turra T8 homogenizer, Ika, Germany) at 3000 RPM for 3 min. The resulting pre-emulsion was ultrasonicated under ice using a Qsonica ultrasonic processor Q700 (Qsonica, Newtown, CT, USA) with a microtip at an amplitude of 80 for 10 min. The sonication was at a pulsation mode of 20 s ON followed by 10 s OFF, repeated for 10 min. The resulting nanoemulsion was cooled in an ice bath for flash crystallization, leading to the formation of the nanoparticles. The optimum formulation was chosen according to the acceptable particle size, zeta potential, and PDI.

4.2. Transferrin NLC Conjugation

In order to conjugate Tf (Sigma-Aldrich St. Louis, MO, USA) with NLC formulations, stearyl amine (SA) (Sigma-Aldrich St. Louis, MO, USA) was added to the above-prepared lipid phase. The primary terminal amine of SA conjugates with Tf and forms an amide bond. The SA was added in the lipid phase with Compritol®, Labrafil®, and lecithin, followed by the same preparation method. Transferrin was conjugated to the NLC through a modified coupling reaction reported by Gupta et al. [32]. The carboxyl group of Tf was covalently coupled with SA in the NLC formulations in a ratio of 90:10 w/w (lipids: Tf). EDC (TCI America, Portland, OR, USA) was added as a coupling agent, as well as NHS (Sigma-Aldrich St. Louis, MO, USA), which improves the efficiency of the coupling reaction. First, Tf, EDC, and NHS were dissolved in PBS using a magnetic stirrer for 15 min. Next, either NLC-SA or NLC-ATM-SA was added to the mixture and stirred for 24 h. Ultrafiltration (0.22 μm) was used to remove the excess TF, EDC, and NHS. Dialysis of the formulation in distilled water for 1 or 2 h using a dialysis bag was investigated as a method to remove excess Tf, EDC, and NHS. This method was excluded as 15% of the drug was released after 2 h of dialysis.

4.3. Coupling Efficiency

The coupling of Tf to NLC-SA was quantified using Bradford protein assay (Bio-Rad, Hercules, CA, USA). Using ultrafiltration, the unconjugated Tf was separated and analyzed. The average quantity of Tf coupled to the NLC was calculated indirectly by subtracting the unconjugated Tf in the filtrate from the initial amount of Tf added.

4.4. Characterization of Nanostructured Lipid Carrier (NLC) Formulations

4.4.1. Particle Size, Zeta Potential, and Polydispersity

The mean particle size, zeta potential, and polydispersity index of the NLC formulations were determined by dynamic light scattering using a Brookhaven 90 plus size analyzer (Brookhaven instruments, Holtsville, NY, USA) (n = 3).

4.4.2. Encapsulation Efficiency

The encapsulation efficiency was calculated by determining the free drug in the medium using ultrafiltration. Samples from NLC-ATM and NLC-TF-ATM of 500 µL were transferred to a centrifugal filter with a pore size of 0.1 µm for centrifugation. The samples were centrifuged at 1844 g for 12 min to separate the free ATM using 0.1 µm ultrafilters and then analyzed by HPLC. The entrapment efficiency was calculated from the following equation:

$$\text{Encapsulation efficiency} EE(\%) = \frac{(\text{W initial drug} - \text{W unbound drug})}{\text{W initial drug}} \times 100 \quad (1)$$

4.5. FT-IR Analysis

FT-IR confirmed the conjugation of Tf with SA in the NLC formulations.

The spectra of the lyophilized samples of NLC, NLC-ATM, NLC-TF, NLC-TF-ATM, Compritol®, SA, and Tf were collected using Nicolet FT-IR spectrometer iS series from Thermo Scientific (Thermo Scientific, Waltham, MA, USA).

4.6. In Vitro Cytotoxicity of NLC-TF-ATM and Its Components

The cytotoxicity of ATM in DMSO, NLC-ATM, NLC-TF-ATM, Labrafil®, tween80, and lecithin was assessed against skin melanoma cell line (SK-MEL-19) (Sloan Kettering Institute, New York, NY, USA) using MTS assay (Promega, Madison, WI, USA). Two other cell lines, A375 and PC3, were also investigated at the beginning of the study. The cells were seeded at 5×10^3 cells/well in 96-well culture plates for 24 h. Next, the cells were treated with the different NLC formulations as well as the NLC components at different concentrations. The plates were incubated for 24 and 48 h, and then 20 µL of the MTS reagent was added. The plates were incubated for four hours, and the absorbance was measured at 490 nm by a microplate reader (Synergy HT, BioTek, Winooski, VT, USA), where the untreated cells were used as a control with 100% cell viability. The following equation was used to calculate cell viability:

$$\text{Cell viability\%} = \frac{\text{Mean absorbance of the sample}}{\text{absorbance of the control}} \times 100 \quad (2)$$

4.7. Lyophilization

Lyophilization of different NLC samples was performed using a Labconco FreeZone triad cascade benchtop freeze-dryer (Labconco, Kansas City, MO, USA). The samples were prefrozen at −80 °C for 24 h and then freeze-dried at −25 °C for 48 h.

4.8. Stability Studies

Samples of NLC, NLC-ATM, NLC-TF, and NLC-TF-ATM were stored at different storage conditions, room temperature, and 4 °C. Characterization of the size, PDI, and zeta potential was performed on diluted samples on days 0, 7, 14, and 30 using a particle size analyzer (n = 3).

4.9. Statistical Analysis

Statistical analysis was accomplished using analysis of variance (ANOVA) and *t*-test followed by Tukey's test. Differences are considered statistically significant at a level of p-value < 0.01.

5. Conclusions

We conclude that transferrin-conjugated nanostructured lipid carriers (NLC-Tfs) are a promising delivery system for Artemisone (ATM), a repurposed antimalaria drug, to melanoma cells. The medication's cytotoxicity significantly increased with the conjugation of Tf to NLCs, suggesting that this carrier might deliver various chemical compounds to cancer cells. NLCs are an efficient drug delivery system capable of addressing challenging drug properties and have the potential for scalable industrial production. Hence, additional research is necessary to evaluate this delivery system's safety and effectiveness in animal models. These results highlight the value of repurposing medications and the potential of nanotechnology to enhance medication efficacy and delivery.

Author Contributions: Conceptualization, E.A. and N.A.; methodology, N.A.; software, N.A.; validation, E.A. and N.A.; formal analysis, N.A.; investigation, N.A.; resources, E.A. and N.A.; data curation, N.A.; writing—original draft preparation, E.A. and N.A.; writing—review and editing, E.A. and N.A.; visualization, N.A.; supervision, E.A.; project administration, E.A.; funding acquisition, E.A. and N.A. All authors have read and agreed to the published version of the manuscript.

Funding: This research received no external funding.

Institutional Review Board Statement: Not applicable.

Informed Consent Statement: Not applicable.

Data Availability Statement: The data that support the findings of this study are available from the corresponding author upon reasonable request.

Acknowledgments: The authors would like to thank the King Abdullah Scholarship Program, King Saud University, and the Saudi Arabia Cultural Mission (SACM) for the financial support of this project.

Conflicts of Interest: The authors declare no conflict of interest.

Abbreviations

ATM	Artemisone
EDC	1-(3-Dimethylaminopropyl)-3-ethylcarbodiimide Hydrochloride
EE	Encapsulation efficiency
EPR	Enhanced permeability and retention effect
MTS	3-(4:5-Dimethylthiazol-2-yl)-5-(3-carboxymethoxyphenyl)-2-(4-sulfophenyl)-2H-tetrazolium)
NHS	N-Hydroxy succinimide
NLC	Nanostructured lipid carriers
PDI	Polydispersity index
ROS	Reactive oxygen species
SA	Stearylamine
SLN	Solid lipid nanoparticle
Tf	Transferrin
TfR	Transferrin receptor

References

1. Ng, Y.L.; Salim, C.K.; Chu, J.J.H. Drug repurposing for COVID-19: Approaches, challenges and promising candidates. *Pharmacol. Ther.* **2021**, *228*, 107930. [CrossRef] [PubMed]
2. Kumar, S.; Roy, V. Repurposing Drugs: An Empowering Approach to Drug Discovery and Development. *Drug Res.* **2023**, *73*, 481–490. [CrossRef] [PubMed]
3. Pushpakom, S.; Iorio, F.; Eyers, P.A.; Escott, K.J.; Hopper, S.; Wells, A.; Doig, A.; Guilliams, T.; Latimer, J.; McNamee, C.; et al. Drug repurposing: Progress, challenges and recommendations. *Nat. Rev. Drug Discov.* **2018**, *18*, 41. [CrossRef]
4. Haynes, R.K. From artemisinin to new artemisinin antimalarials: Biosynthesis, extraction, old and new derivatives, stereochemistry and medicinal chemistry requirements. *Curr. Top. Med. Chem.* **2006**, *6*, 509–537. [CrossRef]
5. Zhou, X.Y.; Suo, F.Z.; Haslinger, K.; Quax, W.J. Artemisinin-Type Drugs in Tumor Cell Death: Mechanisms, Combination Treatment with Biologics and Nanoparticle Delivery. *Pharmaceutics* **2022**, *14*, 395. [CrossRef]
6. Van Huijsduijnen, R.H.; Guy, R.K.; Chibale, K.; Haynes, R.K.; Peitz, I.; Kelter, G.; Phillips, M.A.; Vennerstrom, J.L.; Yuthavong, Y.; Wells, T.N.C. Anticancer properties of distinct antimalarial drug classes. *PLoS ONE* **2013**, *8*, e82962. [CrossRef]
7. Beekman, A.C.; Wierenga, P.K.; Woerdenbag, H.J.; Van Uden, W.; Pras, N.; Konings, A.W.T.; El-Feraly, F.S.; Galal, A.M.; Wikström, H.V. Artemisinin-derived sesquiterpene lactones as potential antitumour compounds: Cytotoxic action against bone marrow and tumour cells. *Planta Medica* **1998**, *64*, 615–619. [CrossRef]
8. Bhaw-Luximon, A.; Jhurry, D. Artemisinin and its derivatives in cancer therapy: Status of progress, mechanism of action, and future perspectives. *Cancer Chemother. Pharmacol.* **2017**, *79*, 451–466. [CrossRef]

9. Crespo-Ortiz, M.P.; Wei, M.Q. Antitumor activity of artemisinin and its derivatives: From a well-known antimalarial agent to a potential anticancer drug. *J. Biomed. Biotechnol.* **2012**, *2012*, 247597. [CrossRef] [PubMed]
10. Firestone, G.L.; Sundar, S.N. Anticancer activities of artemisinin and its bioactive derivatives. *Expert Rev. Mol. Med.* **2009**, *11*, e32. [CrossRef]
11. Golenser, J.; Hunt, N.H.; Birman, I.; Jaffe, C.L.; Zech, J.; Mäder, K.; Gold, D. Applicability of Redirecting Artemisinins for New Targets. *Glob. Chall.* **2023**, *7*, 2300030. [CrossRef] [PubMed]
12. Nagelschmitz, J.; Voith, B.; Wensing, G.; Roemer, A.; Fugmann, B.; Haynes, R.K.; Kotecka, B.M.; Rieckmann, K.H.; Edstein, M.D. First assessment in humans of the safety, tolerability, pharmacokinetics, and ex vivo pharmacodynamic antimalarial activity of the new artemisinin derivative artemisone. *Antimicrob. Agents Chemother.* **2008**, *52*, 3085–3091. [CrossRef]
13. Müller, R.H.; Radtke, M.; Wissing, S.A. Nanostructured lipid matrices for improved microencapsulation of drugs. *Int. J. Pharm.* **2002**, *242*, 121–128. [CrossRef]
14. Muchow, M.; Maincent, P.; Müller, R.H. Lipid nanoparticles with a solid matrix (SLN®, NLC®, LDC®) for oral drug delivery. *Drug Dev. Ind. Pharm.* **2008**, *34*, 1394–1405. [CrossRef]
15. Shidhaye, S.S.; Vaidya, R.; Sutar, S.; Patwardhan, A.; Kadam, V.J. Solid lipid nanoparticles and nanostructured lipid carriers—Innovative generations of solid lipid carriers. *Curr. Drug Deliv.* **2008**, *5*, 324–331. [CrossRef] [PubMed]
16. Thiruchenthooran, V.; Espina, M.; Świtalska, M.; Bonilla-Vidal, L.; Wietrzyk, J.; Garcia, M.L.; Souto, E.B.; Sánchez-López, E.; Gliszczyńska, A. Combination of Indomethacin with Nanostructured Lipid Carriers for Effective Anticancer Therapy. *Int. J. Nanomed.* **2024**, *19*, 7033–7048. [CrossRef] [PubMed]
17. Jeitler, R.; Glader, C.; König, G.; Kaplan, J.; Tetyczka, C.; Remmelgas, J.; Mußbacher, M.; Fröhlich, E.; Roblegg, E. On the Structure, Stability, and Cell Uptake of Nanostructured Lipid Carriers for Drug Delivery. *Mol. Pharm.* **2024**, *21*, 3674–3683. [CrossRef] [PubMed]
18. Müller, R.H.; Shegokar, R.; Keck, C.M. 20 years of lipid nanoparticles (SLN & NLC): Present state of development & industrial applications. *Curr. Drug Discov. Technol.* **2011**, *8*, 207–227. [CrossRef]
19. Das, S.; Chaudhury, A. Recent advances in lipid nanoparticle formulations with solid matrix for oral drug delivery. *AAPS PharmSciTech* **2011**, *12*, 62–76. [CrossRef]
20. Selvamuthukumar, S.; Velmurugan, R. Nanostructured Lipid Carriers: A potential drug carrier for cancer chemotherapy. *Lipids Health Dis.* **2012**, *11*, 159. [CrossRef]
21. Estanqueiro, M.; Amaral, M.H.; Conceição, J.; Lobo, J.M.S. Nanotechnological carriers for cancer chemotherapy: The state of the art. *Colloids Surf. B Biointerfaces* **2015**, *126*, 631–648. [CrossRef]
22. Wicki, A.; Witzigmann, D.; Balasubramanian, V.; Huwyler, J. Nanomedicine in cancer therapy: Challenges, opportunities, and clinical applications. *J. Control. Release* **2015**, *200*, 138–157. [CrossRef]
23. Sun, T.; Zhang, Y.S.; Pang, B.; Hyun, D.C.; Yang, M.; Xia, Y. Engineered nanoparticles for drug delivery in cancer therapy. *Angew. Chem.-Int. Ed.* **2014**, *53*, 12320–12364. [CrossRef] [PubMed]
24. Hare, J.I.; Lammers, T.; Ashford, M.B.; Puri, S.; Storm, G.; Barry, S.T. Challenges and strategies in anti-cancer nanomedicine development: An industry perspective. *Adv. Drug Deliv. Rev.* **2017**, *108*, 25–38. [CrossRef] [PubMed]
25. Nakase, I.; Gallis, B.; Takatani-Nakase, T.; Oh, S.; Lacoste, E.; Singh, N.P.; Goodlett, D.R.; Tanaka, S.; Futaki, S.; Lai, H.; et al. Transferrin receptor-dependent cytotoxicity of artemisinin-transferrin conjugates on prostate cancer cells and induction of apoptosis. *Cancer Lett.* **2009**, *274*, 290–298. [CrossRef]
26. Daniels, T.R.; Delgado, T.; Rodriguez, J.A.; Helguera, G.; Penichet, M.L. The transferrin receptor part I: Biology and targeting with cytotoxic antibodies for the treatment of cancer. *Clin. Immunol.* **2006**, *121*, 144–158. [CrossRef]
27. Daniels, T.R.; Delgado, T.; Helguera, G.; Penichet, M.L. The transferrin receptor part II: Targeted delivery of therapeutic agents into cancer cells. *Clin. Immunol.* **2006**, *121*, 159–176. [CrossRef]
28. Nakase, I.; Lai, H.; Singh, N.P.; Sasaki, T. Anticancer properties of artemisinin derivatives and their targeted delivery by transferrin conjugation. *Int. J. Pharm.* **2008**, *354*, 28–33. [CrossRef] [PubMed]
29. Lai, H.; Sasaki, T.; Singh, N.P.; Messay, A. Effects of artemisinin-tagged holotransferrin on cancer cells. *Life Sci.* **2005**, *76*, 1267–1279. [CrossRef]
30. Leto, I.; Coronnello, M.; Righeschi, C.; Bergonzi, M.C.; Mini, E.; Bilia, A.R. Enhanced Efficacy of Artemisinin Loaded in Transferrin-Conjugated Liposomes versus Stealth Liposomes against HCT-8 Colon Cancer Cells. *ChemMedChem* **2016**, *11*, 1745–1751. [CrossRef]
31. Zhang, H.; Ji, Y.; Chen, Q.; Jiao, X.; Hou, L.; Zhu, X.; Zhang, Z. Enhancement of cytotoxicity of artemisinin toward cancer cells by transferrin-mediated carbon nanotubes nanoparticles. *J. Drug Target.* **2015**, *23*, 552–567. [CrossRef]
32. Gupta, Y.; Jain, A.; Jain, S.K. Transferrin-conjugated solid lipid nanoparticles for enhanced delivery of quinine dihydrochloride to the brain. *J. Pharm. Pharmacol.* **2007**, *59*, 935–940. [CrossRef] [PubMed]
33. Jain, S.K.; Chaurasiya, A.; Gupta, Y.; Jain, A.; Dagur, P.; Joshi, B.; Katoch, V.M. Development and characterization of 5-FU bearing ferritin appended solid lipid nanoparticles for tumour targeting. *J. Microencapsul.* **2008**, *25*, 289–297. [CrossRef] [PubMed]
34. Gomaa, E.; Fathi, H.A.; Eissa, N.G.; Elsabahy, M. Methods for preparation of nanostructured lipid carriers. *Methods* **2022**, *199*, 3–8. [CrossRef]
35. Fang, J.; Islam, W.; Maeda, H. Exploiting the dynamics of the EPR effect and strategies to improve the therapeutic effects of nanomedicines by using EPR effect enhancers. *Adv. Drug Deliv. Rev.* **2020**, *157*, 142–160. [CrossRef]

36. Shen, X.; Pan, D.; Gong, Q.; Gu, Z.; Luo, K. Enhancing drug penetration in solid tumors via nanomedicine: Evaluation models, strategies and perspectives. *Bioact. Mater.* **2024**, *32*, 445–472. [CrossRef]
37. Bhattacharjee, S. DLS and zeta potential—What they are and what they are not? *J. Control. Release* **2016**, *235*, 337–351. [CrossRef]
38. Rodriguez-Ruiz, V.; Salatti-Dorado, J.A.; Barzegari, A.; Nicolas-Boluda, A.; Houaoui, A.; Caballo, C.; Caballero-Casero, N.; Sicilia, D.; Venegas, J.B.; Pauthe, E.; et al. Astaxanthin-Loaded Nanostructured Lipid Carriers for Preservation of Antioxidant Activity. *Molecules* **2018**, *23*, 2601. [CrossRef] [PubMed]
39. Lu, G.W.; Gao, P. CHAPTER 3—Emulsions and Microemulsions for Topical and Transdermal Drug Delivery. In *Handbook of Non-Invasive Drug Delivery Systems*; Kulkarni, V.S., Ed.; William Andrew Publishing: Boston, MA, USA, 2010; pp. 59–94.
40. Emami, J.; Rezazadeh, M.; Sadeghi, H.; Khadivar, K. Development and optimization of transferrin-conjugated nanostructured lipid carriers for brain delivery of paclitaxel using Box–Behnken design. *Pharm. Dev. Technol.* **2017**, *22*, 370–382. [CrossRef]
41. Mulik, R.S.; Mönkkönen, J.; Juvonen, R.O.; Mahadik, K.R.; Paradkar, A.R. Apoptosis-induced anticancer effect of transferrin-conjugated solid lipid nanoparticles of curcumin. *Cancer Nanotechnol.* **2012**, *3*, 65–81. [CrossRef]
42. Sahoo, S.K.; Ma, W.; Labhasetwar, V. Efficacy of transferrin-conjugated paclitaxel-loaded nanoparticles in a murine model of prostate cancer. *Int. J. Cancer* **2004**, *112*, 335–340. [CrossRef] [PubMed]
43. Khajavinia, A.; Varshosaz, J.; Dehkordi, A.J. Targeting etoposide to acute myelogenous leukaemia cells using nanostructured lipid carriers coated with transferrin. *Nanotechnology* **2012**, *23*, 405101. [CrossRef] [PubMed]
44. Grabarek, Z.; Gergely, J. Zero-length crosslinking procedure with the use of active esters. *Anal. Biochem.* **1990**, *185*, 131–135. [CrossRef]
45. Staros, J.V.; Wright, R.W.; Swingle, D.M. Enhancement by N-hydroxysulfosuccinimide of water-soluble carbodiimide-mediated coupling reactions. *Anal. Biochem.* **1986**, *156*, 220–222. [CrossRef]
46. Badland, M.; Crook, R.; Delayre, B.; Fussell, S.J.; Gladwell, I.; Hawksworth, M.; Howard, R.M.; Walton, R.; Weisenburger, G.A. A comparative study of amide-bond forming reagents in aqueous media—Substrate scope and reagent compatibility. *Tetrahedron Lett.* **2017**, *58*, 4391–4394. [CrossRef]
47. Kowalczyk, A.; Matysiak-Brynda, E.; Bystrzejewski, M.; Sutherland, D.S.; Stojek, Z.; Nowicka, A.M. Conformational control of human transferrin covalently anchored to carbon-coated iron nanoparticles in presence of a magnetic field. *Acta Biomater.* **2016**, *45*, 367–374. [CrossRef]
48. Sahoo, S.K.; Labhasetwar, V. Enhanced Antiproliferative Activity of Transferrin-Conjugated Paclitaxel-Loaded Nanoparticles Is Mediated via Sustained Intracellular Drug Retention. *Mol. Pharm.* **2005**, *2*, 373–383. [CrossRef]
49. Shao, Z.; Shao, J.; Tan, B.; Guan, S.; Liu, Z.; Zhao, Z.; He, F.; Zhao, J. Targeted lung cancer therapy: Preparation and optimization of transferrin-decorated nanostructured lipid carriers as novel nanomedicine for co-delivery of anticancer drugs and DNA. *Int. J. Nanomed.* **2015**, *10*, 1223–1233. [CrossRef]
50. Lin, X.; Yan, S.-Z.; Qi, S.-S.; Xu, Q.; Han, S.-S.; Guo, L.-Y.; Zhao, N.; Chen, S.-L.; Yu, S.-Q. Transferrin-modified nanoparticles for photodynamic therapy enhance the antitumor efficacy of hypocrellin a. *Front. Pharmacol.* **2017**, *8*, 815. [CrossRef]
51. Guo, Y.; Wang, L.; Lv, P.; Zhang, P. Transferrin-conjugated doxorubicin-loaded lipid-coated nanoparticles for the targeting and therapy of lung cancer. *Oncol. Lett.* **2015**, *9*, 1065–1072. [CrossRef]
52. Liu, K.; Dai, L.; Li, C.; Liu, J.; Wang, L.; Lei, J. Self-assembled targeted nanoparticles based on transferrin-modified eight-arm-polyethylene glycol-dihydroartemisinin conjugate. *Sci. Rep.* **2016**, *6*, 29461. [CrossRef] [PubMed]

International Journal of *Molecular Sciences*

MDPI

Article

Targeting CD36-Mediated Lipid Metabolism by Selective Inhibitor-Augmented Antitumor Immune Responses in Oral Cancer

Mayu Takaichi [1], Hidetake Tachinami [1], Danki Takatsuka [1], Amirmoezz Yonesi [1], Kotaro Sakurai [1], Muhammad Irfan Rasul [2], Shuichi Imaue [1], Shin-Ichi Yamada [1], Muhammad Ruslin [2], Manabu Yamazaki [3], Jun-Ichi Tanuma [3], Makoto Noguchi [1] and Kei Tomihara [3,4,*]

1 Department of Oral and Maxillofacial Surgery, Faculty of Medicine, Academic Assembly, University of Toyama, Toyama 930-0194, Japan; takaichi@med.u-toyama.ac.jp (M.T.); hidetake0506@gmail.com (H.T.); tdanki@med.u-toyama.ac.jp (D.T.); sakurai@med.u-toyama.ac.jp (K.S.); simaue@med.u-toyama.ac.jp (S.I.); shinshin@med.u-toyama.ac.jp (S.-I.Y.); toyama.mnoguchi@gmail.com (M.N.)

2 Department of Oral and Maxillofacial Surgery, Hasanuddin University, Makassar 90245, Indonesia; irfanrasul@unhas.ac.id (M.I.R.); mruslin@unhas.ac.id (M.R.)

3 Division of Oral Pathology, Faculty of Dentistry & Graduate School of Medical and Dental Sciences, Niigata University, Niigata 951-8514, Japan; manyamaz@dent.niigata-u.ac.jp (M.Y.); tanuma@dent.niigata-u.ac.jp (J.-I.T.)

4 Division of Oral and Maxillofacial Surgery, Faculty of Dentistry & Graduate School of Medical and Dental Sciences, Niigata University, Niigata 951-8514, Japan

* Correspondence: tomihara@dent.niigata-u.ac.jp; Tel.: +81-25-227-2883; Fax: +81-25-223-5792

Abstract: The fatty acid receptor CD36 is expressed on various malignant cells and is suggested to contribute to tumor progression. CD36 is also expressed by several immune cells and involved in immune responses and may be a potential target in cancer immunotherapy. In this study, we investigated whether the selective inhibition of CD36 can inhibit tumor progression and facilitate an antitumor immune response in oral squamous carcinoma cells (OSCCs). We assessed the effects of sulfosuccinimidyl oleate sodium (SSO), a CD36 inhibitor, on the proliferation apoptosis and alteration in tumor cell surface expression levels of immune accessory molecules in vitro. We also assessed whether SSO-treated OSCCs could promote a T cell response via a Mixed Lymphocyte Reaction (MLR) assay. We also investigated the direct antitumor effects and immunomodulatory effects of SSO using a mouse oral cancer OSCC model. SSO treatment significantly inhibited OSCC proliferation, increased apoptotic cell death, and upregulated the cell surface expression of several immune accessory molecules, including CD83, MHC-Class II, and PD-L1. SSO-treated OSCCs augmented T cell proliferation following MLR. In vivo SSO administration significantly attenuated mouse tumor growth with an increased proportion of immune cells, including $CD4^+$ T, $CD8^+$ T, and dendritic cells; it also decreased the proportion of immune suppressive cells, such as myeloid-derived suppressor and regulatory T cells. These results suggest that the selective inhibition of CD36 can induce direct and indirect antitumor effects by facilitating host antitumor immune responses in OSCCs.

Keywords: CD36; sulfosuccinimidyl oleate sodium; immunotherapy; oral squamous cell carcinoma

check for updates

Citation: Takaichi, M.; Tachinami, H.; Takatsuka, D.; Yonesi, A.; Sakurai, K.; Rasul, M.I.; Imaue, S.; Yamada, S.-I.; Ruslin, M.; Yamazaki, M.; et al. Targeting CD36-Mediated Lipid Metabolism by Selective Inhibitor-Augmented Antitumor Immune Responses in Oral Cancer. *Int. J. Mol. Sci.* **2024**, *25*, 9438. https://doi.org/10.3390/ijms25179438

Academic Editor: Kang-Nan Wang

Received: 5 July 2024
Revised: 15 August 2024
Accepted: 23 August 2024
Published: 30 August 2024

1. Introduction

Oral squamous cell carcinoma (OSCC) is the most common malignancy of the head and neck, and the incidence of OSCC is predicted to increase by up to 40% worldwide by 2040 [1]. The major risk factors of OSCC include smoking, alcohol, human papillomavirus infection, and a particularly high incidence in Asian countries where chewing tobacco is common [2–5].

Although recent advances in treatment have significantly improved the patient outcomes of OSCC, it remains problematic for the strategy in cases of high-grade tumors that are refractory to conventional therapies. Therefore, novel treatment strategies are desired to be developed.

Cancer immunotherapy with immune checkpoint inhibitors (ICIs) has been clinically successful in many cancers, including OSCCs. Particularly, the potent antitumor effects of ICIs for recurrence and/or metastasis cancers are expected [6,7]. However, the clinical efficacies of ICIs are low and novel drugs in combination that improve the efficacy of ICIs are required.

The mechanism of tumor growth mediated by lipid metabolism has been elucidated in the tumor microenvironment (TME). Cancer cells proliferate by fatty acids (FAs) uptake via utilizing plasma membrane FA protein transporters, such as CD36, the family of FA transport proteins (SLC27), and plasma membrane FA-binding proteins (FABPs) [8]. These transporters have been shown to be highly expressed in cells of multiple cancer types [8].

Originally, it was well known that CD36 is homeostatically expressed in various types of cells (adipocytes, monocytes, macrophages, platelets, erythrocytes, dendritic cells (DCs), endothelial cells, cardiomyocytes, and epithelial cells) and involved in functions such as angiogenesis, cell adhesion, apoptosis, inflammatory mechanisms, and the lipid metabolism of these cells [9–12]. However, recent studies have also focused on the pro-tumor roles of CD36. Several studies have revealed its role in the proliferation or metastasis of multiple types of cancer cells, including breast cancer, hepatocellular carcinoma, colorectal cancer, ovarian cancer, bladder cancer, esophageal cancer, glioblastoma, and leukemia [13–25]. Our study and a study from another group also revealed that CD36 could promote the proliferation and migration of OSCCs [26,27].

Recently, the functional role of CD36 in tumor immunity has generated much interest, particularly in immune evasion in cancer. It has been demonstrated that CD36 contributed to the suppression in immune effector cells such as $CD8^+$ T cells and $CD4^+$ T cells [28,29]. Moreover, it has also been demonstrated that CD36 is expressed by immunosuppressive cells, such as regulatory T cells (Tregs) and myeloid-derived suppressor cells (MDSCs), and is associated with immunosuppression in cancer [30,31]. In dendritic cells (DCs), the overexpression of CD36 has also been reported to reduce the priming capacity of $CD4^+$ T cells [32]. Therefore, as a therapeutic target in cancer immunotherapy, strategies targeting CD36 have garnered great interest [22,28–30]. However, little is known about the role of CD36 in the host immune responses in OSCC.

In the present study, we investigated whether the selective inhibition of CD36 could exert the direct antitumor effect and the immunomodulatory effects in OSCCs to elucidate the role of CD36-mediated lipid metabolism in OSCC tumor-bearing hosts. This is the first report to demonstrate the regulatory function of CD36 in the host immune responses in OSCCs.

2. Result

2.1. The Selective Inhibition of the CD36-Exerted Inhibitory Effect on the Proliferation of OSCCs, Accompanied by Increased Apoptotic Cell Death

To confirm the direct antitumor effect of CD36 inhibitors on OSCCs, we investigated the effects of SSO, a CD36 inhibitor, on the cell proliferation and apoptosis of NR-S1K cells and SCCVII cells using in vitro assays. As shown in Figure 1A, the viability of NR-S1K cells and SCCVII cells was significantly reduced after SSO treatment in a dose-dependent manner.

Furthermore, SSO treatment also significantly increased the proportion of apoptotic cell death of NR-S1K cells and SCCVII cells (Figure 1B).

These results indicated that the selective inhibition of CD36 has a direct antitumor effect on OSCCs.

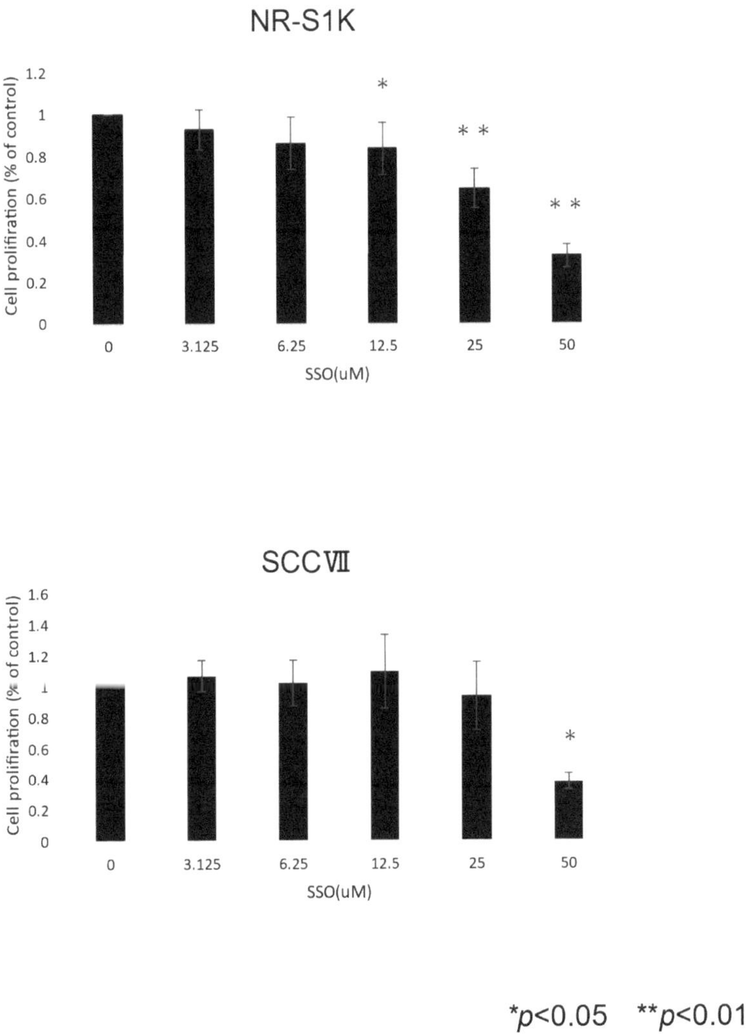

(A)

Figure 1. *Cont.*

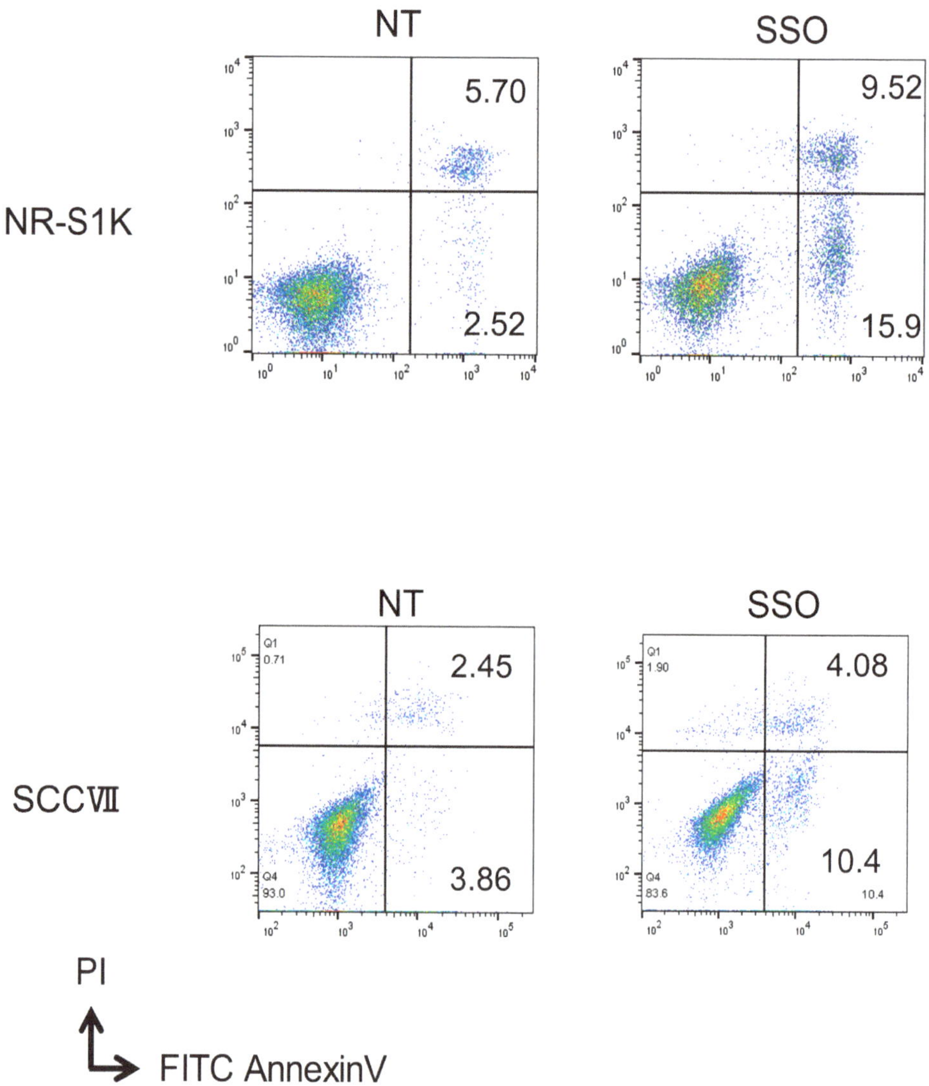

(**B**)

Figure 1. The growth inhibitory effect on OSCCs by SSO. (**A**) OSCCs were cultured in the presence or absence of various concentrations (3.125–50 μM) of SSO for 48 h, and the cell proliferation was measured by the WST-1 cell proliferation assay. The results are presented as mean ± standard deviation from quadruplet determinations. An asterisk indicates a significant difference between the two groups ($p < 0.05$). Experiments were performed three times and similar results were obtained. Representative data from one experiment are shown. (**B**) OSCCs were cultured in the presence or absence of SSO (50 μM) for 48 h, and then the cells were stained with propidium iodide (PI) and annexin V for quantification. Apoptotic cell death was determined by double staining with Annexin V and propidium iodide. The X-axis indicates Annexin V fluorescence and the Y-axis indicates propidium iodide fluorescence. Cells in the lower right and upper right quadrants represent early and late apoptotic cells. Experiments were performed in triplicate and similar results were obtained. Representative data from one experiment are shown.

2.2. The Selective Inhibition of CD36-Altered Surface Antigen Expression in OSCCs

To confirm the effect of the selective inhibition of CD36 on the surface antigen expression changes of OSCCs, we compared the expression level of various T cell costimulatory and adhesion molecules on OSCCs with or without SSO treatment. As shown in Figure 2, the OSCCs treated with SSO showed increased levels of costimulatory molecules, including CD83, MHC class I, and PD-L1, compared to non-treated cells. Conversely, there were no changes in the expression levels of CD86, CD40, MHC-class II, orPD-L2 molecules.

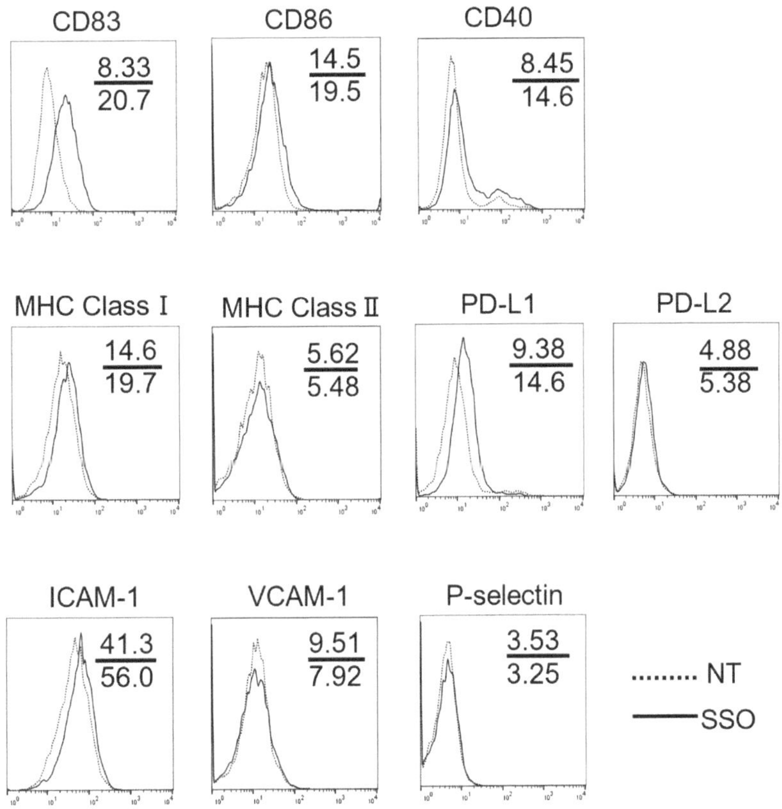

Figure 2. Phenotypic alteration of OSCCs by SSO. OSCCs were cultured in the presence or absence of 50 μM SSO for 48 h, and the cell surface expression of several immune accessary molecules and adhesion molecules was analyzed using flow cytometry. Experiments were performed in triplicate and similar results were obtained. Representative histograms from one experiment are shown. Dashed lines and solid lines indicate results for non-treated (NT) cells and SSO-treated cells. Numbers in each panel indicate the mean fluorescence intensity of each molecule in NT cells (upper) and SSO-treated cells (lower).

Furthermore, OSCCs treated with SSO showed increased levels of ICAM-1, but not VCAM-1 and P-selectin, compared to non-treated cells.

OSCCs treated with SSO showed increased levels of CD36 compared to non-treated cells (Supplementary Figure S1).

These results indicated that the selective inhibition of CD36 induces phenotype alterations of OSCCs, which may facilitate antitumor T cell responses.

2.3. The Selective Inhibition of CD36-Induced Phenotype Alteration of OSCCs That Facilitate T Cell Responses

To confirm changes in the T cell immune response for OSCCs with increased levels of costimulatory molecules due to the selective inhibition of CD36, we compared the capacity to stimulate T cells by a mixed-lymphocyte reaction (MLR) using OSCCs with or without SSO treatment [33]. As shown in Figure 3, IFN-γ-producing T cells were increased when cultured with the OSCCs that were treated with SSO compared with the OSCCs that were not treated. Furthermore, we compared T cell proliferation by coculturing with OSCC with or without SSO treatment. The CFSE-labeled T cells cocultured with the SSO-pretreated OSCCs had more proliferation than did the T cells cocultured with non-treated OSCCs (Figure 3). These results indicated that the selective inhibition of CD36 alters the phenotype of the OSCCs that facilitate T cell responses.

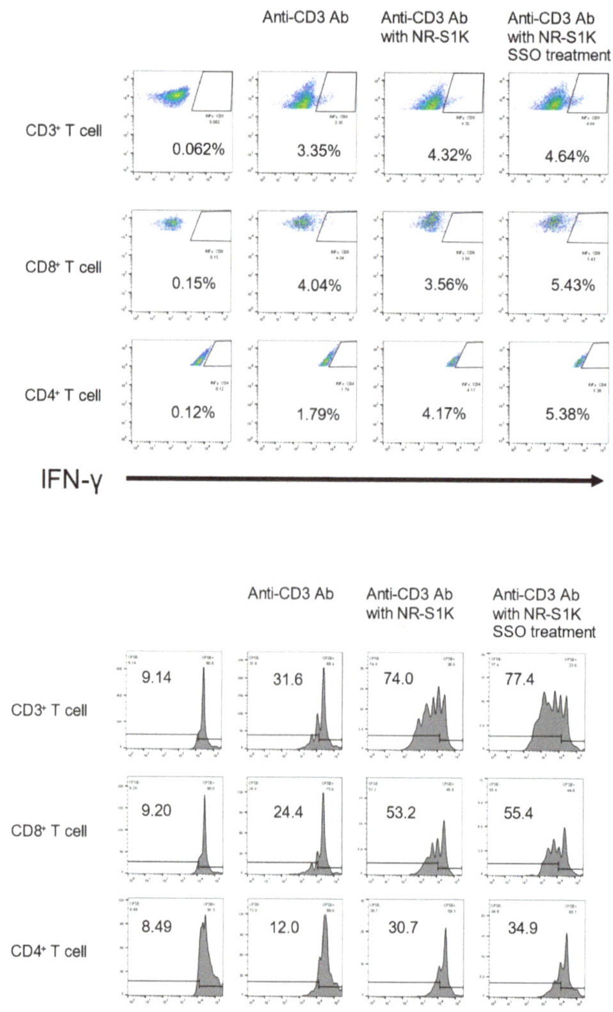

(**A**)

Figure 3. *Cont.*

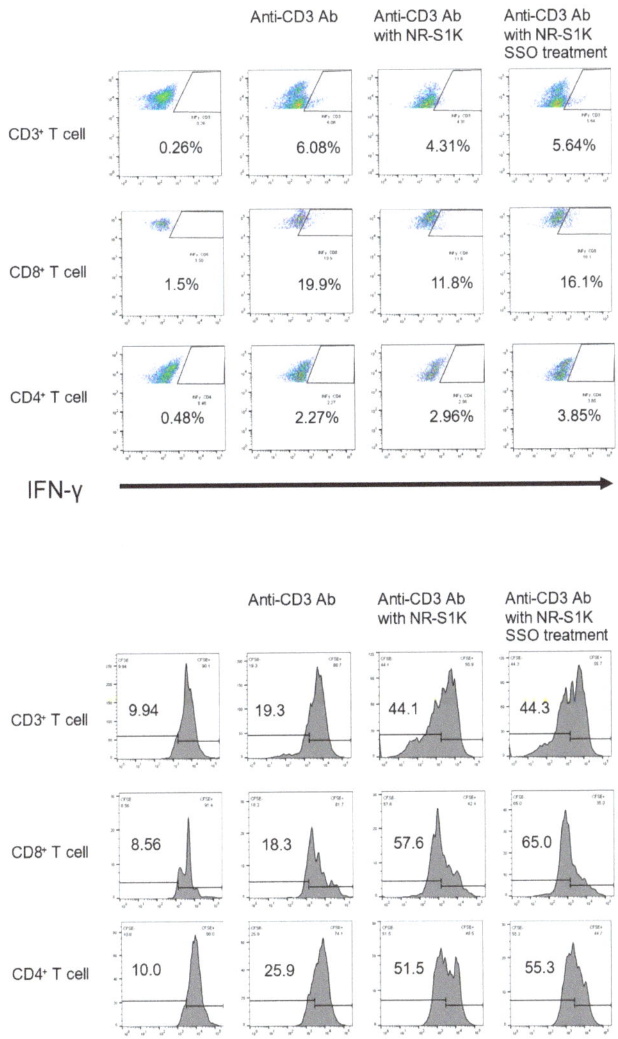

(**B**)

Figure 3. Augmented T cell responses by SSO-treated OSCCs. An in vitro Mixed Lymphocyte Reaction (MLR) assay was performed. NR-S1K cells were pretreated with 50μM SSO for 48 h. Then, lymph node cells (**A**) or spleen cells (**B**) from naive mice and NR-S1K cells (1×10^4) with or without SSO pretreatment were cocultured in a 96-well U-bottom culture plate in the presence of 0.5 μg/mL anti-CD3 for 72 h. Cells were restimulated with 50 ng/mL PMA, 500 ng/mL ionomycin, and 4 μM monensin for 4 h before the end of the culture. The intracellular IFN-γ of T cells was determined using flow cytometry. Experiments were performed three times and similar results were obtained. Representative scatter plots from one experiment are shown. Furthermore, CFSE-labeled T cells from lymph node cells (**A**) or spleen cells (**B**) were cocultured with NR-S1K cells with or without SSO pretreatment. T cells were gated with the same gating as in the experiment for IFN-γ levels. T cell proliferation was measured by CFSE dilution. Experiments were performed in triplicate and similar results were obtained. Representative histograms from one experiment are shown. In both experiments, the control is the figure in the second column from the right.

2.4. The Selective Inhibition of CD36-Exerted Antitumor Effects against an In Vivo Mouse Oral Cancer Model

To confirm the growth inhibitory effect on OSCCs via the CD36 inhibitor in vivo, the tumor growth was assessed during SSO treatment in a mouse model of OSCCs. As shown in Figure 4, SSO-treated OSCC mice showed significantly delayed tumor growth compared with the control mice. These results indicate that the SSO treatment in OSCC model mice may have been effective.

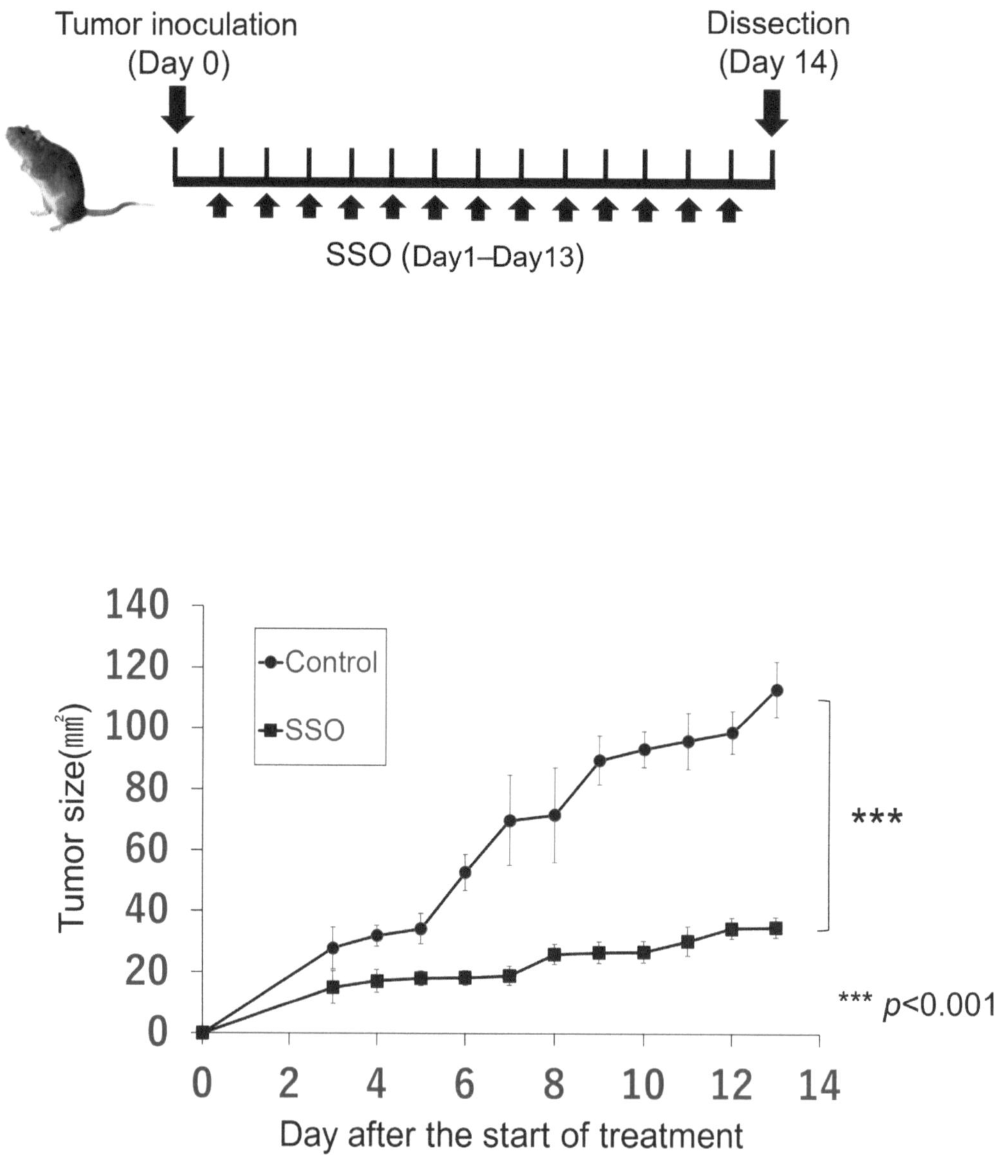

(A)

Figure 4. *Cont.*

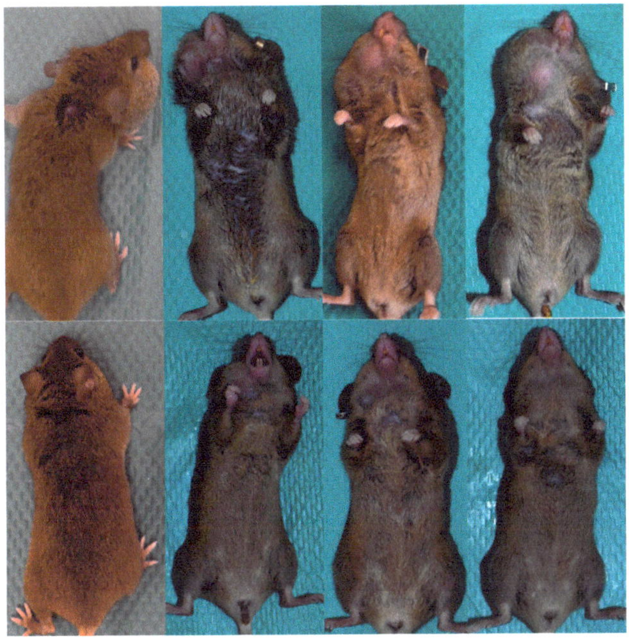

(B)

Figure 4. Growth inhibitory effect by SSO in a mouse model of OSCCs. Mice were challenged with NR-S1K cells. The mice were injected intraperitoneally with a dose of 20 mg/kg of SSO or saline every day from day 1 to day 13. (**A**) Tumor sizes in mice were measured at the indicated time points ($n = 4$/group); *** $p < 0.01$, NT vs. SSO. (**B**) Photos of tumor-bearing mice at the time of dissection. The upper images are of saline-treated control mice and the lower images are of SSO-treated mice.

2.5. The In Vivo Administration of CD36 Inhibitor and the Modulated Distribution of Immune Cell Populations in a Mouse OSCC Model

To evaluate the immunological alterations on OSCCs via CD36 inhibitor in vivo, we compared immune cell populations between SSO-treated OSCC mice and control mice. Tumors, spleens, peripheral blood, peripheral lymph nodes, and cervical lymph nodes were harvested and the percentages of different immune cell types in each organ were determined after 14 days of SSO treatment.

As shown in Figure 5A, the total proportion of $CD8^+$ T cells was significantly increased in the tumors, spleens, peripheral blood, and cervical lymph nodes of the SSO-treated mice compared to the control mice. Furthermore, the total proportion of IFN-γ-producing $CD8^+$ T cells was significantly increased in the tumors, peripheral lymph nodes, and cervical lymph nodes of the SSO-treated mice compared to the control mice (Figure 5B). The total proportion of $CD4^+$ T cells was significantly increased in the tumors of the SSO-treated mice compared to the control mice (Figure 5A). The total proportion of IFN-γ-producing $CD4^+$ T cells was significantly increased in the spleens, peripheral lymph nodes, and cervical lymph nodes of the SSO-treated mice compared to the control mice (Figure 5C). Figure 5D shows that the total proportion of dendritic cells (DCs) was significantly increased in the tumors, spleens, peripheral blood, and peripheral lymph nodes of the SSO-treated mice compared to the control mice. However, as shown in Figure 5E, the total proportion of regulatory T cells (Tregs) was significantly decreased in the tumors and spleens of the SSO-treated mice compared to the control mice. Furthermore, the total proportion of myeloid-derived suppressor cells (MDSCs) was significantly decreased in the tumors and peripheral blood of the SSO-treated mice compared to the control mice (Figure 5F).

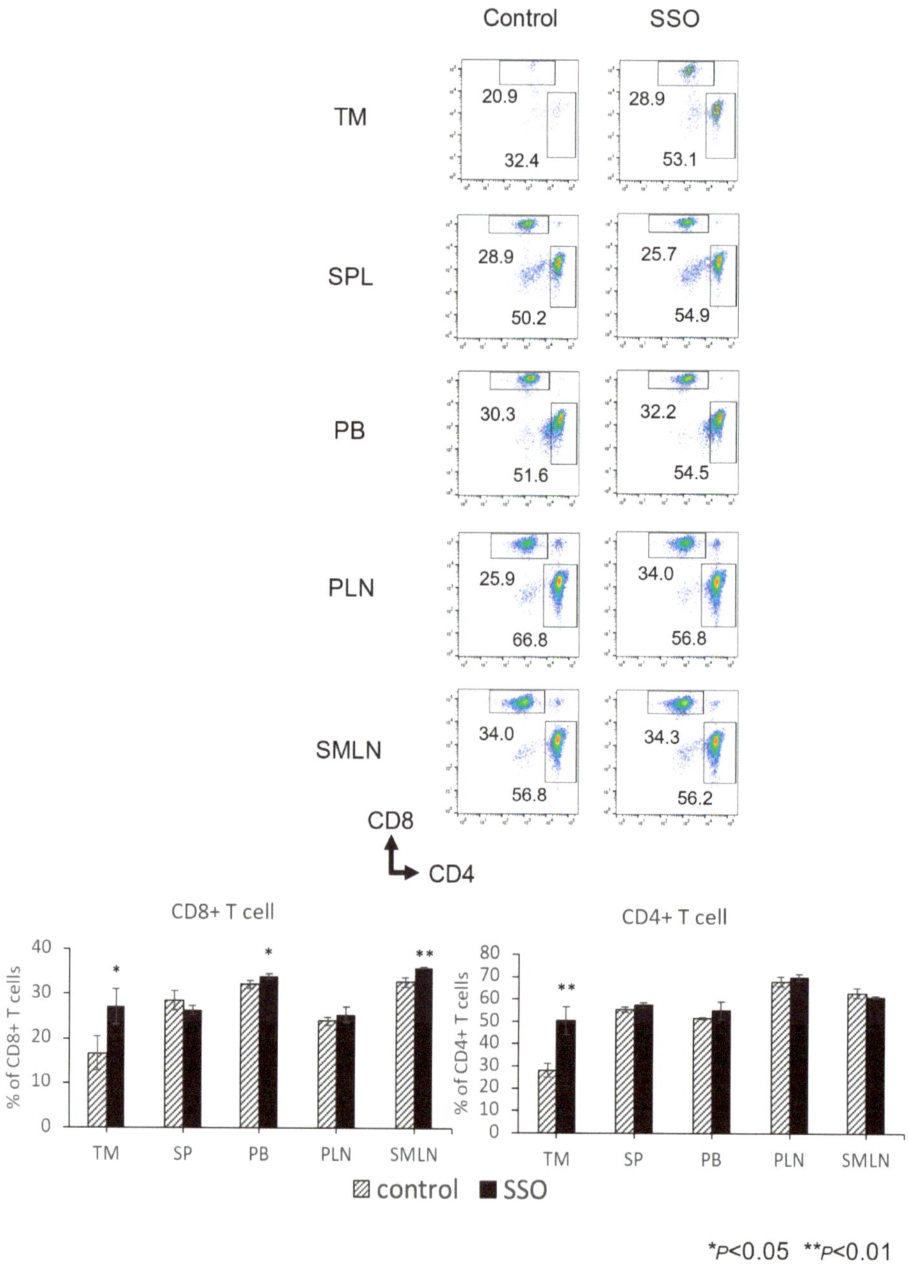

(A)

Figure 5. *Cont.*

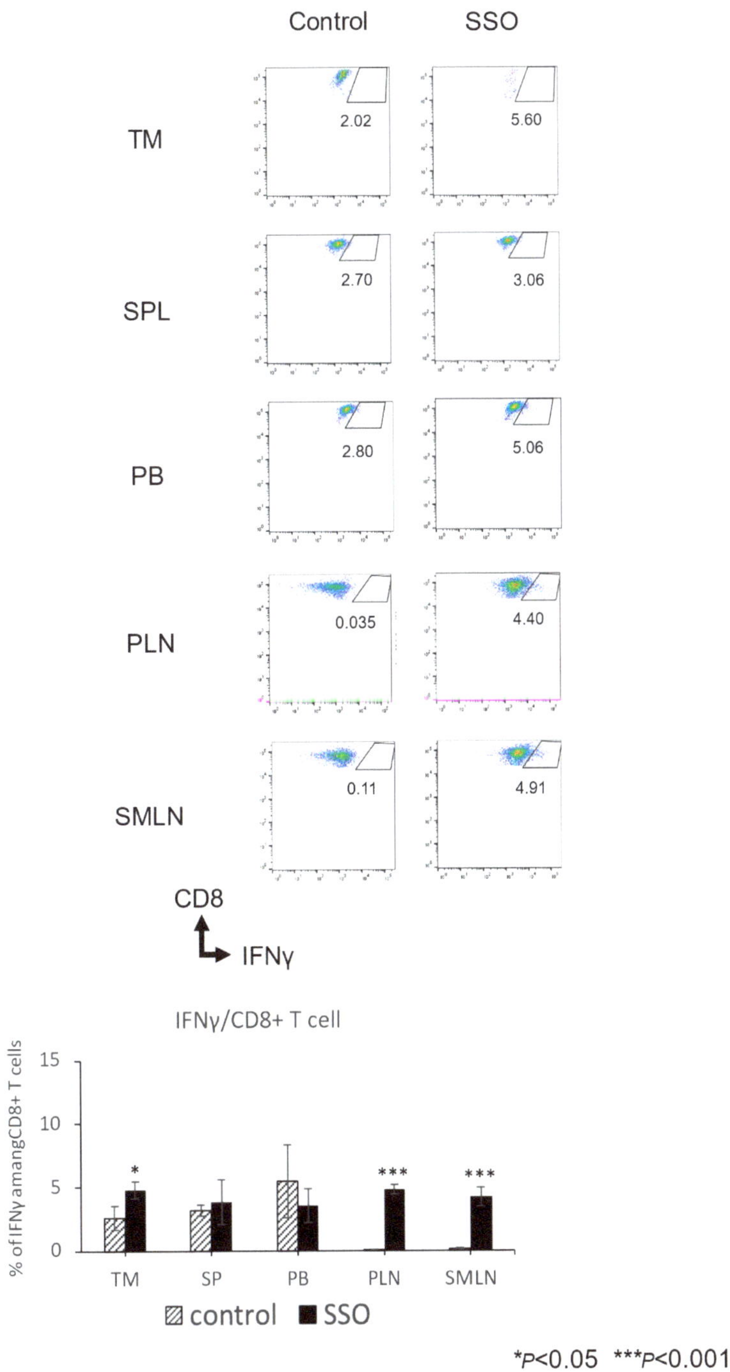

(B)

Figure 5. *Cont.*

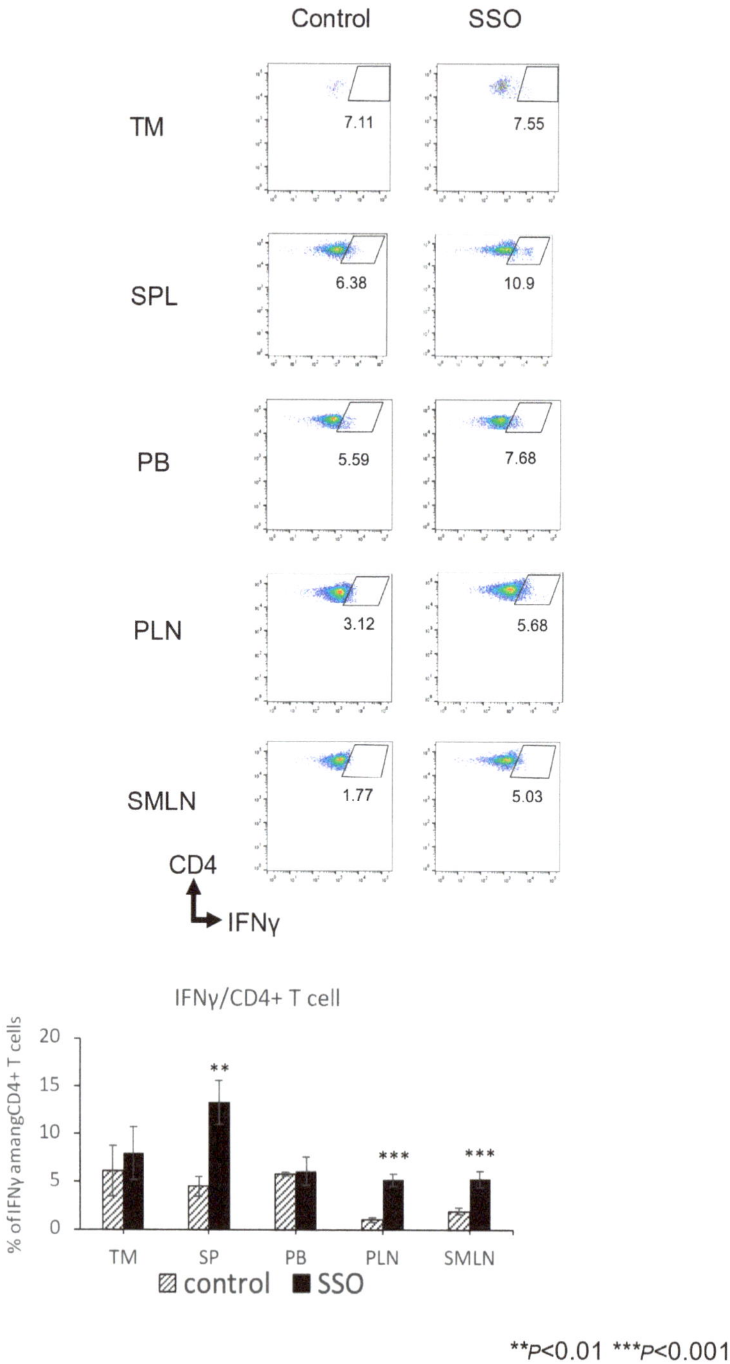

(C)

Figure 5. *Cont.*

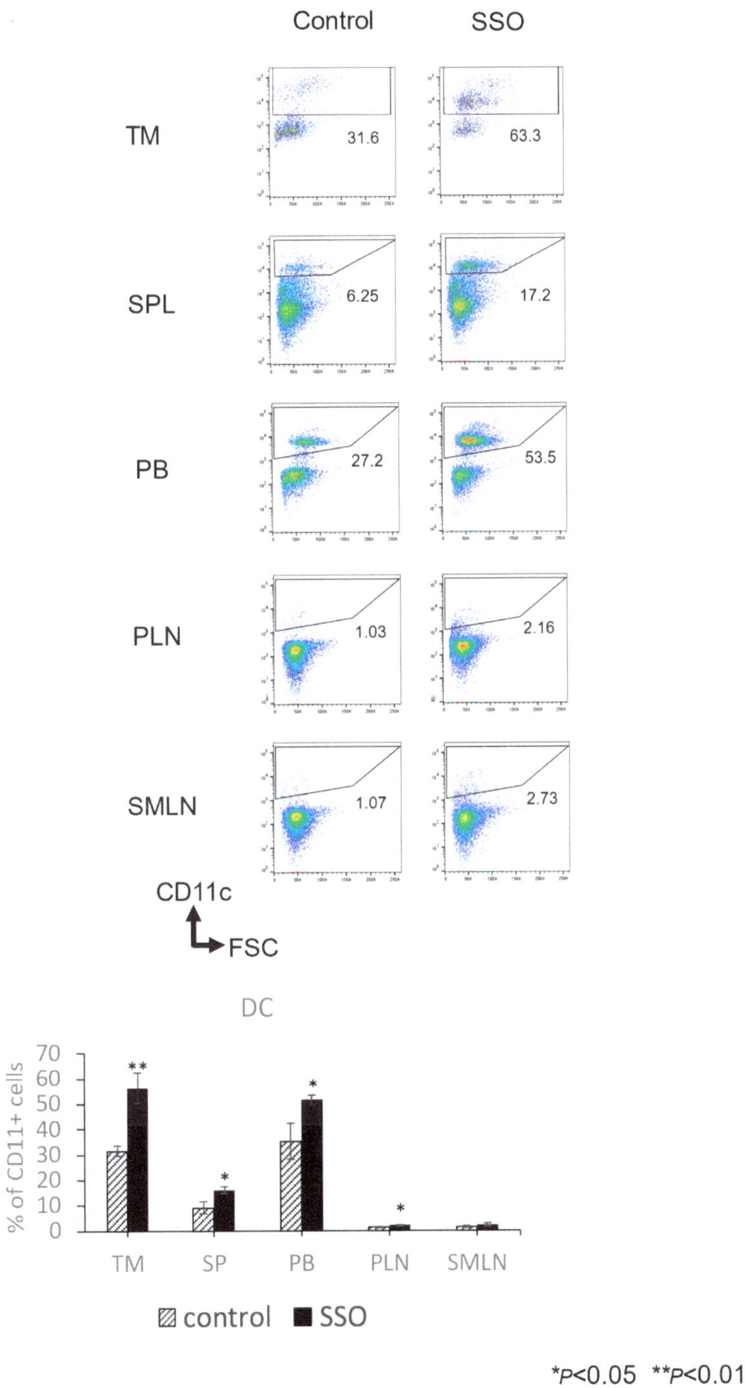

(**D**)

Figure 5. *Cont.*

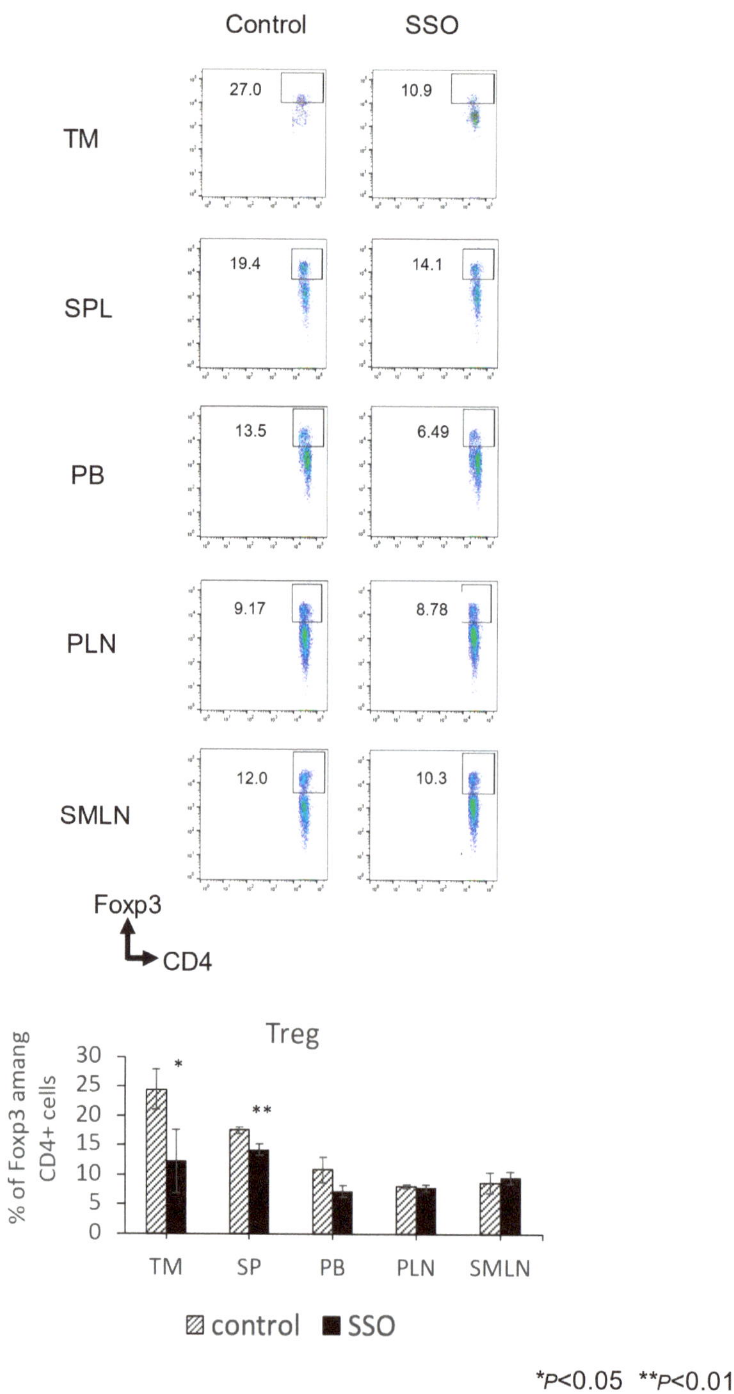

(E)

Figure 5. *Cont.*

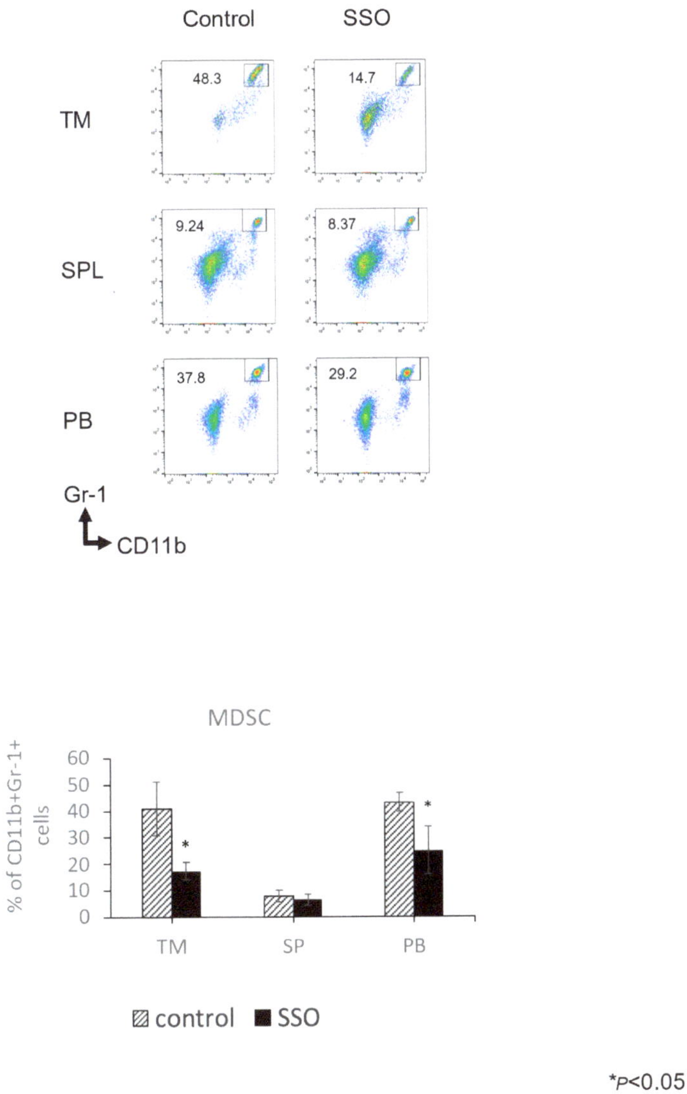

(F)

Figure 5. The comparison of the proportions of various immune cells in OSCC-bearing mice between the SSO-treated and non-treated populations. Mice were challenged with NR-S1K cells. The mice were injected intraperitoneally with a dose of 20 mg/kg of SSO every day from day 1 to day 13. After 14 days of tumor inoculation, the tumor, spleen, peripheral blood, peripheral lymph nodes, and submandibular lymph nodes were harvested and the percentages of various immune cells in each organ were determined using flow cytometry: (**A**) $CD8^+$ T cells, $CD4^+$ T cells, (**B**) IFN-γ-producing $CD8^+$ T cells, (**C**) IFN-γ-producing $CD4^+$ T cells, (**D**) Dendritic cells (DCs), (**E**) Regulatory T cells (Tregs), and (**F**) Myeloid-derived suppressor cells (MDSCs). Representative scatter plots and a summary of these results are shown (n = 3/group); * $p < 0.05$, ** $p < 0.01$, *** $p < 0.001$, NT vs. SSO. (TM: Tumor, SP: spleen, PB: peripheral blood, PLN: peripheral lymph nodes, and SMLN: submandibular lymph nodes).

These results indicated that the selective inhibition of CD36 could alter the distribution of immune cell populations, which may facilitate antitumor immune responses in oral cancer-bearing hosts.

3. Discussion

Recently, the mechanism of lipid metabolism-mediated tumor progression has been elucidated in the TME [8]. It was elucidated that fatty acid synthesis is more abundant in tumor cells compared to normal cells, and fatty acids are used in the proliferation of tumor cells [34,35]. Particularly, fatty acid synthases (FASN) play an important role in the fatty acid-mediated proliferation of tumor cells [35]. It was demonstrated that the selective inhibition of FASN attenuated the progression of prostate cancer, non-small-cell lung cancer, and metastasized breast cancer [36–38]. Moreover, the growth-inhibitory effects of cholesterol-lowering drugs have been demonstrated in OSCCs, suggesting a relationship between tumor growth and lipid metabolism [39]. Lipid metabolism is mediated by FA protein transporters in the plasma membrane. Recent studies have revealed that the CD36 expressed by various tumor cells particularly affects tumor growth and metastasis [13–25].

Furthermore, CD36 is expressed by various immune cells. Particularly, the CD36 expression by immunosuppressive cells such as Tregs and MDSCs in the tumor suggests its contribution to the immunosuppressive function of these cell types in the tumor-bearing hosts [28–31].

Our previous study also revealed that CD36 was abundantly expressed by OSCCs and involved in the proliferation and migration of the tumor cells, whereas little is known about the role of CD36 in the host immune responses in OSCCs [26]. Therefore, we evaluated the role of CD36 in the OSCC host's immune responses.

In the present study, a CD36 inhibitor was used to target lipid metabolism in OSCCs, and it showed not only direct antitumor effects but also the enhancement of antitumor immune responses, both in vitro and in vivo. Previous studies have revealed that CD36 inhibition enhances T cell immune responses and dendritic cell function in the tumor-bearing hosts of various types of cancers [29–32,40]. In contrast, the immunosuppressive functions of Tregs and MDSCs were diminished [28,29]. This is due to the energy metabolism of each immune cell: Tregs and MDSCs have enhanced immunosuppressive functions due to enhanced lipid metabolism via CD36 [22,28]. In $CD8^+$ T cells, increased fatty acid uptake via CD36 leads to ferroptosis and attenuated effector function [30]. Our results also suggested that CD36 was involved in the regulation of T cell immune responses, dendritic cell functions, and the deregulation of immunosuppressive cell functions in OSCCs, suggesting that lipid metabolism is important in the regulation of antitumour immune responses in OSCCs. Future experiments on the gene modification of CD36 would be desirable, to investigate whether these effects directly attribute to the CD36 inhibition.

In recent years, the emergence of immune ICIs has revolutionized the treatment of recurrent and/or metastatic oral cancer, resulting in a better overall survival rate than conventional therapies [6,7]. However, the response rate of ICIs is limited, and it is expected that combination drugs will be developed to enhance the efficacy of ICIs. It was suggested that one of the causes of poor responses to ICIs is the accumulation of immunosuppressive cells such as Tregs and MDSCs in tumor-bearing hosts, which acquire resistance to ICIs [36–38]. Our results and others' studies suggested that the function of immunosuppressive cells could be decreased by CD36 inhibition, and the effect of ICIs may be increased by their combination with CD36 inhibitors [28,29]. The combination of ICIs and CD36 inhibitor in a hepatocellular carcinoma (HCC) in vivo model significantly improved the growth inhibitory effect and survival rate compared to ICIs alone [41]. Furthermore, antitumor immunomodulatory effects in tumor-bearing hosts have been demonstrated; a decrease in Tregs and MDSCs and an increase in $CD8^+$ T cell-producing IFN-γ and granzyme B was shown, suggesting that the combination of ICIs and CD36 inhibitors may have synergistic effects [42]. In the present study, the treatment with CD36 inhibitors of OSCC-bearing mice decreased immunosuppressive cells and increased T cell-producing IFN-γ. Further study would be necessary to elucidate the synergistic effects of ICIs and CD36 inhibitors.

The optimization of combined immunotherapy would require the exploration of therapeutic efficacy and the identification of predictive biomarkers for the therapeutic efficacy simultaneously. CD36 expression is higher in poorly differentiated OSCCs than in well-differentiated OSCCs, and the metastatic ability of CD36-positive OSCCs is higher than in CD36-negative OSCCs, suggesting that CD36 may be an excellent predictive biomarker for the aggressiveness of tumors [22,26]. Moreover, our present study revealed that CD36 inhibition attenuated the immune suppressive properties in OSCC-bearing hosts, suggesting that CD36 may be a potential biomarker for antitumour immune responses in oral cancer. For the development of more personalized medicine for ICI refractory oral cancer patients, further studies are necessary to elucidate the synergistic effects of CD36 inhibitors in combination with ICIs and to evaluate the utility of CD36 expression as a predictive biomarker for the response to immunotherapy.

4. Materials and Methods

4.1. Mice and Cell Lines

The C3H/HeN mice were purchased from Sankyo Laboratory Services, Inc. All the mice were maintained under specific pathogen-free conditions under the institutional guidelines of the University of Toyama.

The mouse OSCC line NR-S1K was established from the NR-S1 cell line [43]. This cell line was kindly provided by Dr. Masato Azuma, Department of Molecular Immunology, Graduate School, Tokyo Medical and Dental University. The mouse OSCC line SCCVII is an established cell line from the C3H/HeN mouse line [33].

4.2. Cell Proliferation Assay

Cell proliferation assays were performed, as described previously [43]. The OSCCs were seeded in 96-well plates at a density of 1.5×10^4 cells/well and cultured for 72 h in the presence or absence of various concentrations of SSO. The viability of adherent cells was measured by adding (WST-1) premix (Takara Bio, Kusatsu, Japan), according to the manufacturer's instructions.

4.3. Assessment of Cellular Apoptosis

The OSCCs were seeded in 6-well plates at a density of 1.0×10^5 cells/well and cultured for 48 h in the presence or absence of 50 μM SSO. Apoptotic cell death was detected using Annexin V (BD Pharmingen, San Diego, CA, USA) staining, according to the manufacturer's instructions, as described previously [44].

4.4. Assessment of Phenotypic Alteration of Cells

The OSCCs were seeded in 6-well plates at a density of 1.0×10^5 cells/well and cultured in the presence or absence of 50 μM SSO for 48 h. Cells were collected and stained with antibodies against each cell surface molecule. The samples were analyzed by flow cytometry.

4.5. Mixed Lymphocyte Reaction (MLR) and Intracellular Cytokine Staining

Mixed Lymphocyte Reaction (MLR) assays were performed, as previously described [45]. The OSCCs were seeded in 6-well plates at a density of 1.0×10^5 cells/well and cultured in the presence or absence of 50 μM SSO for 72 h. The total spleen cells from naive mice and the OSCCs were cocultured for 72 h in the presence of 1 μg/mL anti-CD3 antibodies (spleen cells:OSCCs = 10:1).

4.6. In Vivo SSO Treatment

The NR-S1K cells (2.0×10^6) were administered subcutaneously into the right masseter of C3H/HeN mice. The mice were injected intraperitoneally with 20 mg/kg SSO every day from day 1 to day 13. The SSO was dissolved in a minimal amount of DMSO and then dissolved in the saline for in vivo administration. The same amount of saline was administered to the control mice. After 14 days of tumor inoculation, mice were humanely sacrificed, and the tumor, peripheral blood, lymph nodes, and spleen were extracted, and the cells were analyzed by flow cytometry. The tumor size (length × width) was measured daily from the first i.p. injection to monitor the tumor's size progression.

4.7. Antibodies and Reagents

The following antibodies were purchased from eBioscience (San Diego, CA, USA): FITC-conjugated anti-mouse CD40, CD54(I-CAM), CD83, CD273(B7-DC, PD-L2); PE-conjugated anti-mouse CD80, MHC Class, PD-L1, CD62P(P-Selectin); PerCP-Cy5.5-conjugated anti-mouse; PE-Cy7-conjugated anti-mouse IFN-γ, CD8, CD19; APC-conjugated anti-mouse CD106(VCAM-1), CD8, Foxp3; and APC-Cy7 conjugated anti-mouse Ly-6G/Ly-6C(Gr-1). The following antibodies were purchased from TONBO biosciences (Osaka, Japan): PE-conjugated anti-mouse CD11c and APC-conjugated anti-mouse CD86, MHC class II. The following antibodies were purchased from Biolegend (San Diego, CA, USA): FITC-conjugated anti-mouse CD11b; PE-conjugated anti-mouse IFN-γ; PerCP-Cy5.5-conjugated anti-mouse CD4, CD45R(B220); APC-Cy7 conjugated anti-mouse CD3; and BV510-conjugated anti-mouse CD45. Carboxyfluorescein diacetate succinimidyl ester (CFSE) was purchased from Invitrogen (Carlsbad, CA, USA). Sulfosuccinimidyl oleate sodium (SSO) was purchased from Abcam (Cambridge, UK).

4.8. Flow Cytometry

A flow cytometry analysis was performed as described previously [45]. Samples were assessed using FACS Celesta (Becton Dickinson, San Jose, CA, USA).

4.9. Statistical Analysis

Group comparisons were made using Student's *t*-test or ANOVA with the statistical software OriginPro 2018 (OriginLab Corporation, Northampton, MA, USA), and a value of $p < 0.05$ was considered statistically significant.

5. Conclusions

Overall, our data suggest that CD36 is indispensable in the proliferation and survival of oral cancer cells and regulates various immunological functions in oral tumor-bearing hosts. Because the synergistic effect of CD36 inhibitors and immunotherapy has generated much interest as a new therapeutic strategy, CD36 may be a potential therapeutic target for cancer therapy as a central regulator for tumor progression and immune evasion in tumor-bearing hosts. Further studies are necessary to elucidate an effective and safe clinical setting for CD36 inhibitors and to develop a more personalized medicine for oral cancer treatment.

Supplementary Materials: The following supporting information can be downloaded at: https://www.mdpi.com/article/10.3390/ijms25179438/s1.

Author Contributions: Conceptualization, K.T.; Data curation, K.T. and M.T.; Funding acquisition, K.T. and H.T.; Investigation, M.T., H.T., D.T., A.Y., M.Y. and K.T.; Methodology, M.T., H.T., D.T., K.S., A.Y., M.I.R., S.I., S.-I.Y., M.R., M.Y., J.-I.T. and M.N.; Software, S.I.; Supervision, K.T.; Validation, M.T., H.T., D.T., M.I.R., M.R. and M.N.; Writing—original draft, K.T.; Writing—review & editing, K.T., M.T., H.T., D.T., A.Y., K.S., M.I.R., S.I., S.-I.Y., M.R., M.Y., J.-I.T. and M.N. All authors have read and agreed to the published version of the manuscript.

Funding: This research was funded by the Japan Society for the Promotion of Science (JSPS) Grants-in-Aid for Scientific Research (KAKENHI) (JP22K10214 to K.T. and JP24K20027 to H.T.).

Institutional Review Board Statement: The animal study protocol was approved by the University of Toyama Animal Welfare Committee (Approval number: A2022 MED-31).

Informed Consent Statement: Not applicable.

Data Availability Statement: Any personal or patient data are unavailable due to privacy or ethical restrictions. All other data are available from the corresponding author upon reasonable request.

Acknowledgments: All individuals included in this section have consented to the acknowledgement.

Conflicts of Interest: The authors declare no conflicts of interest.

References

1. Sung, H.; Ferlay, J.; Laversanne, M.; Soerjomataram, I.; Jemal, A.; Bray, F. Global cancer statistics 2020: Globocan estimates of incidence and mortality worldwide for 36 cancers in 185 countries. *CA Cancer J. Clin.* **2021**, *71*, 209–249. [CrossRef]
2. McCoy, G.D.; Wynder, E.L. Etiological and preventive implications in alcohol carcinogenesis. *Cancer Res.* **1979**, *39 Pt 2*, 2844–2850. [PubMed]
3. Blot, W.L.; McLaughlin, J.K.; Winn, D.M.; Austin, D.F.; Greenberg, R.S.; Preston-Martin, S.; Bernstein, L.; Schoenberg, J.B.; Stemhagen, A.; Fraumeni, J.F., Jr. Smoking and drinking in relation to oral and pharyngeal cancer. *Cancer Res.* **1988**, *48*, 3282–3287. [PubMed]
4. van Wyk, C.W.; Stander, I.; Padayachee, A.; Grobler-Rabie, A.F. The areca nut chewing habit and oral squamous cell carcinoma in South African Indians. A retrospective study. *S. Afr. Med. J.* **1993**, *83*, 425–429.
5. Sturgis, E.M.; Cinciripini, P.M. Trends in head and neck cancer incidence in relation to smoking prevalence: An emerging epidemic of human papillomavirus-associated cancers? *Cancer* **2007**, *110*, 1429–1435. [CrossRef] [PubMed]
6. Ferris, R.L.; Blumenschein, G., Jr.; Fayette, J.; Guigay, J.; Colevas, A.D.; Licitra, L.; Harrington, K.; Kasper, S.; Vokes, E.E.; Even, C.; et al. Nivolumab for recurrent squamous-cell carcinoma of the head and neck. *N. Engl. J. Med.* **2016**, *375*, 1856–1867. [CrossRef]
7. Mehra, R.; Seiwert, T.Y.; Gupta, S.; Weiss, J.; Gluck, I.; Eder, J.P.; Burtness, B.; Tahara, M.; Keam, B.; Kang, H.; et al. Efficacy and safety of pembrolizumab in recurrent/metastatic head and neck squamous cell carcinoma: Pooled analyses after long-term follow-up in KEYNOTE-012. *Br. J. Cancer* **2018**, *119*, 153–159. [CrossRef]
8. Su, X.; Abumrad, N.A. Cellular fatty acid uptake: A pathway under construction. *Trends Endocrinol. Metab.* **2009**, *20*, 72–77. [CrossRef] [PubMed]
9. Choromańska, B.; Myśliwiec, P.; Choromańska, K.; Dadan, J.; Chabowski, A. The role of CD36 receptor in the pathogenesis of atherosclerosis. *Adv. Clin. Exp. Med.* **2017**, *26*, 717–722. [CrossRef] [PubMed]
10. Pepino, M.Y.; Kuda, O.; Samovski, D. Structure-function of CD36 and importance of fatty acid signal transduction in fat metabolism. *Annu. Rev. Nutr.* **2014**, *34*, 281–303. [CrossRef]
11. Zhao, L.; Zhang, C.; Luo, X.; Wang, P.; Zhou, W.; Zhong, S.; Xie, Y.; Jiang, Y.; Yang, P.; Tang, R.; et al. CD36 palmitoylation disrupts free fatty acid metabolism and promotes tissue inflammation in non-alcoholic steatohepatitis. *J. Hepatol.* **2018**, *69*, 705–717. [CrossRef]
12. Hao, J.W.; Wang, J.; Guo, H.; Zhao, Y.Y.; Sun, H.H.; Li, Y.F.; Lai, X.Y.; Zhao, N.; Wang, X.; Xie, C.; et al. CD36 facilitates fatty acid uptake by dynamic palmitoylation-regulated endocytosis. *Nat. Commun.* **2020**, *11*, 4765. [CrossRef]
13. El-Aziz, A.; Sherin, M.; Dalia, A. Clinical relevance of thrombospondin receptor (CD36) expression in Egyptian de novo adult acute myeloid leukemia. *Egypt. J. Haematol.* **2013**, *38*, 1–6. [CrossRef]
14. Hale, J.S.; Otvos, B.; Sinyuk, M.; Alvarado, A.G.; Hitomi, M.; Stoltz, K.; Wu, Q.; Flavahan, W.; Levison, B.; Johansen, M.L.; et al. Cancer stem cell-specific scavenger receptor CD36 drives glioblastoma progression. *Stem Cells* **2014**, *32*, 1746–1758. [CrossRef]
15. Ladanyi, A.; Mukherjee, A.; Kenny, H.A.; Johnson, A.; Mitra, A.K.; Sundaresan, S.; Nieman, K.M.; Pascual, G.; Benitah, S.A.; Montag, A.; et al. Adipocyte-induced CD36 expression drives ovarian cancer progression and metastasis. *Oncogene* **2018**, *37*, 2285–2301. [CrossRef]
16. Liang, Y.; Han, H.; Liu, L.; Duan, Y.; Yang, X.; Ma, C.; Zhu, Y.; Han, J.; Li, X.; Chen, Y. CD36 plays a critical role in proliferation, migration and tamoxifen-inhibited growth of ER-positive breast cancer cells. *Oncogenesis* **2018**, *7*, 98. [CrossRef]
17. Drury, J.; Rychahou, P.G.; He, D.; Jafari, N.; Wang, C.; Lee, E.Y.; Weiss, H.L.; Evers, B.M.; Zaytseva, Y.Y. Inhibition of fatty acid synthase upregulates expression of CD36 to sustain proliferation of colorectal cancer cells. *Front. Oncol.* **2020**, *10*, 1185. [CrossRef]
18. Gyamfi, J.; Yeo, J.H.; Kwon, D.; Min, B.S.; Cha, Y.J.; Koo, J.S.; Jeong, J.; Lee, J.; Choi, J. Interaction between CD36 and FABP4 modulates adipocyte-induced fatty acid import and metabolism in breast cancer. *NPJ Breast Cancer* **2021**, *7*, 129. [CrossRef]
19. Luo, X.; Zheng, E.; Wei, L.; Zeng, H.; Qin, H.; Zhang, X.; Liao, M.; Chen, L.; Zhao, L.; Ruan, X.Z.; et al. The fatty acid receptor CD36 promotes HCC progression through activating Src/PI3K/AKT axis-dependent aerobic glycolysis. *Cell Death Dis.* **2021**, *12*, 328. [CrossRef] [PubMed]
20. Yoshida, T.; Yokobori, T.; Saito, H.; Kuriyama, K.; Kumakura, Y.; Honjo, H.; Hara, K.; Sakai, M.; Miyazaki, T.; Obinata, H.; et al. CD36 expression is associated with cancer aggressiveness and energy source in esophageal squamous cell carcinoma. *Ann. Surg. Oncol.* **2021**, *28*, 1217–1227. [CrossRef] [PubMed]
21. Pardo, J.C.; Sanhueza, T.; de Porras, V.R.; Etxaniz, O.; Rodriguez, H.; Martinez-Cardús, A.; Grande, E.; Castellano, D.; Climent, M.A.; Lobato, T.; et al. Prognostic impact of CD36 immunohistochemical expression in patients with muscle-invasive bladder cancer treated with cystectomy and adjuvant chemotherapy. *J. Clin. Med.* **2022**, *11*, 497. [CrossRef] [PubMed]

22. Pascual, G.; Avgustinova, A.; Mejetta, S.; Martín, M.; Castellanos, A.; Attolini, C.S.; Berenguer, A.; Prats, N.; Toll, A.; Hueto, J.A.; et al. Targeting metastasis-initiating cells through the fatty acid receptor CD36. *Nature* **2017**, *541*, 41–45. [CrossRef] [PubMed]
23. Deng, M.; Cai, X.; Long, L.; Xie, L.; Ma, H.; Zhou, Y.; Liu, S.; Zeng, C. CD36 promotes the epithelial-mesenchymal transition and metastasis in cervical cancer by interacting with TGF-β. *J. Transl. Med.* **2019**, *17*, 352. [CrossRef] [PubMed]
24. Jiang, M.; Wu, N.; Xu, B.; Chu, Y.; Li, X.; Su, S.; Chen, D.; Li, W.; Shi, Y.; Gao, X.; et al. Fatty acid-induced CD36 expression via O-GlcNAcylation drives gastric cancer metastasis. *Theranostics* **2019**, *9*, 5359–5373. [CrossRef] [PubMed]
25. Aoki, T.; Kinoshita, J.; Munesue, S.; Hamabe-Horiike, T.; Yamaguchi, T.; Nakamura, Y.; Okamoto, K.; Moriyama, H.; Nakamura, K.; Harada, S.; et al. Hypoxia-induced CD36 expression in gastric cancer cells promotes peritoneal metastasis via fatty acid uptake. *Ann. Surg. Oncol.* **2022**, *30*, 3125–3136. [CrossRef] [PubMed]
26. Sakurai, K.; Tomihara, K.; Yamazaki, M.; Heshiki, W.; Moniruzzaman, R.; Sekido, K.; Tachinami, H.; Ikeda, A.; Imaue, S.; Fujiwara, K.; et al. CD36 expression on oral squamous cell carcinoma cells correlates with enhanced proliferation and migratory activity. *Oral. Dis.* **2020**, *26*, 745–755. [CrossRef]
27. Haidari, S.; Tröltzsch, M.; Knösel, T.; Liokatis, P.; Kasintsova, A.; Eberl, M.; Ortner, F.; Otto, S.; Fegg, F.; Boskov, M.; et al. Fatty acid receptor CD36 functions as a surrogate parameter for lymph node metastasis in oral squamous cell carcinoma. *Cancers* **2021**, *13*, 4125. [CrossRef]
28. Al-Khami, A.A.; Zheng, L.; Del Valle, L.; Hossain, F.; Wyczechowska, D.; Zabaleta, J.; Sanchez, M.D.; Dean, M.J.; Rodriguez, P.C.; Ochoa, A.C. Exogenous lipid uptake induces metabolic and functional reprogramming of tumor-associated myeloid-derived suppressor cells. *Oncoimmunology* **2017**, *6*, e1344804. [CrossRef]
29. Wang, H.; Franco, F.; Tsui, Y.C.; Xie, X.; Trefny, M.P.; Zappasodi, R.; Mohmood, S.R.; Fernández-García, J.; Tsai, C.H.; Schulze, I.; et al. CD36-mediated metabolic Ma X, adaptation supports regulatory T cell survival and function in tumors. *Nat. Immunol.* **2020**, *21*, 298–308. [CrossRef]
30. Ma, X.; Xiao, L.; Liu, L.; Ye, L.; Su, P.; Bi, E.; Wang, Q.; Yang, M.; Qian, J.; Yi, Q. CD36-mediated ferroptosis dampens intratumoral $CD8^+$ T cell effector function and impairs their antitumor ability. *Cell Metab.* **2021**, *33*, 1001–1012.e5. [CrossRef]
31. Xu, S.; Chaudhary, O.; Rodríguez-Morales, P.; Sun, X.; Chen, D.; Zappasodi, R.; Xu, Z.; Pinto, A.F.; Williams, A.; Schulze, I.; et al. Uptake of oxidized lipids by the scavenger receptor CD36 promotes lipid peroxidation and dysfunction in $CD8^+$ T cells in tumors. *Immunity* **2021**, *54*, 1561–1577.e7. [CrossRef] [PubMed]
32. Oh, D.S.; Lee, H.K. Autophagy protein ATG5 regulates CD36 expression and antitumour MHC class II antigen presentation in dendritic cells. *Autophagy* **2019**, *15*, 2091–2106. [CrossRef] [PubMed]
33. Tomihara, K.; Fuse, H.; Heshiki, W.; Takei, R.; Zhang, B.; Arai, N.; Nakamori, K.; Noguchi, M. Gemcitabine chemotherapy induces phenotypic alterations of tumor cells that facilitate antitumor T cell responses in a mouse model of oral cancer. *Oral Oncol.* **2014**, *50*, 457–467. [CrossRef]
34. Kridel, S.J.; Axelrod, F.; Rozenkrantz, N.; Smith, J.W. Orlistat is a novel inhibitor of fatty acid synthase with antitumor activity. *Cancer Res.* **2004**, *64*, 2070–2075. [CrossRef] [PubMed]
35. Currie, E.; Schulze, A.; Zechner, R.; Walther, T.C.; Farese, R.V., Jr. Cellular fatty acid metabolism and cancer. *Cell Metab.* **2013**, *18*, 153–161. [CrossRef] [PubMed]
36. Migita, T.; Ruiz, S.; Fornari, A.; Fiorentino, M.; Priolo, C.; Zadra, G.; Inazuka, F.; Grisanzio, C.; Palescandolo, E.; Shin, E.; et al. Fatty acid synthase: A metabolic enzyme and candidate oncogene in prostate cancer. *J. Natl. Cancer Inst.* **2009**, *101*, 519–532. [CrossRef]
37. Svensson, R.U.; Parker, S.J.; Eichner, L.J.; Kolar, M.J.; Wallace, M.; Brun, S.N.; Lombardo, P.S.; Van Nostrand, J.L.; Hutchins, A.; Vera, L.; et al. Inhibition of acetyl-CoA carboxylase suppresses fatty acid synthesis and tumor growth of non-small-cell lung cancer in preclinical models. *Nat. Med.* **2016**, *22*, 1108–1119. [CrossRef]
38. Ferraro, G.B.; Ali, A.; Luengo, A.; Kodack, D.P.; Deik, A.; Abbott, K.L.; Bezwada, D.; Blanc, L.; Prideaux, B.; Jin, X.; et al. Fatty acid synthesis is required for breast cancer brain metastasis. *Nat. Cancer* **2021**, *2*, 414–428. [CrossRef]
39. Chan, N.N.; Yamazaki, M.; Maruyama, S.; Abé, T.; Haga, K.; Kawaharada, M.; Izumi, K.; Kobayashi, T.; Tanuma, J.I. Cholesterol is a regulator of CAV1 localization and cell migration in oral squamous cell carcinoma. *Int. J. Mol. Sci.* **2023**, *24*, 6035. [CrossRef]
40. Herber, D.L.; Cao, W.; Nefedova, Y.; Novitskiy, S.V.; Nagaraj, S.; Tyurin, V.A.; Corzo, A.; Cho, H.I.; Celis, E.; Lennox, B.; et al. Lipid accumulation and dendritic cell dysfunction in cancer. *Nat. Med.* **2010**, *16*, 880–886. [CrossRef]
41. Zhu, C.Q.; Tang, Z.; Huang, R.; Qu, W.F.; Fang, Y.; Yang, R.; Tao, C.Y.; Gao, J.; Wu, X.L.; Sun, H.X.; et al. $CD36^+$ cancer-associated fibroblasts provide immunosuppressive microenvironment for hepatocellular carcinoma via secretion of macrophage migration inhibitory factor. *Cell Discov.* **2023**, *9*, 25. [CrossRef]
42. Tanaka, K.; Jinhua, P.; Omura, K.; Azuma, M. Multipotency of CD11bhighGr-1+ immature myeloid cells accumulating in oral squamous cell carcinoma-bearing mice. *Oral Oncol.* **2007**, *43*, 586–592. [CrossRef]
43. Takatsuka, D.; Tachinami, H.; Suzuki, N.; Yamazaki, M.; Yonesi, A.; Takaichi, M.; Imaue, S.; Yamada, S.I.; Tanuma, J.I.; Tomihara, K.; et al. PAK4 inhibition augments anti-tumour effect by immunomodulation in oral squamous cell carcinoma. *Sci. Rep.* **2024**, *14*, 14092. [CrossRef] [PubMed]

44. Yamaguchi, M.; Tomihara, K.; Heshiki, W.; Sakurai, K.; Sekido, K.; Tachinami, H.; Rohan, M.; Inoue, S.; Fujiwara, K.; Noguchi, M. Astaxanthin ameliorates cisplatin-induced damage in normal human fibroblasts. *Oral. Sci. Int.* **2019**, *16*, 171–177. [CrossRef]
45. Yonesi, A.; Tomihara, K.; Takatsuka, D.; Tachinami, H.; Yamazaki, M.; Jadidi, A.R.Y.; Takaichi, M.; Imaue, S.; Fujiwara, K.; Yamada, S.I.; et al. Rapamycin Induces Phenotypic Alterations in Oral Cancer Cells That May Facilitate Antitumor T Cell Responses. *Biomedicines* **2024**, *12*, 1078. [CrossRef]

International Journal of
Molecular Sciences

MDPI

Article

Development of Bivalent Aptamer-DNA Carrier-Doxorubicin Conjugates for Targeted Killing of Esophageal Squamous Cell Carcinoma Cells

Tianlu Zhang [1,†], Kai Yin [1,†], Xidong Niu [1], Xue Bai [1], Zhaoting Wang [1], Mengmeng Ji [1] and Baoyin Yuan [1,2,3,*]

1 School of Basic Medical Sciences, Zhengzhou University, Zhengzhou 450001, China; tianlu09062023@163.com (T.Z.); yink2000@163.com (K.Y.); nxd20031118@163.com (X.N.); baixue739808@163.com (X.B.); wangzhaoting7@163.com (Z.W.); jimeng0203@163.com (M.J.)
2 Henan Provincial Cooperative Innovation Center for Cancer Chemoprevention, Zhengzhou 450001, China
3 State Key Laboratory of Esophageal Cancer Prevention & Treatment, Zhengzhou University, Zhengzhou 450001, China
* Correspondence: yuanbaoyin@zzu.edu.cn
† These authors contributed equally to this work.

Abstract: Esophageal cancer ranks the seventh in cancer incidence and the sixth in cancer death. Esophageal squamous cell carcinoma (ESCC) accounts for approximately 90% of the total cases of esophageal cancer. Chemotherapy is the most effective drug-based method for treatment of esophageal cancer. However, severe side effects of traditional chemotherapy limit its treatment efficacy. Targeted chemotherapy can deliver chemotherapeutic drugs to cancer cells and specifically kill these cells with reduced side effects. In the work, the bivalent aptamer-DNA carrier (BAD) was designed by using an ESCC cell-specific aptamer as the recognition molecule and a GC base-rich DNA sequence as the drug carrier. With doxorubicin (Dox) as chemotherapeutic drugs, the bivalent aptamer-DNA-Dox conjugate (BADD) was constructed for targeted killing of ESCC cells. Firstly, the truncated A2(35) aptamer with a retained binding ability was obtained through optimization of an intact A2(80) aptamer and was used to fuse with DNA carrier sequences for constructing the BAD through simple DNA hybridization. The results of gel electrophoresis and flow cytometry analysis showed that the BAD was successfully constructed and had a stronger binding affinity than monovalent A2(35). Then, the BAD was loaded with Dox drugs to construct the BADD through noncovalent intercalation. The results of fluorescence spectra and flow cytometry assays showed that the BADD was successfully constructed and can bind to target cells strongly. Confocal imaging further displayed that the BADD can be specifically internalized into target cells and release Dox. The results of CCK-8 assays, Calcein AM/PI staining, and wound healing assays demonstrated that the BADD can specifically kill target cells, but not control cells. Our results demonstrate that the developed BADD can specifically deliver doxorubicin to target ESCC cells and selectively kill these cells, offering a potentially effective strategy for targeted chemotherapy of ESCC.

Keywords: ESCC; bivalent aptamer; DNA carrier; Dox; targeted chemotherapy

Citation: Zhang, T.; Yin, K.; Niu, X.; Bai, X.; Wang, Z.; Ji, M.; Yuan, B. Development of Bivalent Aptamer-DNA Carrier-Doxorubicin Conjugates for Targeted Killing of Esophageal Squamous Cell Carcinoma Cells. *Int. J. Mol. Sci.* **2024**, *25*, 7959. https://doi.org/10.3390/ijms25147959

Academic Editor: Kang-Nan Wang

Received: 19 June 2024
Revised: 17 July 2024
Accepted: 19 July 2024
Published: 21 July 2024

1. Introduction

Esophageal cancer is the seventh most common cancer and the sixth most common cause of cancer death worldwide [1]. Esophageal cancer mainly includes two subtypes: esophageal squamous cell carcinoma (ESCC) and esophageal adenocarcinoma (EAC). ESCC accounts for approximately 90% of esophageal cancer cases [2,3]. The major clinical treatment methods of esophageal cancer are surgical resection, radiotherapy, and chemotherapy [4,5]. Due to lack of widely overexpressed molecular targets, molecular-targeted therapy for esophageal cancer using small molecular inhibitors and monoclonal antibodies is not ideal [6]. Additionally, reliable specific recognition molecules for ESCC are lacked, which makes targeted delivery of drugs for ESCC difficult. Although immunotherapy is popular in recent years, it is limited in treatment of esophageal cancer because of low remission rates [7]. Traditional chemotherapy remains the most effective drug-based treatment for esophageal cancer, despite its severe side effects.

Targeted chemotherapy can specifically kill cancer cells by delivering chemotherapeutic drugs to target cancer cells but not normal cells, thereby reducing side effects and improving treatment efficacy of chemotherapeutic drugs. Doxorubicin (Dox) is an FDA-approved broad-spectrum chemotherapeutic drug to treat various cancers, including esophageal cancer [8,9]. Dox can nonspecifically enter into normal cells and lead to severe side effects, especially cardiotoxicity, which heavily restricts its clinical application [10–12]. It is a good way to resolve this issue to specifically deliver Dox to cancer cells for targeted chemotherapy of cancers. Because Dox can noncovalently intercalate into GC base-pairs of double-stranded DNA, DNA structures can be used as an ideal drug carrier for targeted delivery of Dox to cancer cells [13–15].

Aptamers are short single-stranded DNA or RNA molecules that can specifically recognize targets with high affinity [16]. Aptamers are usually selected through the systematic evolution of ligands by the exponential enrichment (SELEX) technique [17,18]. Known as novel recognition molecules, aptamers have various advantages over protein antibodies, such as easy synthesis, easy modification, flexible programmability, low cost, low immunogenicity, and low molecular weight [19,20]. Cancer cell-specific aptamers can be obtained via Cell-SELEX and are attracting more and more attention in cancer diagnosis, biomarker discovery, and targeted therapy [21–26]. For example, we have previously developed several aptamers target to ESCC cells and successfully identified potential cancer biomarkers using the aptamers [27–29]. Easy chemical modification and flexible design of aptamers allow them to carry chemotherapeutic drugs through covalent coupling or noncovalent intercalation and specifically deliver drugs to cancer cells via aptamer recognition, thus achieving targeted killing of cancer cells [30–33].

Bivalent aptamers have not only stronger binding ability and biostability than monovalent aptamers, but also a lower molecular weight than multivalent aptamers to penetrate tissues, which are more appropriate to use as recognition molecules for cancer cell targeting [34–36]. In addition, bivalent aptamers are facile to synthesize and construct and are controllable for drug loading. In previous work, we have developed an intact A2 aptamer that specifically binds to ESCC KYSE410 cells through Cell-SELEX and identified integrin β1 that is widely overexpressed in ESCC as the molecular target of A2 aptamer [27]. In this study, we aimed to obtain a truncated A2 aptamer by optimizing an intact A2 aptamer and then construct the bivalent aptamer-DNA carrier (BAD) through simple DNA hybridization by using the truncated A2 aptamer as the targeting element. The drug loading sequences (DNA carrier) located on the hybridization region of the BAD and were rich in GC base-pairs. With Dox as the chemotherapeutic drug, bivalent aptamer-DNA carrier-Dox conjugates (BADD) were constructed by noncovalently intercalating Dox into GC base-pairs of the DNA carrier. The BADD could specifically bind to target KYSE410 cells and then be internalized into target cells, killing target cells by releasing Dox. The proposed BADD may provide a simple and useful strategy for targeted chemotherapy of ESCC.

2. Results and Discussion

2.1. Design Principle of the BADD

To construct bivalent aptamer-DNA carrier-Dox conjugate (BADD), the bivalent aptamer-DNA carrier (BAD) was first designed. As shown in Figure 1, cancer cell-specific aptamer sequences are fused with drug loading sequences 1 (DLS1) and 2 (DLS2) separately, and the fused sequences are termed as Aptamer-DLS1 and Aptamer-DLS2, respectively. DLS1 and DLS2 are complementary to each other and rich in GC base-pairs. By mixing Aptamer-DLS1 with Aptamer-DLS2, the BAD can be simply constructed through base-pairing of DLS1 and DLS2. Doxorubicin (Dox) is a conventional FDA-approved chemotherapeutic drug to effectively treat various cancers. Due to its unique structure, Dox can noncovalently intercalate into GC base-pairs of double-stranded DNA sequences. Using this property, Dox molecules are loaded into the BAD that rich in GC base-pairs, which comprise the BADD. Because of the bivalent aptamer-induced specific and strong binding, the BADD can specifically recognize and enter into the target cells and release Dox for selective killing of cancer cells.

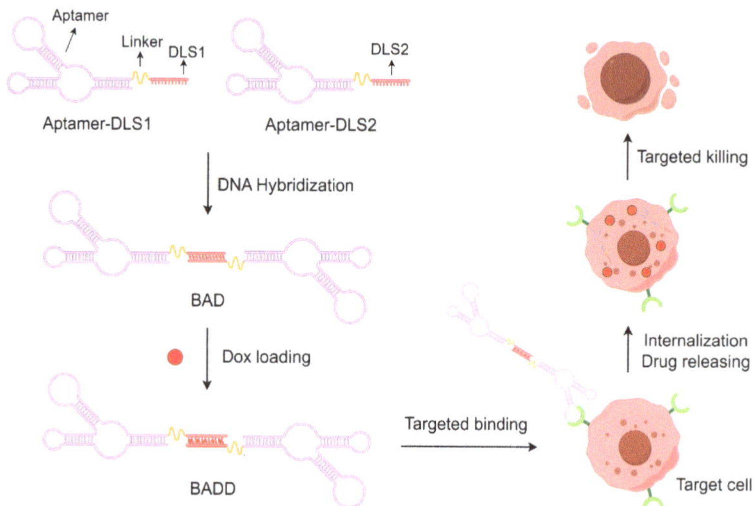

Figure 1. Schematic illustration of the design principle of the BADD for targeted killing of cancer cells.

2.2. Optimization of the Aptamer

In previous work, we have developed an intact A2(80) aptamer that can recognize ESCC cells, but not normal esophageal epithelial cells. The molecular target of A2(80) was identified as integrin β1 that is overexpressed in most ESCC cells, which possesses great potential for diagnosis and therapy of ESCC. Therefore, the A2(80) aptamer was used as a parental recognition molecule for construction of the BADD for targeted chemotherapy of ESCC. Because intact A2(80) is too long (80 bases) and contains redundant sequences that may affect its binding, we needed to optimize A2(80) to achieve a better truncated A2 aptamer. Flow cytometry analysis was used to detect the binding ability of A2(80) to target KYSE410 and control EC109 cells. The results showed that A2(80) had a good binding ability to target KYSE410 cells, but it almost did not bind to control EC109 cells (Figure 2A), verifying that A2 could specifically recognize target KYSE410 cells. The secondary structure of A2(80) was simulated by NUPACK software. As shown in Figure 2B, A2(80) includes two main loop structures, which probably contribute to aptamer binding according to previous reports [37–39]. Four truncated aptamer sequences were obtained by using four optimization methods (Figure 2B). A2(40) was obtained by removing the primer sequences at both the ends of A2(80). A2(35) was obtained by retaining both loops 1 and 2. A2(12) and A2(19) were obtained by retaining loops 1 and 2, respectively. The detailed aptamer sequences are listed in Table 1. These truncated sequences were labeled with a fluorescein amidite (FAM) fluorescent group for target cell binding assays by flow cytometry. As shown in Figure 2C, both A2(12) and A2(19) lost their binding abilities to target cells, while A2(40) and A2(35) retained a similar binding ability compared to A2(80), suggesting that both loops 1 and 2 are necessary in aptamer binding. Considering A2(35) has a good binding ability with a shorter length, it was selected for the subsequent experiments.

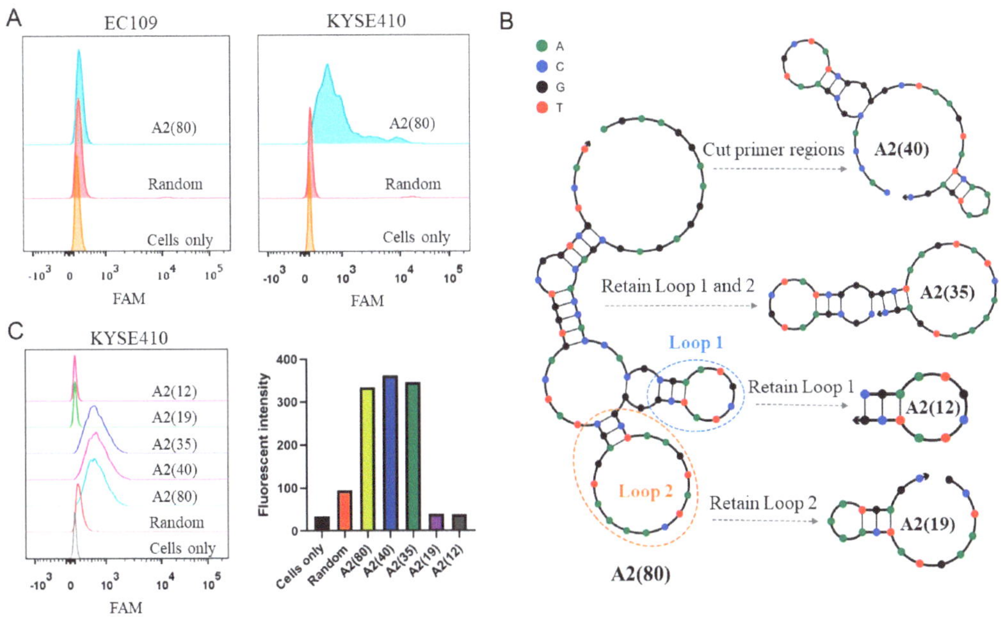

Figure 2. Optimization of the intact A2 aptamer. (**A**) Flow cytometry analysis of target KYSE410 and control EC109 cells incubated with the intact A2 aptamer. (**B**) Optimization strategies of the intact A2 aptamer. (**C**) Flow cytometry analysis of KYSE410 cells incubated with various truncated aptamers of A2. The corresponding quantitative fluorescent intensities were calculated.

Table 1. Detailed DNA sequences used in the work.[a]

Name	Sequences (5′–3′)
A2(80)	AGAAGGAAGGAGAGCGACACCACCACGCGAATGCTATCGGGGCTAAGTATCAAAATGAGCTATCAGTGGTCGGTCGTCAT
A2(40)	CACCACGCGAATGCTATCGGGGCTAAGTATCAAAATGAGC
A2(35)	CGCGAATGCTATCGGGGCTAAGTATCAAAATGAGC
A2(19)	GCTAAGTATCAAAATGAGC
A2(12)	CGAATGCTATCG
A2(35)-DLS1	CGTCGTCGTCGTCGTCGTTTTCGCGAATGCTATCGGGGCTAAGTATCAAAATGAGC
A2(35)-DLS2	ACGACGACGACGACGACGTTTCGCGAATGCTATCGGGGCTAAGTATCAAAATGAGC

[a] Aptamer recognition sequences are in brown, and drug loading sequences are in orange.

2.3. Construction of the BAD

To construct the BAD, a truncated A2(35) aptamer was fused with GC-rich DLS1 and DLS2 sequences separately. The fused A2(35)-DLS1 and A2(35)-DLS2 sequences (Table 1) were mixed in equal proportions to form the BAD through simple base-pairing. The agarose gel electrophoresis experiment was performed to verify the construction of the BAD. As shown in Figure 3A, the molecular weight of the product formed by mixing A2(35)-DLS1 and A2(35)-DLS2 became significantly larger, indicating that A2(35)-DLS1 and A2(35)-DLS2 successfully hybridized to form the BAD.

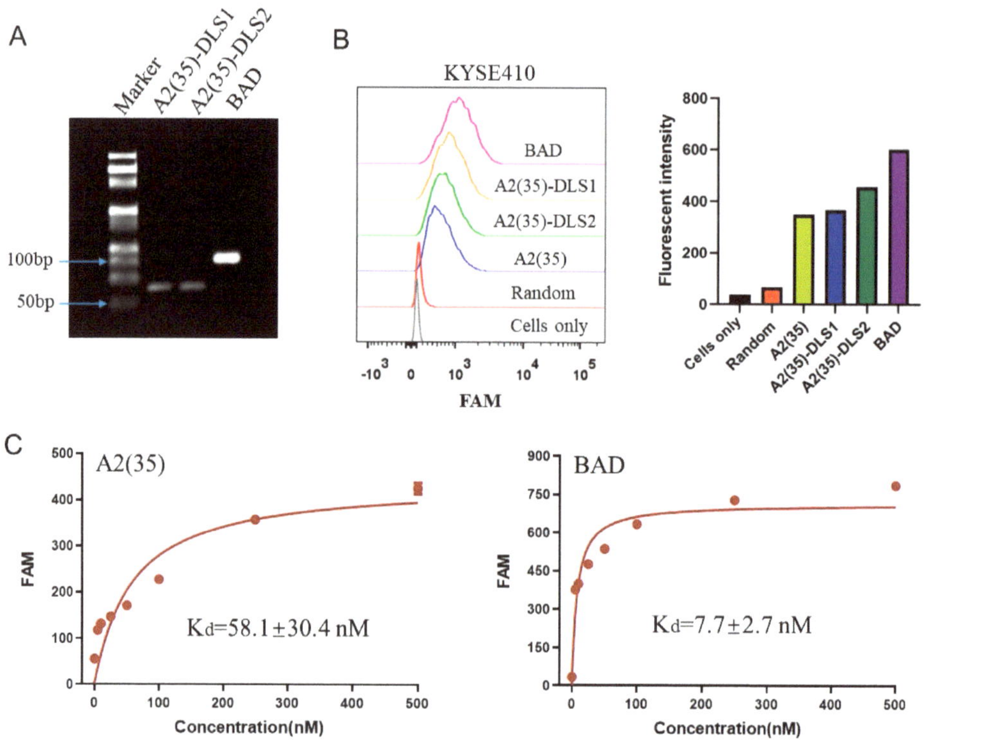

Figure 3. Construction and characterization of the BAD. (**A**) Formation of the BAD was analyzed by agarose gel electrophoresis. (**B**) Binding abilities of 250 nM of A2(35)-DLS1, A2(35)-DLS2, and the BAD to target KYSE410 cells detected by flow cytometry. Quantitative fluorescence intensity was calculated. (**C**) Determination of the dissociation constants (K_d) of A2(35) and the BAD to target KYSE410 cells. Fluorescence intensity values were obtained by flow cytometry, and K_d was calculated by GraphPad Prism 8 software.

2.4. Characterization of the BAD

After the successful construction of the BAD, flow cytometry was used to evaluate its binding ability to target cells. The results showed that the fluorescence signals of A2(35)-DLS1 and A2(35)-DLS2 to target cells were similar to that of A2(35), while the fluorescence signal of the BAD to target cells increased significantly (Figure 3B), suggesting an enhanced binding ability of the bivalent BAD to target cells compared to that of a monovalent aptamer. Moreover, we further evaluated the binding affinity by determining the dissociation constants (K_d) of A2(35) and the BAD binding to target KYSE410 cells. As shown in Figure 3C, the K_d values of A2(35) and the BAD to target cells were in low nanomolar levels, and the K_d of the BAD (7.7 ± 2.7 nM) was significantly lower than that of A2(35) (58.1 ± 30.4 nM), indicating that the BAD can bind to target cells with a higher affinity. The above results demonstrated that the constructed bivalent BAD has a stronger binding ability to target cells than a monovalent aptamer.

2.5. Construction of the BADD

After formation of the BAD, it was used as the DNA carrier to load Dox to construct the BADD for targeted killing of ESCC cells. Because the inherent fluorescence of Dox is quenched when Dox is intercalated in GC base-pairs, the fluorescence spectra of the BADD was recorded to verify the loading of Dox into the BAD. As displayed in Figure 4A, the fluorescence of Dox was gradually decreased with an increasing ratio of the BAD to

Dox, indicating that Dox was successfully loaded into the BAD. When the ratio of the BAD to Dox was 1:5, the fluorescence quenching almost reached saturation, indicating that the maximum drug loading was achieved with five Dox loaded into one BAD. Next, we investigated whether the Dox loading affected the binding of the BADD to target KYSE410 cells. The flow cytometry results showed that the BADD had a similar binding ability to target cells compared to the BAD (Figure 4B), suggesting that Dox loading had no reduced effects on the BADD binding to target cells. Additionally, the stability of the BADD in a serum-free medium was evaluated by detecting the fluorescence of leaching Dox. The results showed that the fluorescence intensity did not change significantly within 4 h (Figure 4C), suggesting that there was almost no leaching of free Dox into the medium and the BADD was stable in the medium.

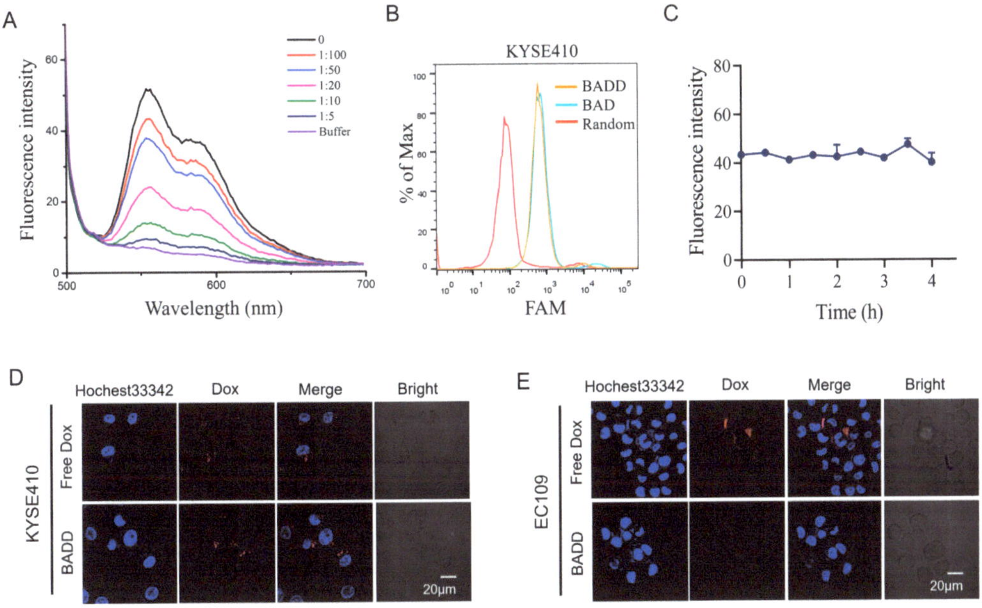

Figure 4. Construction and selective internalization of the BADD. (**A**) Fluorescence spectra of the BADD with different ratios of the BAD to Dox. The Dox concentration was fixed at 10 μM. (**B**) Binding abilities of the BAD and the BADD to target KYSE410 cells analyzed by flow cytometry. 250 nM of the BAD or the BADD was incubated with KYSE410 cells for 30 min. (**C**) Stability analysis of the BADD in a serum-free medium by detecting Dox fluorescence at different times. Internalization of the BADD to target KYSE410 cells (**D**) and control EC109 cells (**E**) were visualized by confocal imaging. 2 μM of the BADD was incubated with the cells for 2 h in a serum-free medium. Dox fluorescence was displayed in red. Cell nuclei were stained with Hochest33342 (blue). The scale bar is 20 μm.

2.6. Selective Internalization of the BADD

Subsequently, selective internalization of the BADD into target cells was evaluated using confocal imaging. The confocal images showed obvious Dox fluorescence in target KYSE410 cells, but not in control EC109 cells, after treatment with the BADD (Figure 4D), and showed similar strong Dox fluorescence in both KYSE410 and EC109 cells after treatment with free Dox (Figure 4E). These results indicated that the BADD can be selectively internalized into target cells and release Dox drugs, while free Dox can non-selectively enter into target and control cells and lead to undesired killing of control cells.

2.7. BADD-Induced Selective Killing of Target Cells

To evaluate BADD-induced selective killing of target cells, CCK-8 analysis was performed to detect cell viability. As shown in Figure 5A, the cell viability of control EC109 cells treated with the BADD was significantly higher than that of EC109 cells treated with free Dox, and the cell viabilities of target KYSE410 cells treated with the BADD and free Dox were similar. These results indicated that the BADD can selectively kill target cells, but not control cells. Calcein AM/PI staining was further performed to verify BADD-induced selective killing of target cells. Target KYSE410 and control EC109 cells were treated with the BAD, the BADD, and free Dox, respectively, and the corresponding cell death ratios were calculated. The results (Figure 5B) showed that the cell death ratios of EC109 cells treated with the BAD and the BADD, respectively, were significantly lower than that of EC109 cells treated with free Dox, while the cell death ratio of KYSE410 cells treated with BADD and free Dox were similar, suggesting that Dox can non-selectively kill KYSE410 and EC109 cells while the BADD can selectively kill target KYSE410 cells. Moreover, wound healing analysis was also performed to validate this effect. After the target KYSE410 and control EC109 cells were treated with free Dox and the BADD for 2 h in a serum-free medium and for 24 and 48 h in a complete medium, the corresponding wounded areas were measured. As shown in Figure 5C, the wounded areas of KYSE410 cells treated with free Dox and BADD were similar, while the wounded area of EC109 cells treated with BADD was lower than that of EC109 cells treated with free Dox, indicating that BADD can specifically inhibit target cell growth and migration. Lamellipodia can promote cell migration, which is usually used as the hallmark of migrating cells [22]. Therefore, lamellipodia imaging was performed to verify BADD-induced changes of cell migration ability. As shown in Figure 5D, almost no lamellipodia was observed on the target KYSE410 cells treated with the BADD compared to that on the control, while obvious lamellipodia could be observed on the EC109 cells treated with the BADD, which further verified that the BADD can specifically inhibit target KYSE410 cell growth and migration, but not control EC109 cells. The above results demonstrated that the BADD can selectively inhibit cell growth and may be used as an effective therapeutic reagent for targeted chemotherapy of ESCC.

In previous work, bivalent aptamers are usually used to deliver siRNA for targeted therapy, but it may need complicated design due to integration of siRNA sequences [40–42]. Circular bivalent aptamers have also been reported to deliver therapeutics with enhanced stability, while enzyme-catalyzed ligation make the synthesis of circular bivalent aptamers not easy [43–45]. Compared to the reported bivalent aptamer complex, the developed BADD can be fast constructed through simple DNA self-assembly and noncovalent intercalation of Dox without difficult sequence design. Because the BADD is composed of the target BAD carrier and Dox, it can specifically enter into cancer cells and release Dox and then kill the cancer cells through Dox. Therefore, the mechanism of the BADD killing ESCC cells is similar to that of Dox killing ESCC cells. As we know, Dox is an FDA-approved drug to treat various cancers, and its mechanism of action for killing tumor cells is well-known. According to the reports, Dox can induce cell death, especially apoptosis, via multiple modes of action, such as intercalate into DNA, inhibit topoisomerase II, produce reactive oxygen species, disrupt mitochondria function and evict histone [8,46,47].

Drug resistance often occurs in chemotherapy. It has been reported that EC109 cells are relatively sensitive to both cisplatin and paclitaxel while KYSE410 cells are relatively resistant to cisplatin [48]. Gene expression profiling is closely associated with these performances; for example, overexpression of MUC4 and MUC20 contributes to drug resistance, while overexpression of MUC13 contributes to drug sensitivity in ESCC cells. The drug resistance of the BADD in ESCC cells remains to be clarified. Although the BADD is feasible to selectively kill cancer cells in vitro, there are some limitations for its in vivo applications. For instance, the BAD carrier is a DNA in nature, so the BADD is probably sensitive to nuclease and may degrade in complex living systems [20]. In the future, improving the biostability of the BADD may be helpful for its in vivo applications. Chemical modifications, such as phosphorothioate, inverted dT, 2′-O-methyl, 2′-fluoro, locked nucleic acid modifications [49–51], or integration with DNA nanostructures [52,53], would be available strategies for improving the biostability of the BADD.

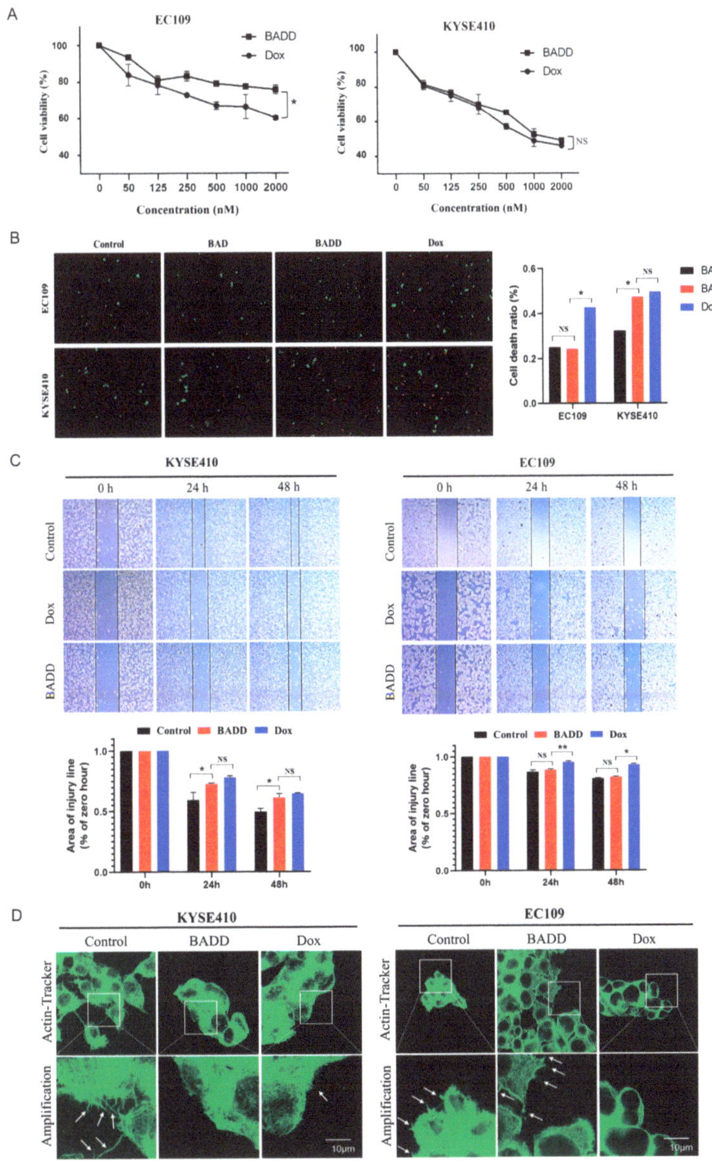

Figure 5. BADD-induced selective killing of target cells. (**A**) CCK-8 analysis of target KYSE410 and control EC109 cells treated with different concentrations (0, 0.05, 0.125, 0.25, 0.5, 1, and 2 μM) of Dox and the BADD for 2 h. (**B**) Cellular activity analysis of KYSE410 and EC109 cells treated with 2 μM of the BAD, Dox, and the BADD for 2 h and stained by Calcein AM/PI. Calcein AM-stained living cells are in green, and PI-stained dead cells are in red. The cell death ratio (number of dead cells/number of all cells) was calculated. (**C**) Wound healing analysis of KYSE410 and EC109 cells treated with 2 μM of Dox and the BADD for 2 h. The ratios of the wound areas at 24 and 48 h to that at 0 h were calculated. The asterisks denote significance (* $p < 0.05$; ** $p < 0.01$), NS denotes no significance. (**D**) Confocal images of lamellipodia at the edge of KYSE410 and EC109 cells treated with 2 μM of Dox and the BADD for 2 h. An actin-Tracker Green-488 dye was used to visualize lamellipodia. The arrow indicates lamellipodia. The scale bar is 10 μm.

3. Conclusion

In summary, we have developed the bivalent aptamer-DNA carrier-Dox conjugate (BADD) for targeted killing of ESCC cells. Firstly, a truncated A2(35) aptamer was obtained through optimization of an intact A2(80) aptamer. The flow cytometry results demonstrated that the A2(35) aptamer retained a strong binding ability to target KYSE410 cells. Then, the A2(35) aptamer was used as the recognition molecule to fuse with GC base-rich DNA sequences for constructing the bivalent aptamer-DNA carrier (BAD) through simple DNA hybridization. The agarose gel electrophoresis results showed that the BAD was successfully constructed, and flow cytometry assays showed that the BAD had a stronger binding ability than monovalent A2(35), and the K_d of the BAD was significantly lower than that of the monovalent A2(35). The BAD was then used to carry the chemotherapeutic drug Dox to construct the BADD. The fluorescence spectrum analysis showed that Dox was successfully loaded into the BAD and formed the BADD with a drug loading ratio of approximately 1:5. The flow cytometry assays showed that the BADD can bind to target cells strongly, and confocal imaging results further displayed that the BADD can be specifically internalized into target cells and release Dox. CCK-8 assays showed that the BADD can specifically kill target cells, but no significant killing effects on control cells, indicating selective cytotoxicity of the BADD. The selective cytotoxicity was further verified by Calcein AM/PI staining and wound healing analysis. The proposed BADD can be simply constructed and has great binding and targeting ability, which is expected to provide a feasible strategy for targeted chemotherapy of ESCC.

4. Materials and Methods

4.1. Chemicals and Materials

An RPMI-1640 culture medium and a fetal bovine serum (FBS) were purchased from Biological Industries (Beit-Haemek, Israel). DNA sequences were synthesized and purified by Sangon Biotech (Shanghai, China). Doxorubicin, phosphate-buffered saline (PBS), and a DNA marker were purchased from BBI (Shanghai, China). Agarose was purchased from Invitrogen (Waltham, MA, USA). A GelRed dye was purchased from Biosharp (Hefei, China). CCK-8 was purchased from Topscience (Shanghai, China). A Hoechst 33,342 dye, a Calcein AM/PI dye, bovine serum albumin (BSA), and a DNA loading buffer were purchased from Solarbio (Beijing, China). An Actin-Tracker Green-488 dye was purchased from Beyotime (Shanghai, China). The washing buffer was 0.01 M PBS with 4.5 g/L glucose and 5 mM $MgCl_2$. The binding buffer was a washing buffer with 1 mg/mL BSA and 0.1 mg/mL yeast tRNA.

4.2. Cells and Culturing

KYSE410 and EC109 ESCC cell lines were obtained from the Department of Pathophysiology, School of Basic Medical Sciences, Zhengzhou University. KYSE410 and EC109 cells were cultured in an RPMI-1640 medium supplied with 10% FBS and 100 μg/mL penicillin-gentamicin. The cells were aseptically cultured at 37 °C in a humidified incubator with 5% CO_2.

4.3. Preparation and Characterization of the BAD and the BADD

To construct the BAD, equal proportions of 1 μM of A2(35)-DLS1 and 1 μM of A2(35)-DLS2 were dissolved in PBS, heated at 95 °C for 5 min and then cooled to room temperature slowly. Agarose gel electrophoresis (2.5%, 100 V, 50 min) was performed to characterize the formation of the BAD. The gel was stained by the GelRed dye and subjected to UV imaging. To construct the BADD, 2 μM of BAD was mixed with 10 μM of Dox in PBS for 30 min at room temperature. To evaluate the Dox loading, the BAD was mixed with 10 μM of Dox in different proportions (BAD:Dox = 0, 1:100, 1:50, 1:20, 1:10, and 1:5). The Dox fluorescence was recorded using a fluorescence spectrometer.

4.4. Flow Cytometry Analysis

A2(80), A2(40), A2(35), A2(19), A2(12), A2(35)-DLS1, A2(35)-DLS2, and random sequences were separately labeled with an FAM fluorescence group at the 5′ end for flow cytometry analysis. Generally, KYSE410 or EC109 cells (approximately 1×10^5) were mixed with different probes (250 nM) in a binding buffer for 30 min on ice. After incubation, cells were washed with the washing buffer for 3 times and resuspended in 400 μL of the binding buffer for flow cytometry analysis. The data were processed by Flowjo 7.6 software. To determine the dissociation constants (K_d) of A2(35) and the BAD to KYSE410 cells, different concentrations (0, 10, 25, 50, 100, 200, and 500 nM) of A2(35) or the BAD were separately mixed with KYSE410 cells, and the fluorescence values of their binding were detected by flow cytometry. K_d was calculated by $Y = B_{max} X/(K_d + X)$ using GraphPad Prism 8 software, where X represents the probe concentration and Y represents the fluorescence value.

4.5. Confocal Imaging

KYSE410 and EC109 cells were seeded in 35 mm confocal dishes (BDD012035, JET BIOFIL, Guangzhou, China) and cultured for 24 h. After the old medium was removed, 2 μM of Dox or the BADD were incubated with the cells for 2 h in a serum-free medium. After washing with PBS for 3 times, the cells were mixed with Hoechst 33,342 for cell nuclei staining. After washing, the medium was added, and the cells were imaged by a laser confocal microscope (Olympus, Tokyo, Japan). Dox fluorescence was excicted at a wavelength of 545 nm and collected at a wavelength of 570 nm with a long-pass filter; Hoechst 33,342 fluorescence was excited at a wavelength of 405 nm and collected at a wavelength range of 435–460 nm. To visualize lamellipodia, KYSE410 and EC109 cells were treated with 2 μM of Dox or the BADD for 2 h. After cultured in the medium for 48 h, the cells were fixed with 3% paraformaldehyde for 15 min and then permeabilized with 0.3% Triton X-100 for 5 min. Afterwards, the cells were stained with the Actin-Tracker Green-488 dye for 1 h and captured by the confocal microscope.

4.6. CCK-8 Analysis

KYSE410 and EC109 cells (6000 cells/well) were seeded in 96-well plates and cultured for 24 h. After removing the old medium, different concentrations (0, 0.05, 0.125, 0.25, 0.5, 1, and 2 μM) of Dox or the BADD were incubated with the cells for 2 h in a serum-free medium. Then, the solution was removed and replaced with a culture medium. The cells were then cultured at 37 °C for 48 h in the incubator. Subsequently, 10 μL of CCK-8 solution was added to each well and incubated at 37 °C for 2 h. The absorbance at 450 nm was detected by a microplate reader.

4.7. Calcein AM/PI Staining

KYSE410 and EC109 cells (1.5×10^4 cells/well) were seeded into 24-well plates and cultured for 24 h. After removing the old medium, 2 μM of Dox or the BADD was added into each well and cultured for 2 h in a serum-free medium. The solution was removed, and cultivation continued at 37 °C for 48 h in the medium. The adherent cells and the detached cells suspended in the culture medium were all collected. The cells were stained with the diluted Calcein AM/PI solution at 37 °C for 30 min in dark. The cell images were captured by a fluorescence microscope.

4.8. Wound Healing Analysis

KYSE410 and EC109 cells (5×10^5 cells/well) were seeded into 6-well plates and cultured at 37 °C for 24 h. A 10 μL pipette tip was used to wound the cells, and the detached cells were washed with PBS. Then, 2 μM of Dox or the BADD was added into each well with a serum-free medium. After incubation for 2 h, the solution was removed, and cultivation continued at 37 °C. Cell images were captured using a microscope at 0, 24, and 48 h from the same field. The wound healing area was measured by ImageJ 2 software.

Author Contributions: Conceptualization, B.Y.; Methodology, T.Z. and K.Y.; Software, K.Y. and X.N.; Validation, T.Z., K.Y., X.B., Z.W. and M.J.; Formal analysis, X.N., X.B., Z.W. and M.J.; Writing—original draft, T.Z. and X.N.; Writing—review & editing, B.Y.; Supervision, B.Y.; Funding acquisition, B.Y. All authors have read and agreed to the published version of the manuscript.

Funding: The work was supported by the National Natural Science Foundations of China (No. 22007084) and the China Postdoctoral Science Foundation (No. 2023M733215).

Institutional Review Board Statement: Not applicable.

Informed Consent Statement: Not applicable.

Data Availability Statement: The data presented in this study are available upon request from the corresponding author.

Acknowledgments: The schematic illustration of the design principle of the BADD in Figure 1 was drawn by using Figdraw 2.0.

Conflicts of Interest: The authors declare no competing financial interest.

References

1. Sung, H.; Ferlay, J.; Siegel, R.L.; Laversanne, M.; Soerjomataram, I.; Jemal, A.; Bray, F. Global Cancer Statistics 2020: GLOBOCAN Estimates of Incidence and Mortality Worldwide for 36 Cancers in 185 Countries. *CA Cancer J. Clin.* **2021**, *71*, 209–249. [CrossRef] [PubMed]
2. Abnet, C.C.; Arnold, M.; Wei, W.Q. Epidemiology of Esophageal Squamous Cell Carcinoma. *Gastroenterology* **2018**, *154*, 360–373. [CrossRef] [PubMed]
3. Smyth, E.C.; Lagergren, J.; Fitzgerald, R.C.; Lordick, F.; Shah, M.A.; Lagergren, P.; Cunningham, D. Oesophageal cancer. *Nat. Rev. Dis. Primers* **2017**, *3*, 17048. [CrossRef] [PubMed]
4. Lagergren, J.; Smyth, E.; Cunningham, D.; Lagergren, P. Oesophageal cancer. *Lancet* **2017**, *390*, 2383–2396. [CrossRef] [PubMed]
5. Ohashi, S.; Miyamoto, S.; Kikuchi, O.; Goto, T.; Amanuma, Y.; Muto, M. Recent Advances from Basic and Clinical Studies of Esophageal Squamous Cell Carcinoma. *Gastroenterology* **2015**, *149*, 1700–1715. [CrossRef] [PubMed]
6. Yang, Y.M.; Hong, P.; Xu, W.W.; He, Q.Y.; Li, B. Advances in targeted therapy for esophageal cancer. *Signal Transduct. Target Ther.* **2020**, *5*, 229. [CrossRef]
7. He, S.; Xu, J.; Liu, X.; Zhen, Y. Advances and challenges in the treatment of esophageal cancer. *Acta Pharm. Sin. B* **2021**, *11*, 3379–3392. [CrossRef] [PubMed]
8. Kciuk, M.; Gielecińska, A.; Mujwar, S.; Kołat, D.; Kałuzińska-Kołat, Ż.; Celik, I.; Kontek, R. Doxorubicin-An Agent with Multiple Mechanisms of Anticancer Activity. *Cells* **2023**, *12*, 659. [CrossRef]
9. Carvalho, C.; Santos, R.X.; Cardoso, S.; Correia, S.; Oliveira, P.J.; Santos, M.S.; Moreira, P.I. Doxorubicin: The Good, the Bad and the Ugly Effect. *Curr. Med. Chem.* **2009**, *16*, 3267–3285. [CrossRef]
10. Kong, C.Y.; Guo, Z.; Song, P.; Zhang, X.; Yuan, Y.P.; Teng, T.; Yan, L.; Tang, Q.Z. Underlying the Mechanisms of Doxorubicin-Induced Acute Cardiotoxicity: Oxidative Stress and Cell Death. *Int. J. Biol. Sci.* **2022**, *18*, 760–770. [CrossRef]
11. Pugazhendhi, A.; Edison, T.; Velmurugan, B.K.; Jacob, J.A.; Karuppusamy, I. Toxicity of Doxorubicin (Dox) to different experimental organ systems. *Life Sci.* **2018**, *200*, 26–30. [CrossRef] [PubMed]
12. Luo, W.; Zou, X.; Wang, Y.; Dong, Z.; Weng, X.; Pei, Z.; Song, S.; Zhao, Y.; Wei, Z.; Gao, R.; et al. Critical Role of the cGAS-STING Pathway in Doxorubicin-Induced Cardiotoxicity. *Circ. Res.* **2023**, *132*, e223–e242. [CrossRef]
13. Ijäs, H.; Shen, B.; Heuer-Jungemann, A.; Keller, A.; Kostiainen, M.A.; Liedl, T.; Ihalainen, J.A.; Linko, V. Unraveling the interaction between doxorubicin and DNA origami nanostructures for customizable chemotherapeutic drug release. *Nucleic Acids Res.* **2021**, *49*, 3048–3062. [CrossRef]
14. Yuan, B.; Xi, Y.; Qi, C.; Zhao, M.; Zhu, X.; Tang, J. A sequentially triggered DNA nanocapsule for targeted drug delivery based on pH-responsive i-motif and tumor cell-specific aptamer. *Front. Bioeng. Biotechnol.* **2022**, *10*, 965337. [CrossRef]
15. Ma, W.; Sun, H.; Chen, B.; Jia, R.; Huang, J.; Cheng, H.; He, X.; Huang, M.; Wang, K. Engineering a Facile Aptamer "Molecule-Doctor" with Hairpin-Contained I-Motif Enables Accurate Imaging and Killing of Cancer Cells. *Anal. Chem.* **2021**, *93*, 14552–14559. [CrossRef] [PubMed]
16. Ellington, A.D.; Szostak, J.W. In vitro selection of RNA molecules that bind specific ligands. *Nature* **1990**, *346*, 818–822. [CrossRef] [PubMed]
17. DeRosa, M.C.; Lin, A.; Mallikaratchy, P.; McConnell, E.M.; McKeague, M.; Patel, R.; Shigdar, S. In vitro selection of aptamers and their applications. *Nat. Rev. Methods Primers* **2023**, *3*, 55. [CrossRef]
18. Tuerk, C.; Gold, L. Systematic evolution of ligands by exponential enrichment: RNA ligands to bacteriophage T4 DNA polymerase. *Science* **1990**, *249*, 505–510. [CrossRef] [PubMed]
19. Li, L.; Xu, S.; Yan, H.; Li, X.; Yazd, H.S.; Li, X.; Huang, T.; Cui, C.; Jiang, J.; Tan, W. Nucleic Acid Aptamers for Molecular Diagnostics and Therapeutics: Advances and Perspectives. *Angew. Chem. Int. Ed.* **2021**, *60*, 2221–2231. [CrossRef]

20. Zhou, J.; Rossi, J. Aptamers as targeted therapeutics: Current potential and challenges. *Nat. Rev. Drug Discov.* **2017**, *16*, 181–202. [CrossRef]
21. Tang, J.; Li, B.; Qi, C.; Wang, Z.; Yin, K.; Guo, L.; Zhang, W.; Yuan, B. Imaging specific cell-surface sialylation using DNA dendrimer-assisted FRET. *Talanta* **2022**, *243*, 123399. [CrossRef] [PubMed]
22. Tang, J.; Qi, C.; Bai, X.; Ji, M.; Wang, Z.; Luo, Y.; Ni, S.; Zhang, T.; Liu, K.; Yuan, B. Cell Membrane-Anchored DNA Nanoinhibitor for Inhibition of Receptor Tyrosine Kinase Signaling Pathways via Steric Hindrance and Lysosome-Induced Protein Degradation. *ACS Pharmacol. Transl. Sci.* **2024**, *7*, 110–119. [CrossRef] [PubMed]
23. Li, Z.; Yuan, B.; Lin, X.; Meng, X.; Wen, X.; Guo, Q.; Li, L.; Jiang, H.; Wang, K. Intramolecular trigger remodeling-induced HCR for amplified detection of protein-specific glycosylation. *Talanta* **2020**, *215*, 120889. [CrossRef] [PubMed]
24. Li, Y.; Tam, W.W.; Yu, Y.; Zhuo, Z.; Xue, Z.; Tsang, C.; Qiao, X.; Wang, X.; Wang, W.; Li, Y.; et al. The application of Aptamer in biomarker discovery. *Biomark. Res.* **2023**, *11*, 70. [CrossRef] [PubMed]
25. Xiong, H.; Liu, L.; Liu, X.; Jiang, H.; Wang, X. Aptamer-Based Immune Drug Systems (AptIDCs) Potentiating Cancer Immunotherapy. *Chemistry* **2023**, *5*, 1656–1680. [CrossRef]
26. Shangguan, D.; Li, Y.; Tang, Z.; Cao, Z.C.; Chen, H.W.; Mallikaratchy, P.; Sefah, K.; Yang, C.J.; Tan, W. Aptamers evolved from live cells as effective molecular probes for cancer study. *Proc. Natl. Acad. Sci. USA* **2006**, *103*, 11838–11843. [CrossRef] [PubMed]
27. Zhang, Y.; Chen, X.; Qiao, Y.; Yang, S.; Wang, Z.; Ji, M.; Yin, K.; Zhao, J.; Liu, K.; Yuan, B. DNA Aptamer Selected against Esophageal Squamous Cell Carcinoma for Tissue Imaging and Targeted Therapy with Integrin beta1 as a Molecular Target. *Anal. Chem.* **2022**, *94*, 17212–17222. [CrossRef] [PubMed]
28. Chen, X.; Zhang, Y.; Shi, Y.; Niu, T.; Li, B.; Guo, L.; Qiao, Y.; Zhao, J.; Yuan, B.; Liu, K. Evolution of DNA aptamers against esophageal squamous cell carcinoma using cell-SELEX. *Analyst* **2021**, *146*, 4180–4187. [CrossRef] [PubMed]
29. Qiao, Y.; Shi, Y.; Ji, M.; Wang, Z.; Bai, X.; Zhang, K.; Yin, K.; Zhang, Y.; Chen, X.; Zhang, Y.; et al. Selection and identification of a prohibitin 2-binding DNA aptamer for tumor tissue imaging and targeted chemotherapy. *Int. J. Biol. Macromol.* **2024**, *259*, 129002. [CrossRef]
30. Zhu, L.; Yang, J.; Ma, Y.; Zhu, X.; Zhang, C. Aptamers Entirely Built from Therapeutic Nucleoside Analogues for Targeted Cancer Therapy. *J. Am. Chem. Soc.* **2022**, *144*, 1493–1497. [CrossRef]
31. Li, F.; Lu, J.; Liu, J.; Liang, C.; Wang, M.; Wang, L.; Li, D.; Yao, H.; Zhang, Q.; Wen, J.; et al. A water-soluble nucleolin aptamer-paclitaxel conjugate for tumor-specific targeting in ovarian cancer. *Nat. Commun.* **2017**, *8*, 1390. [CrossRef] [PubMed]
32. Zhang, H.; Jin, C.; Zhang, L.; Peng, B.; Zhang, Y.; Liu, Y.; Li, L.; Ye, M.; Xiong, W.; Tan, W. CD71-Specific Aptamer Conjugated with Monomethyl Auristatin E for the Treatment of Uveal Melanoma. *ACS Appl. Mater. Interfaces* **2022**, *14*, 32–40. [CrossRef] [PubMed]
33. Li, Y.; Peng, Y.; Tan, Y.; Xuan, W.; Fu, T.; Wang, X.Q.; Tan, W. A new paradigm for artesunate anticancer function: Considerably enhancing the cytotoxicity via conjugating artesunate with aptamer. *Signal Transduct. Target Ther.* **2021**, *6*, 327. [CrossRef] [PubMed]
34. Hu, X.; Tang, L.; Zheng, M.; Liu, J.; Zhang, Z.; Li, Z.; Yang, Q.; Xiang, S.; Fang, L.; Ren, Q.; et al. Structure-Guided Designing Pre-Organization in Bivalent Aptamers. *J. Am. Chem. Soc.* **2022**, *144*, 4507–4514. [CrossRef] [PubMed]
35. Li, X.; Chen, X.; Mao, M.; Peng, C.; Wang, Z. Accelerated CRISPR/Cas12a-based small molecule detection using bivalent aptamer. *Biosens. Bioelectron.* **2022**, *217*, 114725. [CrossRef]
36. Riccardi, C.; Napolitano, E.; Musumeci, D.; Montesarchio, D. Dimeric and Multimeric DNA Aptamers for Highly Effective Protein Recognition. *Molecules* **2020**, *25*, 5227. [CrossRef] [PubMed]
37. Wang, S.; Ma, R.; Li, L.; Wang, L.; Li, J.; Sun, J.; Mao, X.; Tan, W. Engineering Robust Aptamers with High Affinity by Key Fragment Evolution and Terminal Fixation. *Anal. Chem.* **2022**, *94*, 16282–16289. [CrossRef]
38. Gao, S.; Zheng, X.; Jiao, B.; Wang, L. Post-SELEX optimization of aptamers. *Anal. Bioanal. Chem.* **2016**, *408*, 4567–4573. [CrossRef] [PubMed]
39. Yuan, B.; Zhou, Y.; Guo, Q.; Wang, K.; Yang, X.; Meng, X.; Wan, J.; Tan, Y.; Huang, Z.; Xie, Q.; et al. A signal-on split aptasensor for highly sensitive and specific detection of tumor cells based on FRET. *Chem. Commun.* **2016**, *52*, 1590–1593. [CrossRef]
40. Xue, L.; Maihle, N.J.; Yu, X.; Tang, S.C.; Liu, H.Y. Synergistic Targeting HER2 and EGFR with Bivalent Aptamer-siRNA Chimera Efficiently Inhibits HER2-Positive Tumor Growth. *Mol. Pharm.* **2018**, *15*, 4801–4813. [CrossRef]
41. Yang, F.; Li, S.; Yuan, R.; Xiang, Y. A bivalent aptamer and terminus-free siRNA junction nanostructure for targeted gene silencing in cancer cells. *J. Mater. Chem. B* **2022**, *10*, 8315–8321. [CrossRef] [PubMed]
42. Cong, Y.; Liu, K.; Huang, Z.; Lu, J.; Wang, H.-H.; Hong, Y.; Xie, Z.; Li, H. A Bivalent Aptamer-Based DNA Agonist for EGFR Signaling Effectively Alleviates Ulcerative Colitis In Vivo. *ACS Chem. Biol.* **2024**, *19*, 1280–1290. [CrossRef]
43. Kuai, H.; Zhao, Z.; Mo, L.; Liu, H.; Hu, X.; Fu, T.; Zhang, X.; Tan, W. Circular Bivalent Aptamers Enable in Vivo Stability and Recognition. *J. Am. Chem. Soc.* **2017**, *139*, 9128–9131. [CrossRef]
44. Jiang, Y.; Pan, X.; Chang, J.; Niu, W.; Hou, W.; Kuai, H.; Zhao, Z.; Liu, J.; Wang, M.; Tan, W. Supramolecularly Engineered Circular Bivalent Aptamer for Enhanced Functional Protein Delivery. *J. Am. Chem. Soc.* **2018**, *140*, 6780–6784. [CrossRef]
45. Zhou, F.; Wang, P.; Peng, Y.; Zhang, P.; Huang, Q.; Sun, W.; He, N.; Fu, T.; Zhao, Z.; Fang, X.; et al. Molecular Engineering-Based Aptamer-Drug Conjugates with Accurate Tunability of Drug Ratios for Drug Combination Targeted Cancer Therapy. *Angew. Chem. Int. Ed.* **2019**, *58*, 11661–11665. [CrossRef] [PubMed]
46. Zhao, H.; Yu, J.; Zhang, R.; Chen, P.; Jiang, H.; Yu, W. Doxorubicin prodrug-based nanomedicines for the treatment of cancer. *Eur. J. Med. Chem.* **2023**, *258*, 115612. [CrossRef]

47. Agudelo, D.; Bourassa, P.; Bérubé, G.; Tajmir-Riahi, H.A. Intercalation of antitumor drug doxorubicin and its analogue by DNA duplex: Structural features and biological implications. *Int. J. Biol. Macromol.* **2014**, *66*, 144–150. [CrossRef]
48. Shen, L.Y.; Wang, H.; Dong, B.; Yan, W.P.; Lin, Y.; Shi, Q.; Chen, K.N. Possible prediction of the response of esophageal squamous cell carcinoma to neoadjuvant chemotherapy based on gene expression profiling. *Oncotarget* **2016**, *7*, 4531–4541. [CrossRef] [PubMed]
49. Egli, M.; Manoharan, M. Chemistry, structure and function of approved oligonucleotide therapeutics. *Nucleic Acids Res.* **2023**, *51*, 2529–2573. [CrossRef]
50. Wang, R.E.; Wu, H.; Niu, Y.; Cai, J. Improving the stability of aptamers by chemical modification. *Curr. Med. Chem.* **2011**, *18*, 4126–4138. [CrossRef]
51. Ni, S.; Yao, H.; Wang, L.; Lu, J.; Jiang, F.; Lu, A.; Zhang, G. Chemical Modifications of Nucleic Acid Aptamers for Therapeutic Purposes. *Int. J. Mol. Sci.* **2017**, *18*, 1683. [CrossRef] [PubMed]
52. Chandrasekaran, A. R, Nuclease resistance of DNA nanostructures. *Nat. Rev. Chem.* **2021**, *5*, 225–239. [CrossRef] [PubMed]
53. Qi, C.; Li, W.; Luo, Y.; Ni, S.; Ji, M.; Wang, Z.; Zhang, T.; Bai, X.; Tang, J.; Yuan, B.; et al. Selective inhibition of c-Met signaling pathways with a bispecific DNA nanoconnector for the targeted therapy of cancer. *Int. J. Biol. Macromol.* **2024**, *273*, 133134. [CrossRef] [PubMed]

International Journal of
Molecular Sciences

MDPI

Review

Exploring the Key Signaling Pathways and ncRNAs in Colorectal Cancer

Yun Ju Lee [1,2], Woo Ryung Kim [1,2], Eun Gyung Park [1,2], Du Hyeong Lee [1,2], Jung-min Kim [1,2], Hae Jin Shin [1,2], Hyeon-su Jeong [1,2], Hyun-Young Roh [2,3] and Heui-Soo Kim [2,3,*]

1 Department of Integrated Biological Sciences, Pusan National University, Busan 46241, Republic of Korea; lsg5821@naver.com (Y.J.L.); dnfud647@pusan.ac.kr (W.R.K.); ehdtodt@pusan.ac.kr (E.G.P.); doo2080@naver.com (D.H.L.); jmk95@naver.com (J.-m.K.); 0705haejin@naver.com (H.J.S.); tbd97@pusan.ac.kr (H.-s.J.)

2 Institute of Systems Biology, Pusan National University, Busan 46241, Republic of Korea; susan9416@naver.com

3 Department of Biological Sciences, College of Natural Sciences, Pusan National University, Busan 46241, Republic of Korea

* Correspondence: khs307@pusan.ac.kr

Abstract: Colorectal cancer (CRC) is the third most prevalent cancer to be diagnosed, and it has a substantial mortality rate. Despite numerous studies being conducted on CRC, it remains a significant health concern. The disease-free survival rates notably decrease as CRC progresses, emphasizing the urgency for effective diagnostic and therapeutic approaches. CRC development is caused by environmental factors, which mostly lead to the disruption of signaling pathways. Among these pathways, the Wingless/Integrated (Wnt) signaling pathway, Phosphatidylinositol 3-kinase/protein kinase B/mammalian target of rapamycin (PI3K/AKT/mTOR) signaling pathway, Mitogen-Activated Protein Kinase (MAPK) signaling pathway, Transforming Growth Factor-β (TGF-β) signaling pathway, and p53 signaling pathway are considered to be important. These signaling pathways are also regulated by non-coding RNAs (ncRNAs), including microRNAs (miRNAs), long non-coding RNAs (lncRNAs), and circular RNAs (circRNAs). They have emerged as crucial regulators of gene expression in CRC by changing their expression levels. The altered expression patterns of these ncRNAs have been implicated in CRC progression and development, suggesting their potential as diagnostic and therapeutic targets. This review provides an overview of the five key signaling pathways and regulation of ncRNAs involved in CRC pathogenesis that are studied to identify promising avenues for diagnosis and treatment strategies.

Keywords: colorectal cancer; Wnt signaling pathway; PI3K/AKT/mTOR signaling pathway; MAPK signaling pathway; TGF-β signaling pathway; p53 signaling pathway; miRNA; long non-coding RNA; circular RNA; CeRNA

Citation: Lee, Y.J.; Kim, W.R.; Park, E.G.; Lee, D.H.; Kim, J.-m.; Shin, H.J.; Jeong, H.-s.; Roh, H.-Y.; Kim, H.-S. Exploring the Key Signaling Pathways and ncRNAs in Colorectal Cancer. *Int. J. Mol. Sci.* **2024**, *25*, 4548. https://doi.org/10.3390/ijms25084548

Academic Editor: Peter J. K. Kuppen

Received: 29 March 2024
Revised: 19 April 2024
Accepted: 19 April 2024
Published: 21 April 2024

1. Introduction

Colorectal cancer (CRC) is the third most commonly diagnosed cancer in both males and females, with an estimated 81,540 new cases in males and 71,720 new cases in females in 2024, according to the American Cancer Society [1]. In addition, CRC is estimated to cause significant mortality, with approximately 28,700 deaths in males, ranking the top three among males, and 24,310 deaths in females, positioning it as the fourth among females. An examination of the five-year disease-free survival rate revealed a notable trend: as the stage of CRC advances, the survival rates decrease. Specifically, the five-year disease-free survival rates are 91.0% for stage I, 79.8% for stage II, 63.3% for stage III, and 18.9% for stage IV [2]. These statistics highlight the severity of this disease and underscore the need for effective strategies for diagnosing and treating CRC.

CRC develops as a result of abnormal growth cells in the colon or rectum, which over time turn cancerous [3]. It is commonly known that obesity, overweight, a lack of physical activity, smoking, alcohol consumption, and diet are risk factors for CRC [4–6]. Most CRCs are thought to be caused by genetic alterations induced by these environmental factors [7]. Consequently, the genetic alterations can induce the aberrant activation of signaling pathways, leading to tumor progression [8]. Many dysregulated signaling pathways in CRC have been identified, most notably the Wingless/Integrated (Wnt) signaling pathway, Phosphatidylinositol 3-kinase/protein kinase B/mammalian target of rapamycin (PI3K/AKT/mTOR) signaling pathway, Mitogen-Activated Protein Kinase (MAPK) signaling pathway, Transforming Growth Factor-β (TGF-β) signaling pathway, and p53 signaling pathway [9]. Aberrant signaling pathways occur due to abnormalities in the expression of genes that function in the pathway, which also affects genes downstream [8]. As these studies continue to expand, there is increasing interest in non-coding RNAs (ncRNAs) as regulators of gene expression.

Unlike coding RNAs, ncRNAs, which are transcribed from DNA but not translated into protein, have been shown to act as a regulator [10]. Well-known ncRNAs such as microRNAs (miRNAs), long non-coding RNAs (lncRNAs), and circular RNAs (circRNAs) have been extensively studied and shown to directly or indirectly regulate the expression of genes involved in CRC, thereby influencing various biological processes [11]. Due to their altered expression patterns and their significant function in CRC, these ncRNAs can also be used not only for the diagnosis but also as potential targets for the treatment. In this review, we aim to summarize the five major signaling pathways and ncRNAs implicated in CRC, to identify potential targets for diagnosis and treatment.

2. Multiple Major Signaling Pathways in CRC

In the human body, various signaling pathways exist, regulating essential cellular processes. However, when these pathways are disrupted, they can lead to various diseases [12]. CRC is no exception, and through numerous studies, differences in signaling pathways between normal and CRC patients have been extensively identified [8]. By comparing these signaling pathways with KEGG pathways (map05210), five signaling pathways play crucial roles in influencing CRC: the Wnt signaling pathway, PI3K/AKT/mTOR signaling pathway, MAPK signaling pathway, TGF-β signaling pathway, and p53 signaling pathway [9,13]. Although these signaling pathways ultimately regulate similar cellular characteristics, there are differences in the genes involved and the factors that trigger the activation of each gene within these pathways (Figure 1). Therefore, in this section, we focus on the changes in gene expression associated with each of these signaling pathways and discuss how these factors affect CRC.

2.1. Wnt Signaling Pathway

The complex and highly conserved Wnt signaling pathway is essential for embryonic development, tissue homeostasis, and various physiological processes [14–16]. The presence or absence of Wnt, a ligand, influences the expression levels of genes, particularly the transcriptional co-activator β-catenin, associated with the Wnt signaling pathway. In the absence of Wnt, β-catenin is phosphorylated by the Axin/GSK3/APC/CK1 complex and degraded by the ubiquitin-dependent proteasome, maintaining it at a low level [17]. However, in the presence of Wnt, Wnt binds to the Frizzled receptor, activating the Wnt signaling pathway [14,18,19]. The Frizzled receptor dimerizes with LRP5/6 and recruits the Axin/GSK3/APC/CK1 complex to the cell membrane, preventing the phosphorylation and degradation of β-catenin. Accumulated cytoplasmic β-catenin then translocates to the nucleus, where it interacts with TCF/LEF transcription factors, inducing the expression of Wnt target genes.

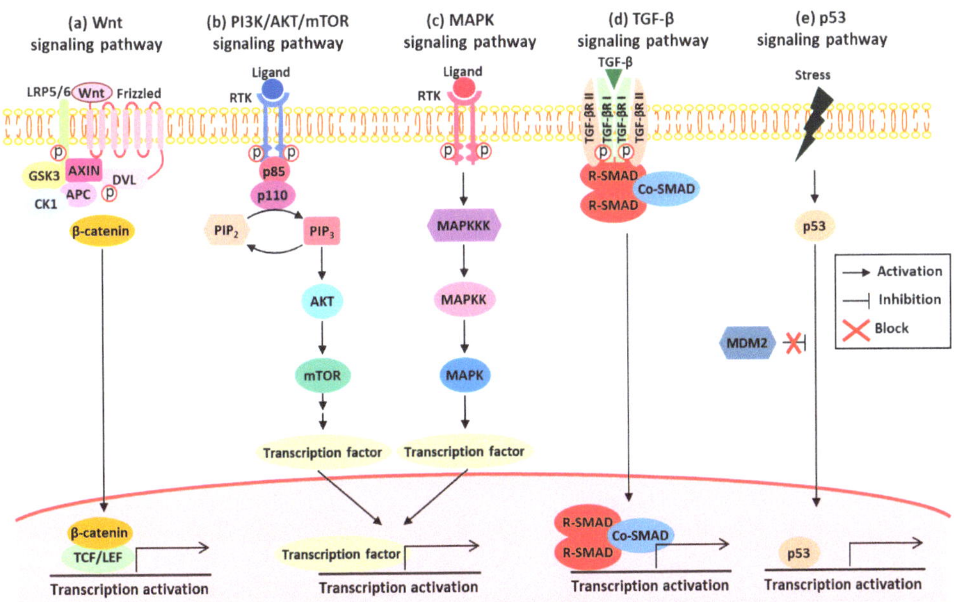

Figure 1. Schematic diagram of activated signaling pathways: (**a**) the Wnt signaling pathway is activated in the presence of Wnt. When Wnt binds to the Frizzled receptor, it dimerizes with LRP5/6, leading to the recruitment of the Axin/GSK3/APC/CK1 complex to the cell membrane. This results in the translocation of β-catenin from the cytoplasm to the nucleus. In the nucleus, β-catenin interacts with TCF/LEF transcription factors, thereby inducing the expression of downstream target genes. (**b**) The PI3K/AKT/mTOR signaling pathway is activated when a ligand binds to receptors such as RTK. Activated PI3K phosphorylates PIP2 to convert it to PIP3. Then, PIP3 activates downstream genes, such as AKT and mTOR. (**c**) The MAPK signaling pathway is initiated when a ligand binds to RTK, leading to the phosphorylation and activation of MAP3K. Activated MAP3K phosphorylates the downstream MAP2K, which in turn phosphorylates MAPK. This phosphorylation cascade transmits signals from the cell surface receptor to the nucleus, influencing gene expression by activating various kinases, enzymes, and transcription factors downstream. (**d**) The TGF-β signaling pathway is activated when TGF-β directly or indirectly binds to TGF-βR II. TGF-βR II recruits and phosphorylates TGF-βR I, which, in turn, phosphorylates R-SAMD. The phosphorylated R-SMAD then forms a trimeric complex with Co-SMAD. This complex translocates to the nucleus, where it acts as a transcription factor, influencing gene expression. (**e**) The p53 signaling pathway is initiated when p53 remains stabilized due to its failure to bind with MDM2, an E3 ubiquitin ligase. This stabilization of p53 leads to the induction of downstream gene expression, resulting in the activation of the pathway. Wnt signaling pathway: Wingless/Integrated signaling pathway; PI3K/AKT/mTOR signaling pathway: Phosphatidylinositol 3-kinase/protein kinase B/mammalian target of rapamycin signaling pathway; RTK: receptor tyrosine kinase; PIP2: phosphatidylinositol 4,5-bisphosphate; PIP3: phosphatidylinositol 3,4,5-trisphosphate; MAPK: Mitogen-Activated Protein Kinase; MAP3K: MAP kinase kinase kinase; MAP2K: MAP kinase kinase; TGF-β: Transforming Growth Factor-β; TGF-βR: Transforming Growth Factor-β receptor; R-SAMD: receptor-regulated SMAD; Co-SMAD: common-mediator SMAD; MDM2: mouse double minute 2 homolog.

The Wnt signaling pathway has been extensively studied to prove its critical role in CRC, and the hyperactivation of this signaling pathway has been observed from early to advanced stages of CRC [20–22]. In this signaling pathway, β-catenin and adenomatous polyposis coli (APC) especially have been identified as representative core genes with frequent genetic changes in CRC, which have been found to significantly impact the development and progression of the disease [23–25]. According to a study, it was confirmed that β-catenin was activated by Wnt1, which plays an important role in CRC progression, and that a high expression of nuclear β-catenin was associated with poor prognosis [26]. Similarly, it was revealed that the activation of β-catenin in CRC contributed to the aggressiveness of CRC by manipulating cell-cell junctions and inducing epithelial-mesenchymal transition (EMT) [27]. In a different study, it was found that β-catenin was activated by HMGB3, leading to the activation of downstream components of the Wnt signaling pathway, including c-myc and MMP7 [28]. This activation promoted cell growth and migration, influencing the carcinogenesis and development of CRC. Unlike β-catenin, APC is known to act as a negative regulator in the Wnt signaling pathway and is recognized as a tumor suppressor [29]. In one study, it was observed that the overexpression of SMYD2, a negative regulator of APC2, was associated with poor prognosis [30]. The knockdown of SMYD2 led to a decrease in cell proliferation, migration, and invasion, along with an upregulation of APC2 and a downregulation of the expression of E-cadherin, inducing EMT. Additionally, inhibiting SMYD2 reduced lung metastasis, confirming that SMYD2 plays an important role in CRC by regulating the expression of APC2 and influencing the Wnt signaling pathway.

2.2. PI3K/AKT/mTOR Signaling Pathway

The PI3K/AKT/mTOR signaling pathway is an important intracellular signaling cascade that is essential for controlling many cellular functions, including proliferation, survival, DNA repair, apoptosis, and gene transcription [12,31,32]. PI3K is known as a conserved lipid kinase family, divided into three classes based on their structure and specific substrates. While classes II and III remain relatively less-explored areas, class I has been extensively researched. Class I PI3K can be further subdivided into classes IA and IB, depending on the types of catalytic and regulatory subunits. Class IA consists of catalytic subunits, namely p110 (p110α, p110β, and p110 δ), and regulatory subunits p85 (p85 α, p55α, p50α, p85β, and p55γ) [33–35]. On the other hand, class IB PI3K is composed of the catalytic subunit p110γ and the regulatory subunit p101. PI3K can be activated downstream of numerous growth factors, including the fibroblast growth factor, vascular endothelial growth factor, and receptors, such as receptor tyrosine kinases [36]. Activated PI3K phosphorylates phosphatidylinositol 4,5-bisphosphate to convert it into phosphatidylinositol 3,4,5-trisphosphate; subsequently, phosphatidylinositol 3,4,5-trisphosphate activates AKT [33,36]. Activated AKT induces the phosphorylation of other substrates in both the cytosol and nucleus, particularly promoting the activation of mTOR, thereby activating the PI3K/AKT/mTOR signaling pathway. In particular, this signaling pathway plays an important role in the early and late stages of CRC, and dysregulation of this signaling pathway leads to carcinogenesis by altering proliferation, angiogenesis, and apoptosis [37–39].

One study has been conducted to investigate the correlation between the expression of PI3K and primary versus metastatic CRC [40]. It was found that metastatic lesions in CRC patients exhibited a higher expression of PI3K compared to primary lesions, suggesting a potential association with metastasis. To examine the actual impact of PI3K on CRC, the PI3K inhibitor LY294002 was used and resulted in the induction of apoptosis in CRC cell lines, indicating the functional significance of PI3K. According to another study, the overexpression of AKT in CRC cell lines was shown to increase their proliferative, migratory, and invasive capacities [41]. Moreover, in a xenograft mouse model of CRC, tumors overexpressing AKT exhibited rapid growth, increased vascularization, and the induction of expression of EMT markers. In a different study, it was observed that a higher expression of mTOR in CRC patients correlated with decreased overall survival and disease-free survival probabilities [42]. Additionally, the treatment of CRC cell lines with mTOR inhibitors, rapamycin, and PP242 resulted in a significant decrease in the number and diameter of spheroid cell models derived from CRC cells.

Besides the mentioned studies, there is extensive research on how changes in the expression of genes involved in the PI3K/AKT/mTOR signaling pathway impact CRC. In a study, the artificial overexpression of LACTB, which is downregulated in both CRC patients and cell lines, led to a decrease in PIK3R3 expression and a reduction in the expression of proteins associated with the PI3K/AKT/mTOR signaling pathway [43]. Furthermore, downregulated PIK3R3 increased autophagy and decreased the proliferation, EMT, and tumor size in CRC cell lines or mouse models. Taken together, these results suggest that LACTB influences CRC development by regulating the PI3K/AKT/mTOR signaling pathway. In another study, it was observed that TGM3 was downregulated in CRC patients and cell lines [44]. Increasing the expression of this gene was found to decrease the phosphorylation of PI3K and AKT, indicating the inactivation of the PI3K/AKT signaling pathway suppressed the proliferation and metastasis of CRC cells. Similarly, it has been revealed that MUC3A, which is overexpressed in CRC and associated with poor prognosis, activates the PI3K/AKT/mTOR signaling pathway [45]. The knockout of MUC3A resulted in the inhibition of the PI3K/AKT/mTOR signaling pathway and induced proliferation, cell cycle arrest, and the invasion of CRC cells.

2.3. MAPK Signaling Pathway

The MAPK signaling pathway is an essential intracellular signaling system that controls multiple cellular processes, such as cell proliferation, differentiation, survival, and apoptosis [46,47]. This signaling pathway is initiated when a ligand binds to receptor tyrosine kinases, leading to the phosphorylation and activation of the MAPK kinase kinase (MAP3K) [48,49]. Activated MAP3K phosphorylates MAPK kinase (MAP2K), which, in turn, phosphorylates MAPK, activating various kinases, enzymes, and transcription factors, thereby transmitting signals from cell surface receptors to the nucleus and impacting gene expression.

In the case of MAPK, it consists of three main families: extracellular signal-regulated kinase 1/2 (ERK1/2), c-Jun N-terminal kinase (JNK), and p38. These families are identified to play important roles in the development, progression, and oncogenic impact of CRC, acting individually or concurrently within the specific cellular context and signaling requirements in CRC [5,50]. There are several studies on the independent action of MAPK in CRC. One study revealed the influence of the ERK/MAPK signaling pathway on the progression of CRC by demonstrating a decrease in CRC cell proliferation and migration when the MAP2K inhibitor U0126 was used to block the ERK/MAPK signaling pathway [51]. Similarly, another study revealed that the knockdown of COPB2, which was overexpressed in CRC, led to a reduction in the proliferation and induction of apoptosis in CRC cells through the JNK signaling pathway [52]. In a different study, activated FOXC1, whose degradation was inhibited by phosphorylation through p38, was overexpressed in CRC patients and correlated with poor prognosis [53]. The knockdown of FOXC1 induced a decreased migration and invasion of CRC cells. Additionally, it reduced the number of metastatic tumors in the CRC mouse model.

In addition to cases where a single family independently affects the MAPK signaling pathway, it has been revealed that two MAPK families show combined actions within the signaling pathway. For instance, investigating both JNK and p38 MAPK signaling pathways simultaneously, it was found that S100A16 was overexpressed in CRC patients and correlated with poor prognosis [54]. The overexpression of S100A16 led to an increase in the expression of mesenchymal markers N-cadherin and vimentin, while decreasing the expression of the epithelial marker E-cadherin, implicating its involvement in EMT. Furthermore, treatment with the JNK inhibitor SP600125 and the p38 inhibitor SB203580 resulted in a decrease in cell migration and invasion. This suggests that S100A16 influences CRC through the JNK and p38 MAPK signaling pathways. Similarly, it has been confirmed that both ERK1/2 and p38 MAPK signaling pathways simultaneously influence CRC. The overexpression of CD24 induced the proliferation of CRC cells and increased the tumor size in a CRC mouse model, promoting tumorigenicity [55]. Additionally, increased CD24 promoted the activation of ERK1/2, Raf-1, and p38 MAPK, confirming that CD24-dependent ERK1/2 and p38 MAPK activation were necessary for CRC proliferation.

2.4. TGF-β Signaling Pathway

The TGF-β signaling pathway is a complex and multifunctional pathway that plays a crucial role in the regulation of cell growth, differentiation, immune responses, and homeostasis [56,57]. This pathway is activated by the binding of TGF-β (TGF-β I, TGF-β II, and TGF-β III) to its receptor, TGF-βR (TGF-βR I, TGF-βR II, and TGF-βR III). TGF-β I and TGF-β III bind directly to TGF-βR II, while TGF-β II can interact with TGF- βR II with the assistance of TGF-βR III [56,58–60]. Subsequently, TGF-βR II recruits and phosphorylates TGF-βR I for activation. Then, receptor-regulated SMADs (R-SMADs, SMAD 1, SMAD 2, SMAD 3, SMAD 5, and SMAD 8) are phosphorylated and activated to initiate the signaling cascade. In turn, the common-mediator SMAD (Co-SMAD, SMAD 4) interacts with R-SMAD to form a trimeric complex. This complex translocates into the nucleus and binds to DNA binding sites as transcription factors, fulfilling their role in gene expression. The activated TGF-β signaling pathway influences tumor progression and development in CRC [56].

A study confirmed that the expression of SMAD 1 was increased in CRC patient tissues [61]. When the expression of SMAD 1 was inhibited, the migration ability of CRC and the expression of N-cadherin were decreased, while the expression of E-cadherin was increased, indicating a reduction in EMT ability. Therefore, it can be anticipated that SMAD1 induces the migration of CRC cells and increases EMT in CRC patients. Another study confirmed the influence of ITGB5 on CRC by regulating the TGF-β signaling pathway [62]. ITGB5 was overexpressed in CRC patients, which correlated with poor survival. The knockdown of ITGB5 resulted in the decreased proliferation and invasion of CRC cells. Additionally, in a CRC mouse model, the suppression of ITGB5 led to reduced tumor size and metastasis. In addition, the repression of ITGB5 resulted in the decreased expression of TGF-βR I and phospho-SMAD 2, while increasing the expression of E-cadherin. This indicates that ITGB5, which was highly expressed in CRC, promotes the TGF-β signaling pathway and induces EMT in CRC. Similarly, it was revealed that the transcription activator EHF was overexpressed in CRC and correlated positively with poor prognosis [63]. EHF knockdown suppressed CRC cell proliferation, viability, and migration while inducing apoptosis. Moreover, in the CRC mouse model, repressed EHF resulted in reduced tumor size and metastasis to the liver. Further experiments revealed a positive correlation between the expressions of EHF and TGF-β1. It was demonstrated that EHF binds to the promoter of TGF-β1, thereby regulating its expression and implicating EHF in the tumorigenesis and metastasis of CRC.

2.5. p53 Signaling Pathway

The p53 signaling pathway is a crucial cellular pathway that plays a central role in regulating cellular responses to stress, including cell cycle, DNA repair, and apoptosis [64,65]. The p53 transcription factor is known to function as a tumor suppressor, aiding in the maintenance of genomic stability [6,66,67]. In unstressed cells, polyubiquitination by the E3 ubiquitin ligase mouse double minute 2 homolog (MDM2) leads to low levels of p53. MDM4 (MDMX), known as both a homology of MDM2 and a negative regulator of p53, interacts with MDM2 to bind to p53. As a result of this interaction, p53 is degraded via MDM2-dependent polyubiquitination, leading to the suppression of its transcriptional activity [6,68]. However, various stress signals, including DNA damage and abnormal growth signals, disrupt the interaction between p53 and MDM2. This disruption results in the stabilization of p53, enabling the activation of the transcription of target genes, which regulate various cellular responses, such as senescence and metabolism [69].

In CRC, mutations in the *TP53* gene, encoding p53, occur frequently, with approximately 43% of cases exhibiting mutations [6]. Mutations in p53 typically impair the function of wild-type p53, leading to a loss of its tumor suppressor properties. These alterations can affect cancer development and progression, such as cancer stem cell formation, cell proliferation, invasion, and metastasis [6,70]. One study found that CRC patients with *TP53* mutations with a complete loss of p53 had significantly different tumor sizes and decreased overall survival compared to CRC patients with wild-type *TP53* [71]. In another study, similar findings were observed where CRC patients with a positive p53 expression had higher overall survival compared to those with a negative p53 expression [72].

In addition to the direct impact of mutations in the *TP53* on CRC, it has been revealed that the expression level of other genes located upstream or downstream of the p53 affects CRC by influencing the p53 signaling pathway. One study confirmed that the p53 signaling pathway influenced cancer progression through the low expression of *MDM2*, a negative regulator of p53, in CRC patients with liver metastasis [73]. Similarly, an analysis using the GEO and GEPIA databases confirmed that *PRDX2* was overexpressed in CRC and correlated with poor prognosis [74]. The downregulation of PRDX2 led to reduced cell proliferation and tumor growth, accompanied by the upregulation of p53 expression. Furthermore, PRDX2 has been shown to interact with MDM2, leading to the increased ubiquitination and degradation of p53. Thus, PRDX2 is implicated in CRC progression by promoting the ubiquitinated degradation of p53 via the p53 signaling pathway. In another study, it was found that HAUS6 was overexpressed in CRC patients and was associated with shorter overall survival [75]. The knockdown of HAUS6 resulted in the activation of the p53 signaling pathway, leading to an increased expression of downstream p21. Consequently, HAUS6 induces tumor growth and proliferation by regulating the p53 signaling pathway. In addition, there is research indicating that GADD45B, overexpressed in CRC patient tissues, plays a role in apoptosis in CRC [76]. Additionally, while the anti-apoptotic member BCL-2 was upregulated, the pro-apoptotic member Bax was downregulated in these tissues. Furthermore, the knockdown of GADD45B disrupted the balance between anti-apoptotic and pro-apoptotic proteins, indicating that GADD45B influenced apoptosis through the p53 signaling pathway. In a different study, the knockdown of RPS15A, which was highly expressed in CRC patient tissues, has been shown to suppress cell proliferation and induce cell cycle arrest at the G0/G1 phase [77]. Additionally, the downregulation of RPS15A increased the expression of *p53* and *p21*, confirming that RPS15A induces CRC malignancy through the p53 signaling pathway.

3. ncRNA Regulation in CRC

In the mid-1900s, non-protein-coding regions of the human genome were initially dismissed as junk DNA, with no apparent function unlike protein-coding genes [78]. However, with the completion of the Human Genome Project (HGP) in 2003, it became clear that protein-coding genes constitute only about 1.5% (21,000 genes) of the human genome, with non-protein-coding regions making up a significant proportion [79,80]. Subsequent projects, such as the Encyclopedia of DNA Elements (ENCODE) and the Functional Annotation of the Mammalian Genome (FANTOM), demonstrated that junk DNA contains essential regulatory elements for transcription, as well as sequences encoding ncRNAs involved in various biological mechanisms and functions [78,81]. Through these projects, sequencing techniques have significantly advanced and have been used to study ncRNAs in the human genome, revealing substantial insights [82]. ncRNAs, not translated to protein, are generally classified into two categories based on RNA transcript length: those shorter than 200 nucleotides are small ncRNAs, while longer ones are called lncRNAs [83]. miRNAs, representative small ncRNAs, and lncRNAs have been extensively studied in various diseases, including CRC, due to their direct and indirect regulation of protein-coding gene expression [84]. In recent years, circRNAs, ncRNA characterized by their circular structure, have gained attention and have been implicated in the regulation of gene expression, similar to lncRNAs [85]. Therefore, in this section, we aim to discuss the regulatory relationships of miRNAs, lncRNAs, and circRNAs, as well as their impact on CRC (Figure 2).

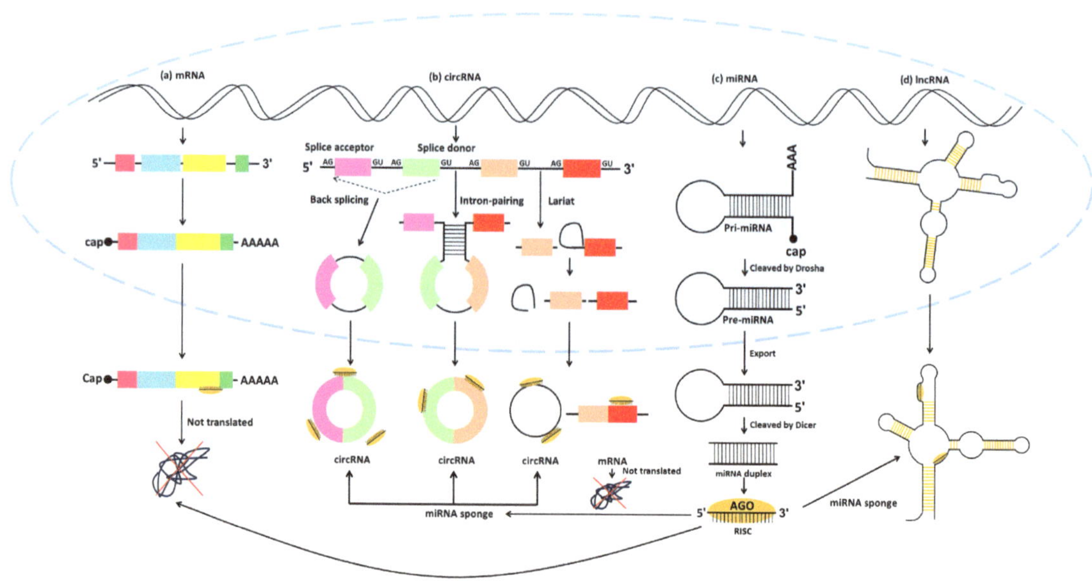

Figure 2. Biogenesis pathways and mutual regulatory relationships of mRNA, miRNA, lncRNA, and circRNA. (**a**) mRNA generated from a gene undergoes capping, tailing, and splicing to produce mature mRNA. This mature mRNA is regulated directly by miRNA or indirectly by lncRNA or circRNA in the cytoplasm, thereby controlling the translation of the mRNA. (**b**) CircRNA is generated through back-splicing, intron-pairing-driven circularization, and lariat-driven circularization. Back-splicing involves connecting the 5′ splice donor site downstream to the 3′ splice acceptor site upstream, forming a circular structure. In intron-pairing-driven circularization, circRNA is formed by the binding of inverted complementary sequences on the sides of long intron base pairs. Lariat-driven circularization occurs when the excision process of regular splicing fails, resulting in the generation of a circRNA. CircRNA binds to miRNA, indirectly influencing gene expression by preventing miRNA from binding to target genes. (**c**) Pri-miRNA is transcribed from the miRNA gene and is cleaved by Drosha to produce pre-miRNA. The pre-miRNA is then transported to the cytoplasm and processed to form a miRNA duplex by Dicer. One of the strands binds with AGO to form the RISC, which subsequently interacts with the target mRNA, leading to an inhibition of translation. (**d**) LncRNA is generated from DNA similarly to mRNA, undergoing capping, splicing, and tailing. Once generated, lncRNA in the cytoplasm binds to miRNA, indirectly affecting gene expression by inhibiting miRNA from binding to target genes. mRNA: messenger RNA; miRNA: microRNA; lncRNA: long non-coding RNA; circRNA: circular RNA; pri-miRNA: primary miRNA; pre-miRNA: precursor miRNA; AGO: Argonaute; RISC: RNA-induced silencing complex.

3.1. miRNA Regulation

miRNAs are a class of small RNA molecules that play a crucial role in the regulation of gene expression in various organisms [86]. They are typically short and single-stranded RNA molecules composed of approximately 18 to 24 nucleotides [87,88]. miRNAs regulate gene expression by binding to the miRNA response element (MRE) of messenger RNA (mRNA) transcribed from genes, thereby modulating post-transcriptional processes [89,90]. By regulating the levels of proteins produced by specific genes, miRNAs play important roles in various biological processes, including development, cell proliferation, differentiation, apoptosis, and immune response [91,92]. The biogenesis of mature miRNAs involves a series of steps (Figure 2). Initially, long primary miRNA transcripts (pri-miRNAs) are transcribed by RNA polymerase II from the miRNA gene [93]. These pri-miRNAs are cleaved in the cell nucleus by an enzyme called Drosha and processed into precursor miRNAs (pre-miRNAs) [94]. Pre-miRNAs are transported to the cytoplasm, where they are further processed by an enzyme called Dicer, resulting in a double-stranded mature, functional miRNA composed of approximately 22 nucleotides. Subsequently, Argonaute binds to the double-stranded mature miRNA [95]. One of the two strands, known as the guide strand, remains in Argonaute, forming the RNA-induced silencing complex, while the other strand, referred to as the passenger strand, is degraded. The guide strand within the RNA-induced silencing complex binds complementarily to the target mRNA, leading to its degradation or the inhibition of translation, thereby directly regulating the expression of the target gene [96].

Research on miRNAs has significantly expanded our understanding of gene regulation and their involvement in various diseases, including cancer, neurodegenerative disorders, and cardiovascular diseases [95,97,98]. Among these, miRNAs exhibit aberrant expression patterns and play a significant role in gene regulation, making them potential targets for diagnosis and treatment (Table 1). In particular, miRNAs influence major signaling pathways in CRC, as discussed in Section 2, resulting in changes in tumor phenotype [99]. For example, miR-590-5p was overexpressed in CRC patient tissues and cell lines and was verified to directly regulate the expression of PDCD4, thereby reducing its expression [100]. It was confirmed that the decreased miR-590-5p induced the expression of PDCD4, resulting in modulating the expression of TGF-β and SMAD2/3, genes involved in the TGF-β signaling pathway. In addition, the downregulation of miR-590-5p induced declined viability, migration, and invasion, suggesting its potential as a therapeutic target for CRC treatment. Similarly, miR-150 was downregulated in both CRC patient tissues and cell lines, exhibiting an inverse expression pattern with β-catenin [101]. Further experiments confirmed *CTNNB1*, encoding β-catenin, as a target gene of miR-150. The overexpression of miR-150 resulted in decreased cell proliferation, inhibiting CRC progression in mouse models, indicating its involvement in the Wnt signaling pathway and potential as a therapeutic target. In another study, miR-130a-3p was downregulated in CRC patient tissues and various cell lines [102]. The overexpression of miR-130a-3p led to decreased proliferation, viability, and tumor size, implicating its role in the Wnt signaling pathway by reducing the expression of its target gene, *Wnt1*. These findings suggest the therapeutic potential of targeting miR-130a-3p in CRC treatment.

In addition, numerous studies have shown that miRNAs affect CRC by regulating the expression of genes for which the signaling mechanisms involved have not yet been exactly identified. For instance, RNF187, known for its tumor-suppressive role in other cancers, was overexpressed in CRC patient tissues and cell lines, correlating negatively with poor prognosis [103]. Furthermore, miR-144-5p was decreased in patient tissues and directly regulated the expression of RNF187. Artificial knockdown of RNF187 resulted in suppressed proliferation, migration, and invasion while increasing apoptosis. This suggests that miR-144-5p regulates the expression of RNF187, impacting the progression of CRC and highlighting its potential as a biomarker for CRC diagnosis and therapy. Similarly, research on miR-1-3p suggests its potential as a novel therapeutic strategy for CRC [104]. Downregulated in CRC patient tissues and cell lines, miR-1-3p directly regulated overexpressed YWHAZ. The overexpression of miR-1-3p led to reduced proliferation and invasion, as well as the inhibition of EMT, impacting the development and progression of CRC.

Table 1. The functions of miRNAs with differential expression relating to the main signaling pathways implicated in CRC, as well as experimental information for validation.

miRNA Expression	miRNAs	Target Genes	Clinical Value	Impact on CRC	Function	Signaling Pathway	Human Samples	Study Model	References
	miR-590-5p	*PDCD4*	therapeutic target	pathogenesis	induce cell viability, migration, and invasion	TGF-β signaling pathway	30 tumor and adjacent normal tissues	in vitro	[100]
	miR-21	*VMP1*	therapeutic target	development	promote migration and invasion, while repressing autophagy and drug sensitivity	PTEN/AKT/TFEB signaling pathway	4 tumor and adjacent normal tissues	Oncomine dataset, in vitro	[105]
	miR-21-5p	*TGF-β1*	therapeutic target	NA	induce pyroptosis while inhibiting cell viability	NA	5 tumor and normal tissues	in vitro	[106]
	miR-452-5p	*PKN2* *DUSP6*	therapeutic target	progression	enhance cell proliferation, cell cycle transition, and chemoresistance, while suppressing apoptosis	MAPK/ERK signaling pathway	87 tumor and matched para-carcinoma mucosal tissues	TCGA, in vitro, in vivo	[107]
	miR-106-5p	*FAT4*	NA	carcinogenic	promote proliferation, migration, invasion, and angiogenesis	NA	tumor and adjacent normal tissues	TCGA, in vitro	[108]
UP	miR-496	*RASSF6*	therapeutic target	progression	induce cell motility, migration, invasion, and EMT	Wnt signaling pathway	28 tumor and adjacent normal tissues	TCGA, in vitro	[109]
	miR-125b	*CFTR* *CGN*	diagnosis biomarker and therapeutic target	metastasis	enhance migration, invasion, metastasis, and EMT	RhoA/ROCK signaling pathway	58 tumor and adjacent normal tissues	human tissue microarray chips, in vitro, in vivo	[110]
	miR-298	*PTEN*	therapeutic target	development	facilitate cell metabolic activity, cell cycle progression, migration, and invasion, while inhibiting apoptosis	AKT/ERK and AKT/mTOR/P70 S6K signaling pathways	100 tumors and matched adjacent normal tissues, 100 tumor and normal plasma	in vitro	[111]
	miR-429	*LATS2*	diagnosis biomarker and therapeutic target	tumorigenesis	stimulate proliferation and tumor growth	YAP/TAZ signaling pathway	21 tumor and normal tissues	in vitro, in vivo	[112]
	miR-645	*EFNA5*	therapeutic target	metastasis	enhance migration, invasion, metastasis, and tumor growth	NA	28 tumor and adjacent normal tissues	in vitro, in vivo	[113]
	miR-150	*CTNNB1*	therapeutic target	progression	inhibit cell proliferation	Wnt signaling pathway	30 tumor and adjacent normal tissues	in vitro, in vivo	[101]
	miR-130a-3p	*WNT1*	biomarker, therapeutic target	development	repress cell proliferation and tumor growth	Wnt signaling pathway	30 tumor and adjacent normal tissues	TCGA, in vitro, in vivo	[102]
	miR-144-5p	*RNF187*	diagnosis biomarker and therapeutic target	progression	suppress migration and invasion	NA	83 tumor and adjacent normal tissues	in vitro	[103]
	miR-1-3p	*YWHAZ*	therapeutic target	progression	inhibit cell proliferation and EMT	NA	20 tumor and adjacent normal tissues	GEO database, in vitro	[104]
	miR-144-3p	*BCL6*	prognosis biomarker and therapeutic target	progression	repress cell proliferation and cell cycle progression	Wnt signaling pathway	20 tumor and adjacent normal tissues	in vitro	[114]
DOWN	miR-450a-5p	*SOX2*		progression	suppress stemness, vasculature, and tumor growth	NA	90 glass-slide tumor tissues for array	in vitro, in vivo	[115]
	miR-215-5p	*CTNNBIP1*	therapeutic target	progression and metastasis	inhibit clonogenic potential, cell cycle, migration, invasion, metastasis, and tumor growth, while inducing apoptosis	Wnt signaling pathway	primary tumor, paired liver metastatic, and adjacent normal tissues	in vitro, in vivo	[116]
	miR-148b	*p55PIK*	therapeutic target	progression	suppress proliferation, tumor growth	p53 signaling pathway	10 tumor and normal tissues	in vitro, in vivo	[117]
	miR-16	*Survivin*	therapeutic target	NA	repress proliferation and induce apoptosis	p53 signaling pathway	52 tumor and adjacent normal tissues	in vitro	[118]
	miR-139-3p	*KRT80*	therapeutic target	NA	inhibit proliferation, migration, and invasion	NA	27 tumor and normal tissues	in vitro	[119]

Table 1. *Cont.*

miRNA Expression	miRNAs	Target Genes	Clinical Value	Impact on CRC	Function	Signaling Pathway	Human Samples	Study Model	References
DOWN	miR-125a-5p miR-217	*FNDC3B*	prognosis biomarker and therapeutic target	progression	reduce proliferation	PI3K/mTOR signaling pathway	36 tumor and adjacent normal tissues	in vitro	[120]
	miR-654-3p	*SRC*	diagnosis and prognosis biomarker, and therapeutic target	development	repress proliferation, migration, and invasion, while promoting apoptosis	NA	103 tumor and adjacent normal tissues	in vitro, in vivo	[121]
	miR-539	*TIPE*	therapeutic target	progression	suppress proliferation and tumor growth, while enhancing ferroptosis	SAPK/JNK signaling pathway	26 tumor and adjacent tissues	in vitro, in vivo	[122]
	miR-3622a-3p	*SALL4*	therapeutic target	progression and metastasis	reduce proliferation and EMT	Wnt signaling pathway	80 tumor and adjacent normal tissues	TCGA, in vitro, in vivo	[123]

3.2. Competing Endogenous RNA (ceRNA) Regulation in CRC

CeRNAs refer to various RNA molecules containing MREs that compete for miRNA binding, of which lncRNAs and circRNAs are well known [124,125]. This dynamic interaction establishes the ceRNA network, where miRNAs compete for interactions with lncRNAs or circRNAs as well as mRNAs. Consequently, it is known that ceRNAs function as a miRNA sponge because they are involved in various physiological processes, such as development, by binding to miRNAs and regulating the availability and activity of miRNAs [126–128]. Recent studies have extensively investigated ceRNAs in numerous diseases, particularly CRC [129]. CeRNAs play a pivotal role in CRC initiation and progression by interfering with the binding between miRNAs and mRNAs, thereby serving as potential therapeutic targets or diagnostic biomarkers [130].

3.2.1. LncRNA/miRNA/mRNA

LncRNAs are a class of RNA molecules longer than 200 nucleotides that do not encode proteins but play diverse roles in many biological processes, such as transcriptional control, post-transcriptional processing, and cellular organization, exerting a crucial influence on gene expression regulation [131–136]. LncRNAs are transcribed from lncRNA genes by RNA polymerase II (Pol II), synthesizing the nascent lncRNA transcript (Figure 2) [137–139]. Similar to mRNA, the nascent lncRNA transcript undergoes processes, including 5′ capping, splicing, and 3′ poly-A tailing, to be stabilized. Some of the produced lncRNAs function in the nucleus, while others transport to the cytoplasm for regulatory roles, where they interact with other RNA molecules, such as miRNAs [140,141]. This interaction influences cellular processes associated with cancer, particularly revealing significant impacts on CRC, suggesting their potential as therapeutic targets for CRC (Table 2) [141–144].

Accumulated studies have indicated that abnormal expression levels of lncRNAs play an important role in the development and progression of CRC acting as miRNA sponges. In particular, genes implicated in CRC-related signaling pathways, as highlighted in Section 2, are regulated through the interaction of lncRNAs and miRNAs. For example, the overexpression of CTBP1-AS2 in CRC has been identified as a miRNA sponge, influencing CRC development [145]. Increased CTBP1-AS2 activated TGF-β, thereby promoting the proliferation and metastasis of CRC cells while concurrently suppressing apoptosis. Experimental evidence further supported the notion that upregulated miR-93-5p led to a reduction in the expression of both CTBP1-AS2 and TGF-β1, establishing miR-93-5p as a common target for these two factors. Similarly, the increase in SNHG16 contributed to the proliferation of CRC cells, promoting tumor development in CRC [146]. This study revealed that the overexpressed SNHG16 interacted with miR-302-3p, influencing the expression of AKT, a target gene of miR-302-3p. Additionally, the upregulation of miR-302-3p resulted in the increased proliferation of CRC cells, providing evidence that SHNG16 acts as a sponge for miR-302-3p.

In addition to the previously mentioned genes related to important CRC pathways, lncRNAs also play a significant role in CRC by regulating the expression of genes involved in various signaling pathways. In a study, both *HIF1A*, a target gene of miR-20b-5p, and COL4A2-AS1 expression were overexpressed in CRC patient tissues and cell lines, while miR-20b-5p was decreased [147]. COL4A2-AS1 was observed to compete with *HIF1A* and acted as a miRNA sponge for miR-20b-3p. Furthermore, reducing COL4A2-AS1 expression in cells and mouse models resulted in decreased cell viability, proliferation, and aerobic glycolysis, thus inhibiting tumorigenesis. These findings suggest the potential of COL4A2-AS1 as a novel target for CRC therapy. In another study, MIR503HG was upregulated in CRC patients, with a more pronounced decrease observed in stages III-IV compared to stages I-II [148]. Its expression was downregulated in patients with lymph node metastasis compared to those without. Increased MIR503HG led to the inhibition of cell proliferation, migration, and invasion, and the induction of apoptosis. Further investigations revealed that MIR503HG inversely regulated miR-107, and PAR4 was negatively controlled by miR-107, establishing a positive relationship between PAR4 and MIR503HG. It was demonstrated that MIR503HG acts as a tumor suppressor by inhibiting CRC development and metastasis, suggesting its potential as a target for CRC treatment.

Table 2. The functions of lncRNAs with differential expression relating to the main signaling pathways implicated in CRC, as well as experimental information for validation.

LncRNA Expression	LncRNAs	miRNAs	Genes	Clinical Value	Impact on CRC	Function	Signaling Pathway	Human Samples	Study Model	References
UP	EGFR-AS1	miR-133b	*STAT3*	early diagnosis biomarker	development and progression	induce cell proliferation, migration, and invasion	NA	130 tumor and 30 normal tissues	in vitro	[142]
	CTBP1-AS2	miR-93-5p	*TGF-β1*	NA	progression and metastasis	promote proliferation, invasion, and metastasis, while inhibiting apoptosis	TGF-β signaling pathway	50 tumor and normal tissues	GEPIA, in vitro, in vivo	[145]
	SNHG16	miR-302a-3p	*AKT*	therapeutic target	development	enhance proliferation	AKT signaling pathway	NA	in vitro	[146]
	COL4A2-AS1	miR-20b-5p	*HIF1A*	biomarker and therapeutic target	progression	facilitate proliferation and aerobic glycolysis	NA	55 tumor and adjacent normal tissues	in vitro, in vivo	[147]
	MIR4435-2HG	miR-206	*YAP1*	prognosis biomarker and therapeutic target	metastasis	stimulate invasion, migration, EMT, metastasis, and growth	Hippo signaling pathway	90 tumor and normal tissues	in vitro, in vivo	[149]
	NEAT1	miR-205-5p	*VEGFA*	diagnosis biomarker and therapeutic target	development	induce proliferation, migration, and invasion	NA	30 tumor and adjacent normal tissues	in vitro	[150]
		miR-34a	*SIRT1*	prognosis biomarker and therapeutic target	metastasis	enhance proliferation, invasion, and tumor growth	Wnt signaling pathway	100 tumor and normal tissues	GEO database, in vitro, in vivo	[151]
	HOTAIR	miR-206	*CCL2*	therapeutic target	progression	stimulate proliferation and invasion, while inhibiting apoptosis	NA	32 tumor and normal tissues	TCGA, GEPIA, UALCAN, SurvExpress, in vitro	[152]
	HCG18	miR-1271	*MTDH*	therapeutic target	development	promote proliferation and invasion	Wnt signaling pathway	20 tumor and adjacent normal tissues	StarBase, in vitro	[153]
	RoR	miR-6833-3p	*SMC4*	NA	tumorigenesis	facilitate proliferation and viability, while inhibiting apoptosis	NA	24 tumor and normal tissues	in vitro	[154]
	SNHG8	miR-588	*ATG7*	therapeutic target	development	induce proliferation and autophagy	NA	NA	TCGA, in vitro	[155]
	MAFG-AS1	miR-149-3p	*HOXB8*	prognosis biomarker and therapeutic target	progression	stimulate proliferation, migration, and invasion	NA	30 tumor and adjacent normal tissues	in vitro, in vivo	[156]
	SNHG6	miR-181b-5p miR-181c-5p	*JAK2*	biomarker and therapeutic target	progression	enhance proliferation, while repressing apoptosis	NA	40 tumor and adjacent normal tissues	in vitro	[157]
	RHPN1-AS1	miR-7-5p	*OGT*	therapeutic target	progression	promote proliferation, migration, and invasion, while suppressing apoptosis	NA	NA	in vitro, in vivo	[158]

Table 2. *Cont.*

LncRNA Expression	LncRNAs	miRNAs	Genes	Clinical Value	Impact on CRC	Function	Signaling Pathway	Human Samples	Study Model	References
UP	CASC21	miR-7-5p	*YAP1*	therapeutic target	progression	facilitate migration, invasion, and EMT, while inhibiting apoptosis	NA	NA	GEPIA, in vitro	[159]
	HOXD-AS1	miR-526b-3p	*CCND1*	NA	progression	stimulate proliferation, migration, and invasion	NA	54 tumor and adjacent normal tissues	in vitro	[160]
	RP11-757G1.5	miR-139-5p	*YAP1*	prognosis biomarker and therapeutic target	progression and metastasis	enhance proliferation, cell cycle progression, migration, invasion, tumor growth, and metastasis	NA	43 tumor and 6 adjacent normal tissues	in vitro, in vivo	[161]
	MCF2L-AS1	miR-874-3p	*CCNE1*	diagnosis and prognosis biomarkers and therapeutic target	initiation and progression	induce proliferation, migration, invasion, and EMT, while inhibiting apoptosis	NA	130 tumor and normal tissues	in vitro	[162]
	DANCR	miR-185-5p	*HMGA2*	NA	progression	promote proliferation, migration, and invasion	NA	50 tumor and adjacent normal tissues	in vitro	[163]
	RNCR3	miR-1301-3p	*AKT1*	therapeutic target	progression	facilitate proliferation and invasion, while suppressing apoptosis	NA	76 tumor and adjacent normal tissues	in vitro, in vivo	[164]
DOWN	MIR503HG	miR-107	*PAR4*	therapeutic target		suppress migration and invasion	NA	80 tumor and adjacent normal tissues	in vitro, in vivo	[148]
	LINC00485	miR-581	*EDEM1*	therapeutic target	progression and metastasis	repress proliferation, migration, invasion, tumor growth, and metastasis	NA	52 tumor and adjacent normal tissues	TCGA, GEO database, in vitro, in vivo	[165]
	DPP10-AS1	miR-127-3p	*ADCY1*	therapeutic target	progression	inhibit stemness, sphere formation, proliferation, migration, invasion, and tumor growth, while enhancing apoptosis	NA	54 rumor and adjacent normal tissues	in vitro, in vivo	[166]
	MBNL1-AS1	miR-412-3p	*MYL9*	therapeutic target	progression, occurrence	repress proliferation, invasion, migration, and tumor formation, while increasing apoptosis	NA	NA	TCGA, GEO database, in vitro, in vivo	[167]
	FENDRR	miR-18a-5p	*ING4*	NA	progression and metastasis	suppress proliferation, migration, invasion, tumor growth, and metastasis	NA	42 tumor and adjacent normal tissues	GEPIA, in vitro, in vivo	[168]
	MCM3AP-AS1	miR-19a-3p	*FOXF2*	biomarker and therapeutic target	progression	inhibit proliferation and migration	NA	53 tumor and adjacent normal tissues	GEPIA, in vitro	[169]

3.2.2. CircRNA/miRNA/mRNA

CircRNAs are endogenous ncRNAs found in various species, including mammals [170,171]. These molecules exhibit a continuous circular structure, connecting the 3′ and 5′ ends through a phosphodiester bond, and notably lack both a 5′ cap or a 3′ poly-A tail [172,173]. This unique characteristic renders them considerably more stable compared to linear RNAs, providing resistance to exonuclease activity. While circRNAs are not fully understood, they are known to be produced through various synthetic mechanisms, including back-splicing, intron-pairing-driven circularization, and lariat-driven circularization (Figure 2). Back-splicing, a non-canonical splicing event, deviates from the typical splicing pathway that leads to the formation of linear mRNA from precursor mRNA (pre-mRNA) [174,175]. Here, a canonical splice site (5′-GU and 3′-AG at intron) is used, where the 5′ splice donor site downstream of the pre-mRNA connects with the 3′ splice acceptor site upstream, forming a 3′, 5′-phosphodiester bond [173,176,177]. This process results in circRNAs, as the end of the downstream exon is ligated to the beginning of the upstream exon. In intron-pairing-driven circularization, circRNAs are formed by the base pairing of reverse complementary sequences, such as Alu elements, within the flanking long introns of pre-mRNA [178,179]. The third biogenesis type, lariat-driven circularization, occurs through canonical splicing events [178,180,181]. In canonical splicing events, the intronic region of pre-mRNA is debranched, leading to the excision of the intron, and exons are ligated to generate mature mRNA. However, if the debranching process escapes, the lariat structure persists due to the 2′, 5′-phosphodiester bond between the splice donor and branch point. This failure to dismantle the lariat structure maintains the circular form, making it a precursor for circRNAs. Similar to lncRNAs, circRNAs function in the nucleus and play crucial roles in the cytoplasm. Residing in the cytoplasm and containing MRE, circRNAs function as sponges for miRNA, allowing them to act as transcriptional regulators [173,182]. Ultimately, the ability of circRNAs to indirectly regulate mRNA expression has revealed diverse biological functions. Specifically, it has been demonstrated that changes in circRNA expression in CRC impact the initiation, growth, and metastasis of CRC (Table 3) [183–185]. Consequently, the altered expression of circRNAs is recognized as a potential biomarker for the diagnosis or treatment of CRC [186,187].

In particular, there are studies on circRNAs that indirectly regulate the expression of genes involved in the key signaling pathways mentioned in Section 2, which play a significant role in CRC. In a study, circ_0008285 expression was downregulated in CRC patient tissues and cell lines [188]. The expression level of circ_0008285 was inversely correlated with tumor size, lymph node metastasis, and tumor–node–metastasis stage. Furthermore, the knockdown of circ_0008285 resulted in the enhanced proliferation and migration of cell lines, revealing its role as a sponge for miR-382-5p. Since miR-382-5p acts as a regulator of PTEN involved in the PI3K/AKT signaling pathway, circ_0008285 affected CRC by indirectly regulating PTEN and is suggested as a potential target for CRC treatment. In a different investigation, the downregulated circ_0009361 in both colorectal cancer patient tissues and cell lines was associated with the promotion of proliferation, EMT, migration, and invasion [189]. Conversely, the overexpression of hsa_circ_0009361 resulted in reduced tumor size and decreased metastasis. Further experiments revealed that hsa_circ_0009361 acted as a miRNA sponge by binding to miR-582, thereby regulating the expression of APC2. Consequently, hsa_circ_0009361 influences colorectal cancer by interacting with miR-582 to regulate APC2 involved in the Wnt signaling pathway.

In addition, it has been revealed that circRNAs can compete with genes unrelated to the main signaling pathways known to have a significant impact on CRC, thereby regulating miRNA. In a study, the expression of circCTNNA1 was upregulated in CRC tissues and associated with poor prognosis, especially in advanced tumor stages showing node metastasis [190]. It has been revealed that circCTNNA1 competed with FOXM1, a target gene of miR-149-5p, thus binding to miR-149-5p. The knockdown of circCTNNA1 resulted in decreased proliferation and tumor size. Therefore, circCTNNA1 plays an oncogenic role, highlighting its potential as a target for CRC therapy and diagnosis. Similarly, circ_GLG1 exhibited high expression in CRC patients, and the downregulation of circ_GLG1 led to the suppression of tumor cell viability, proliferation, migration, and invasion [191]. Circ_GLG1 was identified to act as a sponge for miR-622, influencing the expression of *KRAS*, a target gene of miR-622. This suggests that circ_GLG1 holds potential as a novel biomarker in CRC diagnosis. In another study, it was observed that circ0065378 expression was decreased in CRC tissues and cell lines [192]. Artificially increasing circ0065378 expression led to reduced proliferation, migration, and invasion of CRC cells. Circ0065378 was found to interact with miR-4701-5p, indirectly regulating the target gene *TUSC1* of miR-4701-5p, thereby influencing malignant behavior. These findings suggest that circ-PLXNB1 can serve as a novel target for the diagnosis and treatment of CRC.

Table 3. The functions of circRNAs with differential expression relating to the main signaling pathways implicated in CRC, as well as experimental information for validation.

CircRNA Expression	CircRNAs	miRNAs	Genes	Clinical Value	Impact on CRC	Function	Signaling Pathway	Human Samples	Study Model	References
UP	CircVAPA	miR-125a	*CREB5*	therapeutic target	progression	induce cycle progression, glycolysis, migration, and invasion	NA	42 tumor and normal tissues	in vitro	[187]
	CircCTNNA1	miR-149-5p	*FOXM1*	diagnosis biomarker and therapeutic target	progression	facilitate proliferation, DNA synthesis, migration, invasion, and tumorigenesis	NA	60 tumor and adjacent normal tissues	TCGA, in vitro, in vivo	[190]
	CircGLG1	miR-622	*KRAS*	diagnosis biomarker and therapeutic target	progression	promote proliferation, invasion, and migration	NA	40 tumor and normal tissues	in vitro	[191]
	CircSPARC	miR-485-3p	*JAK2*	diagnosis and prognosis biomarkers and therapeutic target	progression and metastasis	enhance proliferation, migration, invasion, tumor growth, and metastasis	JAK/STAT signaling pathway	84 tumor and adjacent normal tissues, 40 plasmas of patients	GEO database, in vitro, in vivo	[193]
	Circ0007142	miR-122-5p	*CDC25A*	diagnosis biomarker and therapeutic target	progression	stimulate proliferation, migration, invasion, and tumor growth	NA	31 tumor and adjacent normal tissues	in vitro, in vivo	[194]
		miR-455-5p	*SGK1*	therapeutic target	progression	induce proliferation, migration, invasion, and tumor growth while repressing apoptosis	NA	45 tumor and para-carcinoma tissues	in vitro, in vivo	[195]
	Circ0000467	miR-382-5p	*EN2*	therapeutic target	development	enhance proliferation, migration, invasion, and EMT	NA	69 tumor and adjacent normal tissues	GEO database, in vitro	[196]
	Circ0001178	miR-382 miR-587 miR-616	*ZEB1*	therapeutic target	metastasis	stimulate migration, invasion, EMT, and metastasis	NA	102 tumor tissues	in vitro, in vivo	[197]
	Circ0053277	miR-2467-3p	*MMP14*	therapeutic target	development and progression	promote proliferation, migration, and EMT	NA	3 tumor and normal tissues	in vitro	[198]
	Circ0060745	miR-4736	*CSE1L*	therapeutic target	NA	accelerate proliferation, migration, and invasion	NA	28 tumor and para-tumor tissues	in vitro	[199]
	Circ102209	miR-761	*RIN1*	therapeutic target	progression	enhance proliferation, cell cycle progression, migration, invasion, EMT, and tumor growth, while suppressing apoptosis	NA	56 tumor and para-carcinoma tissues	human circRNA array v2, in vitro, in vivo	[200]
	Circ100146	miR-149	*HMGA2*	early diagnosis biomarker and therapeutic target	progression and metastasis	facilitate proliferation, migration, invasion, tumor growth, and metastasis, while inhibiting apoptosis	NA	58 tumor and normal tissues	GEO database, in vitro, in vivo	[201]

Table 3. *Cont.*

CircRNA Expression	CircRNAs	miRNAs	Genes	Clinical Value	Impact on CRC	Function	Signaling Pathway	Human Samples	Study Model	References
UP	Circ0004277	miR-512-5p	*PTMA*	NA	progression	induce proliferation and tumor growth, while inhibiting apoptosis	NA	50 tumor and para-carcinoma tissues	GEO database, in vitro, in vivo	[202]
	CircFARSA	miR-330-5p	*LASP1*	therapeutic target	progression	promote proliferation, migration, invasion, and tumor growth	NA	40 tumor and adjacent normal tissues	in vitro, in vivo	[203]
	Circ000166	miR-326		diagnosis biomarker and therapeutic target	progression	stimulate proliferation, while repressing apoptosis	NA	40 tumor and adjacent normal tissues	GEO database, in vitro	[204]
	CircPRMT5	miR-377	*E2F3*	therapeutic target	progression	facilitate proliferation and tumor growth	NA	30 tumor and adjacent normal tissues	in vitro, in vivo	[205]
	CircERBIN	miR-125a-5p miR-138-5p	*4EBP-1*	therapeutic target	progression	enhance proliferation, migration, invasion, tumor growth, angiogenesis, and metastasis	NA	59 tumor and adjacent normal tissues	in vitro, in vivo	[206]
	CircMAT2B	miR-610	*E2F1*	therapeutic target	progression	induce proliferation	NA	70 tumor and adjacent normal tissues	in vitro	[207]
	Circ0084615	miR-599	*DNMT3A*	biomarker and therapeutic target	progression and metastasis	promote proliferation, migration, invasion, and metastasis	NA	50 tumor and adjacent normal tissues	GEO database, in vitro, in vivo	[208]
	CircDENND4C	miR-760	*GLUT1*	diagnosis biomarker and therapeutic target	progression	enhance proliferation, migration, glycolysis, and tumor growth	NA	tumor and normal tissues	sequencing, in vitro, in vivo	[209]
	CircUBAP2	miR-582-5p	*FOXO1*	biomarker and therapeutic target	progression and metastasis	stimulate migration, invasion, proliferation, autophagy, tumor growth, and metastasis	NA	3 tumor and normal tissues	in vitro, in vivo	[210]
	Circ0089153	miR-198	*SENP1*	therapeutic target	progression	facilitate proliferation, sphere formation, tube formation, and tumor growth, while suppressing apoptosis	NA	50 tumor and adjacent noncancerous tissues	in vitro, in vivo	[211]
	CircNOX4	miR-485-5p	*CKS1B*	NA	progression	induce proliferation, migration, invasion, glycolysis, and tumor growth	NA	46 tumor and adjacent normal tissues	in vitro, in vivo	[212]
	CircHERC4	miR-556-5p	*CTBP2*	prognosis biomarker and therapeutic target	progression and metastasis	promote proliferation, migration, invasion, tumor growth, and metastasis	Notch signaling pathway	120 tumor and adjacent normal tissues	sequencing, in vitro, in vivo	[213]
	Circ0000372	miR-495	*IL6*	prognosis biomarker and therapeutic target	progression	enhance proliferation, migration, invasion, and tumor growth	JAK/STAT signaling pathway	60 tumor and adjacent normal tissues	in vitro, in vivo	[214]
	CircLDLR	miR-30a-3p	*SOAT1*	biomarker and therapeutic target	progression and metastasis	stimulate proliferation, DNA synthesis, cholesterol, migration, invasion, tumor growth, and metastasis	NA	80 tumor and 15 normal tissues	in vitro, in vivo	[215]

Table 3. *Cont.*

CircRNA Expression	CircRNAs	miRNAs	Genes	Clinical Value	Impact on CRC	Function	Signaling Pathway	Human Samples	Study Model	References
DOWN	Circ0008285	miR-382-5p	*PTEN*	therapeutic target	progression	repress proliferation and migration	PI3K/AKT signaling pathway	56 tumor and noncancerous colorectal mucosa tissues	in vitro	[188]
	Circ_0009361	miR-582	*APC2*	therapeutic target	progression and metastasis	inhibit proliferation, EMT, migration, invasion, tumor growth, and metastasis	Wnt signaling pathway	30 tumor and paracancerous tissues	microarray, in vitro, in vivo	[189]
	Circ0065378	miR-4701-5p	*TUSC1*	diagnosis biomarker and therapeutic target	progression	suppress proliferation, migration, invasion, EMT, and tumor growth	NA	25 tumor and adjacent normal tissues	sequencing, in vitro, in vivo	[192]
	CircCUL2	miR-208a-3p	*PPP6C*	diagnosis biomarker and therapeutic target	progression	repress proliferation, apoptosis, and tumor growth, while increasing apoptosis	NA	30 tumor and adjacent normal tissues	microarray, in vitro, in vivo	[216]
	CircSMARCA5	miR-93-5p	*ARID4B*	biomarker	progression	inhibit proliferation, migration, invasion, and tumor growth	NA	tumor and adjacent tissues	in vitro, in vivo	[217]
	Circ0003266	miR-503-5p	*PDCD4*	therapeutic target	progression	suppress proliferation, migration, and invasion, while accelerating apoptosis	NA	46 tumor and paracancerous tissues	GEO database, in vitro	[218]

4. Conclusions

According to numerous studies and KEGG pathway analyses, the key signaling pathways in CRC are the Wnt, PI3K/AKT/mTOR, MAPK, TGF-β, and p53 signaling pathways. Genes involved in each signaling pathway have been shown to influence behaviors of CRC cells, such as proliferation, migration, and invasion, ultimately impacting the progression and metastasis of CRC. Additionally, extensive research has been conducted on ncRNAs in CRC, as they are known regulators of genes. Overexpressed miRNAs have been found to suppress the expression of tumor suppressor genes, while downregulated miRNAs increased the expression of oncogenes, exerting negative effects on CRC. Furthermore, the discovery of ceRNAs has drawn attention to their indirect regulation of gene expression through the interaction between ceRNAs and miRNAs, thus influencing CRC (Figure 3). However, there are several limitations in the current research. Firstly, some studies solely rely on either in vitro or in vivo experiments, lacking validation in patient samples. Secondly, certain studies are exclusively based on in vitro experiments, lacking confirmation of reproducibility in vivo. Thirdly, the functional roles of genes regulated by ncRNAs remain ambiguous. As discussed in Section 2, although significant signaling pathways in CRC have been identified, Section 3 shows a deficiency in research on ncRNAs regulating genes involved in major signaling pathways. Instead, much research has been performed on gene regulation whose mechanisms have not yet been revealed. Therefore, for the studied ncRNAs to have clinical value, in-depth studies are needed on the regulation of genes involved in key signaling pathways or the mechanisms of genes whose exact functions remain unidentified. As these studies accumulate, it is expected that it will be possible to discover meaningful biomarkers or suitable treatment targets in CRC.

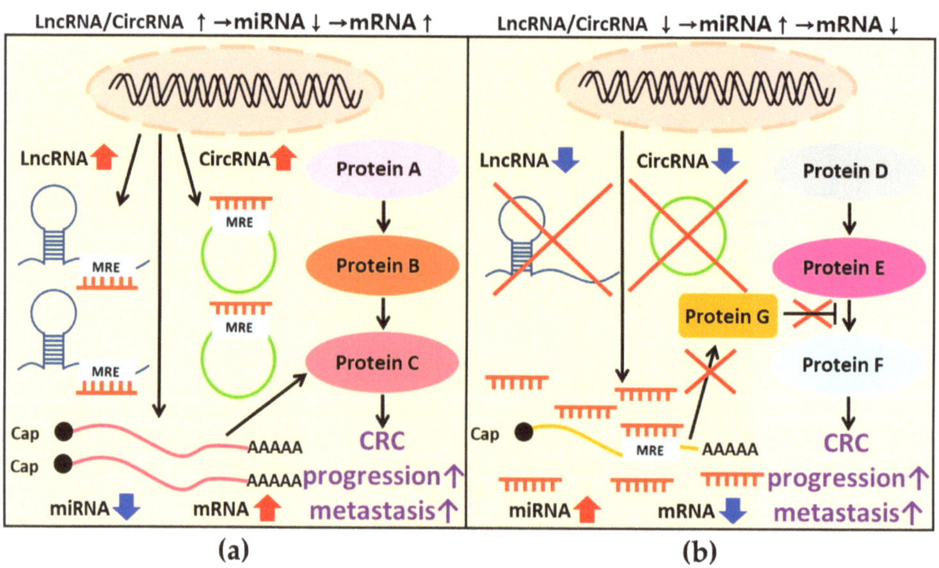

Figure 3. The regulatory mechanisms of ceRNA/miRNA/mRNA to enhance CRC progression or metastasis: (**a**) Upregulated ceRNA binds to miRNA, interfering with the binding between miRNA and target mRNA. This interference results in increased mRNA expression, promoting CRC progression or metastasis. (**b**) Downregulated ceRNA induces an increase in miRNA expression, leading to the binding to mRNA, which acts as an inhibitor of signaling pathways. Consequently, this interaction induces a decrease in mRNA expression, promoting the progression or metastasis of CRC. ceRNA: competing endogenous RNA; miRNA: microRNA; mRNA: messenger RNA; CRC: colorectal cancer.

Author Contributions: Conceptualization, Y.J.L. and H.-S.K.; investigation, Y.J.L.; writing—original draft preparation, Y.J.L.; writing—review and editing, Y.J.L., W.R.K., E.G.P., D.H.L., J.-m.K., H.J.S., H.-s.J., H.-Y.R. and H.-S.K.; supervision, H.-S.K. All authors have read and agreed to the published version of the manuscript.

Funding: This research received no external funding.

Institutional Review Board Statement: Not applicable.

Informed Consent Statement: Not applicable.

Data Availability Statement: Not applicable.

Conflicts of Interest: The authors declare no conflicts of interest.

References

1. Siegel, R.L.; Giaquinto, A.N.; Jemal, A. Cancer statistics, 2024. *CA Cancer J. Clin.* **2024**, *74*, 12–49. [CrossRef]
2. Hong, Y.; Kim, J.; Choi, Y.J.; Kang, J.G. Clinical study of colorectal cancer operation: Survival analysis. *Korean J. Clin. Oncol.* **2020**, *16*, 3. [CrossRef]
3. Vafapour, Z.; Troy, W.; Rashidi, A. Colon cancer detection by designing and analytical evaluation of a water-based THz metamaterial perfect absorber. *IEEE Sens. J.* **2021**, *21*, 19307–19313. [CrossRef]
4. Sawicki, T.; Ruszkowska, M.; Danielewicz, A.; Niedźwiedzka, E.; Arłukowicz, T.; Przybyłowicz, K.E. A review of colorectal cancer in terms of epidemiology, risk factors, development, symptoms and diagnosis. *Cancers* **2021**, *13*, 2025. [CrossRef]
5. Stefani, C.; Miricescu, D.; Stanescu-Spinu, I.-I.; Nica, R.I.; Greabu, M.; Totan, A.R.; Jinga, M. Growth factors, PI3K/AKT/mTOR and MAPK signaling pathways in colorectal cancer pathogenesis: Where are we now? *Int. J. Mol. Sci.* **2021**, *22*, 10260. [CrossRef]
6. Liebl, M.C.; Hofmann, T.G. The role of p53 signaling in colorectal cancer. *Cancers* **2021**, *13*, 2125. [CrossRef]
7. Tomita, N.; Ishida, H.; Tanakaya, K.; Yamaguchi, T.; Kumamoto, K.; Tanaka, T.; Hinoi, T.; Miyakura, Y.; Hasegawa, H.; Takayama, T. Japanese society for cancer of the colon and rectum (JSCCR) guidelines 2020 for the clinical practice of hereditary colorectal cancer. *Int. J. Clin. Oncol.* **2021**, *26*, 1353–1419. [CrossRef]
8. Ahmad, R.; Singh, J.K.; Wunnava, A.; Al-Obeed, O.; Abdulla, M.; Srivastava, S.K. Emerging trends in colorectal cancer: Dysregulated signaling pathways. *Int. J. Mol. Med.* **2021**, *47*, 14. [CrossRef]
9. Wan, M.-l.; Wang, Y.; Zeng, Z.; Deng, B.; Zhu, B.-s.; Cao, T.; Li, Y.-k.; Xiao, J.; Han, Q.; Wu, Q. Colorectal cancer (CRC) as a multifactorial disease and its causal correlations with multiple signaling pathways. *Biosci. Rep.* **2020**, *40*, BSR20200265. [CrossRef] [PubMed]
10. Liu, N.; Wang, Z.-Z.; Zhao, M.; Zhang, Y.; Chen, N.-H. Role of non-coding RNA in the pathogenesis of depression. *Gene* **2020**, *735*, 144276. [CrossRef] [PubMed]
11. Ma, B.; Wang, S.; Wu, W.; Shan, P.; Chen, Y.; Meng, J.; Xing, L.; Yun, J.; Hao, L.; Wang, X. Mechanisms of circRNA/lncRNA-miRNA interactions and applications in disease and drug research. *Biomed. Pharmacother.* **2023**, *162*, 114672. [CrossRef] [PubMed]
12. Nepstad, I.; Hatfield, K.J.; Grønningsæter, I.S.; Reikvam, H. The PI3K-Akt-mTOR signaling pathway in human acute myeloid leukemia (AML) cells. *Int. J. Mol. Sci.* **2020**, *21*, 2907. [CrossRef] [PubMed]
13. Malki, A.; ElRuz, R.A.; Gupta, I.; Allouch, A.; Vranic, S.; Al Moustafa, A.-E. Molecular mechanisms of colon cancer progression and metastasis: Recent insights and advancements. *Int. J. Mol. Sci.* **2020**, *22*, 130. [CrossRef]
14. Zhao, H.; Ming, T.; Tang, S.; Ren, S.; Yang, H.; Liu, M.; Tao, Q.; Xu, H. Wnt signaling in colorectal cancer: Pathogenic role and therapeutic target. *Mol. Cancer* **2022**, *21*, 144. [CrossRef] [PubMed]
15. Steinhart, Z.; Angers, S. Wnt signaling in development and tissue homeostasis. *Development* **2018**, *145*, dev146589. [CrossRef] [PubMed]
16. Hayat, R.; Manzoor, M.; Hussain, A. Wnt signaling pathway: A comprehensive review. *Cell Biol. Int.* **2022**, *46*, 863–877. [CrossRef] [PubMed]
17. Li, X.; Ortiz, M.A.; Kotula, L. The physiological role of Wnt pathway in normal development and cancer. *Medicus. Exp. Biol. Med.* **2020**, *245*, 411–426. [CrossRef] [PubMed]
18. Najafi, S.M.A. Canonical Wnt Signaling (Wnt/β-Catenin Pathway): A Potential Target for Cancer Prevention and Therapy. *Iran. Biomed. J.* **2020**, *24*, 269. [CrossRef]
19. Bian, J.; Dannappel, M.; Wan, C.; Firestein, R. Transcriptional regulation of Wnt/β-catenin pathway in colorectal cancer. *Cells* **2020**, *9*, 2125. [CrossRef]
20. Schatoff, E.M.; Leach, B.I.; Dow, L.E. Wnt signaling and colorectal cancer. *Curr. Color. Cancer Rep.* **2017**, *13*, 101–110. [CrossRef]
21. Fleming-de-Moraes, C.D.; Rocha, M.R.; Tessmann, J.W.; de Araujo, W.M.; Morgado-Diaz, J.A. Crosstalk between PI3K/Akt and Wnt/β-catenin pathways promote colorectal cancer progression regardless of mutational status. *Cancer Biol. Ther.* **2022**, *23*, 1–13. [CrossRef]
22. Basu, S.; Haase, G.; Ben-Ze'ev, A. Wnt signaling in cancer stem cells and colon cancer metastasis. *F1000Research* **2016**, *5*, 699. [CrossRef] [PubMed]

23. Aghabozorgi, A.S.; Ebrahimi, R.; Bahiraee, A.; Tehrani, S.S.; Nabizadeh, F.; Setayesh, L.; Jafarzadeh-Esfehani, R.; Ferns, G.A.; Avan, A.; Rashidi, Z. The genetic factors associated with Wnt signaling pathway in colorectal cancer. *Life Sci.* **2020**, *256*, 118006. [CrossRef]
24. Network, C.G.A. Comprehensive molecular characterization of human colon and rectal cancer. *Nature* **2012**, *487*, 330. [CrossRef] [PubMed]
25. La Vecchia, S.; Sebastián, C. Metabolic pathways regulating colorectal cancer initiation and progression. *Semin. Cell Dev. Biol.* **2020**, *98*, 63–70. [CrossRef]
26. Yoshida, N.; Kinugasa, T.; Ohshima, K.; Yuge, K.; Ohchi, T.; Fujino, S.; Shiraiwa, S.; Katagiri, M.; Akagi, Y. Analysis of Wnt and β-catenin expression in advanced colorectal cancer. *Anticancer Res.* **2015**, *35*, 4403–4410. [PubMed]
27. Kim, W.K.; Kwon, Y.; Jang, M.; Park, M.; Kim, J.; Cho, S.; Jang, D.G.; Lee, W.-B.; Jung, S.H.; Choi, H.J. β-catenin activation down-regulates cell-cell junction-related genes and induces epithelial-to-mesenchymal transition in colorectal cancers. *Sci. Rep.* **2019**, *9*, 18440. [CrossRef]
28. Zhang, Z.; Chang, Y.; Zhang, J.; Lu, Y.; Zheng, L.; Hu, Y.; Zhang, F.; Li, X.; Zhang, W.; Li, X. HMGB3 promotes growth and migration in colorectal cancer by regulating WNT/β-catenin pathway. *PLoS ONE* **2017**, *12*, e0179741. [CrossRef]
29. Wachsmannova, L.; Mego, M.; Stevurkova, V.; Zajac, V.; Ciernikova, S. Novel strategies for comprehensive mutation screening of the APC gene. *Neoplasma* **2017**, *64*, 338–343. [CrossRef]
30. Meng, F.; Liu, X.; Lin, C.; Xu, L.; Liu, J.; Zhang, P.; Zhang, X.; Song, J.; Yan, Y.; Ren, Z. SMYD2 suppresses APC2 expression to activate the Wnt/β-catenin pathway and promotes epithelial-mesenchymal transition in colorectal cancer. *Am. J. Cancer Res.* **2020**, *10*, 997.
31. Miricescu, D.; Totan, A.; Stanescu-Spinu, I.-I.; Badoiu, S.C.; Stefani, C.; Greabu, M. PI3K/AKT/mTOR signaling pathway in breast cancer: From molecular landscape to clinical aspects. *Int. J. Mol. Sci.* **2020**, *22*, 173. [CrossRef] [PubMed]
32. Ruchi Sharma, V.; Kumar Gupta, G.; Sharma, A.K.; Batra, N.; Sharma, D.K.; Joshi, A.; Sharma, A.K. PI3K/Akt/mTOR intracellular pathway and breast cancer: Factors, mechanism and regulation. *Curr. Pharm. Des.* **2017**, *23*, 1633–1638. [CrossRef] [PubMed]
33. Owusu-Brackett, N.; Shariati, M.; Meric-Bernstam, F. Role of PI3K/AKT/mTOR in Cancer Signaling. In *Predictive Biomarkers in Oncology: Applications in Precision Medicine*; Springer: Berlin/Heidelberg, Germany, 2019; pp. 263–270.
34. Wang, Q.; Zhang, P.; Zhang, W.; Zhang, X.; Chen, J.; Ding, P.; Li, L.; Lv, X.; Li, L.; Hu, W. PI3K activation is enhanced by FOXM1D binding to p110 and p85 subunits. *Signal Transduct. Target. Ther.* **2020**, *5*, 105. [CrossRef] [PubMed]
35. Mazloumi Gavgani, F.; Smith Arnesen, V.; Jacobsen, R.G.; Krakstad, C.; Hoivik, E.A.; Lewis, A.E. Class I phosphoinositide 3-Kinase PIK3CA/p110α and PIK3CB/p110β isoforms in endometrial cancer. *Int. J. Mol. Sci.* **2018**, *19*, 3931. [CrossRef] [PubMed]
36. Xue, C.; Li, G.; Lu, J.; Li, L. Crosstalk between circRNAs and the PI3K/AKT signaling pathway in cancer progression. *Signal Transduct. Target. Ther.* **2021**, *6*, 400. [CrossRef] [PubMed]
37. Papadatos-Pastos, D.; Rabbie, R.; Ross, P.; Sarker, D. The role of the PI3K pathway in colorectal cancer. *Crit. Rev. Oncol. Hematol.* **2015**, *94*, 18–30. [CrossRef] [PubMed]
38. Moafian, Z.; Maghrouni, A.; Soltani, A.; Hashemy, S.I. Cross-talk between non-coding RNAs and PI3K/AKT/mTOR pathway in colorectal cancer. *Mol. Biol. Rep.* **2021**, *48*, 4797–4811. [CrossRef] [PubMed]
39. Johnson, S.M.; Gulhati, P.; Rampy, B.A.; Han, Y.; Rychahou, P.G.; Doan, H.Q.; Weiss, H.L.; Evers, B.M. Novel expression patterns of PI3K/Akt/mTOR signaling pathway components in colorectal cancer. *J. Am. Coll. Surg.* **2010**, *210*, 767–776. [CrossRef] [PubMed]
40. Zhu, Y.-F.; Yu, B.-H.; Li, D.-L.; Ke, H.-L.; Guo, X.-Z.; Xiao, X.-Y. PI3K expression and PIK3CA mutations are related to colorectal cancer metastases. *World J. Gastroenterol.* **2012**, *18*, 3745. [CrossRef]
41. Suman, S.; Kurisetty, V.; Das, T.P.; Vadodkar, A.; Ramos, G.; Lakshmanaswamy, R.; Damodaran, C. Activation of AKT signaling promotes epithelial–mesenchymal transition and tumor growth in colorectal cancer cells. *Mol. Carcinog.* **2014**, *53*, E151–E160. [CrossRef]
42. Cai, Z.; Ke, J.; He, X.; Yuan, R.; Chen, Y.; Wu, X.; Wang, L.; Wang, J.; Lan, P.; Wu, X. Significance of mTOR signaling and its inhibitor against cancer stem-like cells in colorectal cancer. *Ann. Surg. Oncol.* **2014**, *21*, 179–188. [CrossRef]
43. Xu, W.; Yu, M.; Qin, J.; Luo, Y.; Zhong, M. LACTB regulates PIK3R3 to promote autophagy and inhibit EMT and proliferation through the PI3K/AKT/mTOR signaling pathway in colorectal cancer. *Cancer Manag. Res.* **2020**, *12*, 5181–5200. [CrossRef]
44. Feng, Y.; Ji, D.; Huang, Y.; Ji, B.; Zhang, Y.; Li, J.; Peng, W.; Zhang, C.; Zhang, D.; Sun, Y. TGM3 functions as a tumor suppressor by repressing epithelial-to-mesenchymal transition and the PI3K/AKT signaling pathway in colorectal cancer. *Oncol. Rep.* **2020**, *43*, 864–876. [CrossRef]
45. Su, W.; Feng, B.; Hu, L.; Guo, X.; Yu, M. MUC3A promotes the progression of colorectal cancer through the PI3K/Akt/mTOR pathway. *BMC Cancer* **2022**, *22*, 602. [CrossRef]
46. Guo, Y.J.; Pan, W.W.; Liu, S.B.; Shen, Z.F.; Xu, Y.; Hu, L.L. ERK/MAPK signalling pathway and tumorigenesis. *Exp. Ther. Med.* **2020**, *19*, 1997–2007. [CrossRef] [PubMed]
47. Sheng, Y.-N.; Luo, Y.-H.; Liu, S.-B.; Xu, W.-T.; Zhang, Y.; Zhang, T.; Xue, H.; Zuo, W.-B.; Li, Y.-N.; Wang, C.-Y. Zeaxanthin induces apoptosis via ROS-regulated MAPK and AKT signaling pathway in human gastric cancer cells. *Onco Targets Ther.* **2020**, *13*, 10995–11006. [CrossRef]
48. Asl, E.R.; Amini, M.; Najafi, S.; Mansoori, B.; Mokhtarzadeh, A.; Mohammadi, A.; Lotfinejad, P.; Bagheri, M.; Shirjang, S.; Lotfi, Z. Interplay between MAPK/ERK signaling pathway and MicroRNAs: A crucial mechanism regulating cancer cell metabolism and tumor progression. *Life Sci.* **2021**, *278*, 119499. [CrossRef]

49. Kciuk, M.; Gielecińska, A.; Budzinska, A.; Mojzych, M.; Kontek, R. Metastasis and MAPK pathways. *Int. J. Mol. Sci.* **2022**, *23*, 3847. [CrossRef] [PubMed]
50. Fang, J.Y.; Richardson, B.C. The MAPK signalling pathways and colorectal cancer. *Lancet Oncol.* **2005**, *6*, 322–327. [CrossRef] [PubMed]
51. Zhou, G.; Yang, J.; Song, P. Correlation of ERK/MAPK signaling pathway with proliferation and apoptosis of colon cancer cells. *Oncol. Lett.* **2019**, *17*, 2266–2270. [CrossRef] [PubMed]
52. Wang, Y.; Xie, G.; Li, M.; Du, J.; Wang, M. COPB2 gene silencing inhibits colorectal cancer cell proliferation and induces apoptosis via the JNK/c-Jun signaling pathway. *PLoS ONE* **2020**, *15*, e0240106. [CrossRef] [PubMed]
53. Zhang, Y.; Liao, Y.; Chen, C.; Sun, W.; Sun, X.; Liu, Y.; Xu, E.; Lai, M.; Zhang, H. p38-regulated FOXC1 stability is required for colorectal cancer metastasis. *J. Pathol. Inform.* **2020**, *250*, 217–230. [CrossRef]
54. Ou, S.; Liao, Y.; Shi, J.; Tang, J.; Ye, Y.; Wu, F.; Wang, W.; Fei, J.; Xie, F.; Bai, L. S100A16 suppresses the proliferation, migration and invasion of colorectal cancer cells in part via the JNK/p38 MAPK pathway. *Mol. Med. Rep.* **2021**, *23*, 164. [CrossRef] [PubMed]
55. Wang, W.; Wang, X.; Peng, L.; Deng, Q.; Liang, Y.; Qing, H.; Jiang, B. CD24-dependent MAPK pathway activation is required for colorectal cancer cell proliferation. *Cancer Sci.* **2010**, *101*, 112–119. [CrossRef]
56. Itatani, Y.; Kawada, K.; Sakai, Y. Transforming growth factor-β signaling pathway in colorectal cancer and its tumor microenvironment. *Int. J. Mol. Sci.* **2019**, *20*, 5822. [CrossRef] [PubMed]
57. Zhao, H.; Wei, J.; Sun, J. Roles of TGF-β signaling pathway in tumor microenvirionment and cancer therapy. *Int. Immunopharmacol.* **2020**, *89*, 107101. [CrossRef]
58. Pawlak, J.B.; Blobe, G.C. TGF-β superfamily co-receptors in cancer. *Dev. Dynam.* **2022**, *251*, 117–143. [CrossRef]
59. Tzavlaki, K.; Moustakas, A. TGF-β Signaling. *Biomolecules* **2020**, *10*, 487. [CrossRef]
60. Yu, X.-Y.; Sun, Q.; Zhang, Y.-M.; Zou, L.; Zhao, Y.-Y. TGF-β/Smad signaling pathway in tubulointerstitial fibrosis. *Front. Pharmacol.* **2022**, *13*, 860588. [CrossRef]
61. Yang, D.; Hou, T.; Li, L.; Chu, Y.; Zhou, F.; Xu, Y.; Hou, X.; Song, H.; Zhu, K.; Hou, Z. Smad1 promotes colorectal cancer cell migration through Ajuba transactivation. *Oncotarget* **2017**, *8*, 110415. [CrossRef]
62. Shi, W.; He, J.; Huang, Y.; Zeng, Z.; Feng, Z.; Xu, H.; Nie, Y. Integrin β5 enhances the malignancy of human colorectal cancer by increasing the TGF-β signaling. *Anti-Cancer Drugs* **2021**, *32*, 717–726. [CrossRef]
63. Wang, L.; Ai, M.; Nie, M.; Zhao, L.; Deng, G.; Hu, S.; Han, Y.; Zeng, W.; Wang, Y.; Yang, M. EHF promotes colorectal carcinoma progression by activating TGF-β1 transcription and canonical TGF-β signaling. *Cancer Sci.* **2020**, *111*, 2310–2324. [CrossRef]
64. Williams, A.B.; Schumacher, B. p53 in the DNA-damage-repair process. *Cold Spring Harb. Perspect. Med.* **2016**, *6*, a026070. [CrossRef]
65. Borrero, L.J.H.; El-Deiry, W.S. Tumor suppressor p53: Biology, signaling pathways, and therapeutic targeting. *Biochim. Biophys. Acta* **2021**, *1876*, 188556.
66. Fu, X.; Wu, S.; Li, B.; Xu, Y.; Liu, J. Functions of p53 in pluripotent stem cells. *Protein Cell* **2020**, *11*, 71–78. [CrossRef] [PubMed]
67. Han, X.; Wei, W.; Sun, Y. PROTAC degraders with ligands recruiting MDM2 E3 ubiquitin ligase: An updated perspective. *Acta Mater. Med.* **2022**, *1*, 244. [CrossRef] [PubMed]
68. Eskandari, M.; Shi, Y.; Liu, J.; Albanese, J.; Goel, S.; Verma, A.; Wang, Y. The expression of MDM2, MDM4, p53 and p21 in myeloid neoplasms and the effect of MDM2/MDM4 dual inhibitor. *Leuk Lymphoma* **2021**, *62*, 167–175. [CrossRef]
69. Wang, P.; Wang, H.-Y.; Gao, X.-J.; Zhu, H.-X.; Zhang, X.-P.; Liu, F.; Wang, W. Encoding and decoding of p53 dynamics in cellular response to stresses. *Cells* **2023**, *12*, 490. [CrossRef]
70. Solomon, H.; Dinowitz, N.; Pateras, I.S.; Cooks, T.; Shetzer, Y.; Molchadsky, A.; Charni, M.; Rabani, S.; Koifman, G.; Tarcic, O. Mutant p53 gain of function underlies high expression levels of colorectal cancer stem cells markers. *Oncogene* **2018**, *37*, 1669–1684. [CrossRef]
71. Nagao, K.; Koshino, A.; Sugimura-Nagata, A.; Nagano, A.; Komura, M.; Ueki, A.; Ebi, M.; Ogasawara, N.; Tsuzuki, T.; Kasai, K. The complete loss of p53 expression uniquely predicts worse prognosis in colorectal cancer. *Int. J. Mol. Sci.* **2022**, *23*, 3252. [CrossRef]
72. Kruschewski, M.; Mueller, K.; Lipka, S.; Budczies, J.; Noske, A.; Buhr, H.J.; Elezkurtaj, S. The prognostic impact of p53 expression on sporadic colorectal cancer is dependent on p21 status. *Cancers* **2011**, *3*, 1274–1284. [CrossRef] [PubMed]
73. Kondo, I.; Iida, S.; Takagi, Y.; Sugihara, K. MDM2 mRNA expression in the p53 pathway may predict the potential of invasion and liver metastasis in colorectal cancer. *Dis. Colon Rectum* **2008**, *51*, 1395–1402. [CrossRef] [PubMed]
74. Wang, W.; Wei, J.; Zhang, H.; Zheng, X.; Zhou, H.; Luo, Y.; Yang, J.; Deng, Q.; Huang, S.; Fu, Z. PRDX2 promotes the proliferation of colorectal cancer cells by increasing the ubiquitinated degradation of p53. *Cell Death Dis.* **2021**, *12*, 605. [CrossRef] [PubMed]
75. Shen, A.; Liu, L.; Huang, Y.; Shen, Z.; Wu, M.; Chen, X.; Wu, X.; Lin, X.; Chen, Y.; Li, L. Down-regulating HAUS6 suppresses cell proliferation by activating the p53/p21 pathway in colorectal cancer. *Front. Cell Dev. Biol.* **2022**, *9*, 772077. [CrossRef] [PubMed]
76. Wang, L.; Xiao, X.; Li, D.; Chi, Y.; Wei, P.; Wang, Y.; Ni, S.; Tan, C.; Zhou, X.; Du, X. Abnormal expression of GADD45B in human colorectal carcinoma. *J. Transl. Med.* **2012**, *10*, 215. [CrossRef] [PubMed]
77. Chen, J.; Wei, Y.; Feng, Q.; Ren, L.; He, G.; Chang, W.; Zhu, D.; Yi, T.; Lin, Q.; Tang, W. Ribosomal protein S15A promotes malignant transformation and predicts poor outcome in colorectal cancer through misregulation of p53 signaling pathway. *Int. J. Oncol.* **2016**, *48*, 1628–1638. [CrossRef] [PubMed]

78. Science in School. Available online: https://www.scienceinschool.org/article/2023/the-importance-of-non-coding-rnas/ (accessed on 1 February 2024).
79. Moraes, F.; Góes, A. A decade of human genome project conclusion: Scientific diffusion about our genome knowledge. *Biochem. Mol. Biol. Educ.* **2016**, *44*, 215–223. [CrossRef] [PubMed]
80. Oyelade, J.; Isewon, I.; Ogunlana, O.; Aworunse, O.; Oyesola, O.; Aromolaran, O.; Dokumu, T.; Ademuwagun, I.; Iheagwam, F.; Babatunde, E. Overview of the human genome. In *Genome Plasticity in Health and Disease*; Academic Press: Cambridge, MA, USA, 2020; pp. 9–26.
81. Santosh, B.; Varshney, A.; Yadava, P.K. Non-coding RNAs: Biological functions and applications. *Cell Biochem. Funct.* **2015**, *33*, 14–22. [CrossRef] [PubMed]
82. Micheel, J.; Safrastyan, A.; Wollny, D. Advances in non-coding RNA sequencing. *Non-Coding RNA* **2021**, *7*, 70. [CrossRef]
83. Bhatti, G.K.; Khullar, N.; Sidhu, I.S.; Navik, U.S.; Reddy, A.P.; Reddy, P.H.; Bhatti, J.S. Emerging role of non-coding RNA in health and disease. *Metab. Brain Dis.* **2021**, *36*, 1119–1134. [CrossRef]
84. Luo, R.; Song, J.; Zhang, W.; Ran, L. Identification of MFI2-AS1, a novel pivotal lncRNA for prognosis of stage III/IV colorectal cancer. *Dig. Dis. Sci.* **2020**, *65*, 3538–3550. [CrossRef] [PubMed]
85. Shao, T.; Pan, Y.-h.; Xiong, X.-d. Circular RNA: An important player with multiple facets to regulate its parental gene expression. *Mol. Ther. Nucleic Acids* **2021**, *23*, 369–376. [CrossRef]
86. Gebremedhn, S.; Ali, A.; Hossain, M.; Hoelker, M.; Salilew-Wondim, D.; Anthony, R.V.; Tesfaye, D. MicroRNA-mediated gene regulatory mechanisms in mammalian female reproductive health. *Int. J. Mol. Sci.* **2021**, *22*, 938. [CrossRef]
87. Wang, L.-L.; Chen, W.-Q.; Wang, Y.-R.; Zeng, L.-P.; Chen, T.-T.; Chen, G.-Y.; Chen, J.-H. Numerous long single-stranded DNAs produced by dual amplification reactions for electrochemical detection of exosomal microRNAs. *Biosens. Bioelectron.* **2020**, *169*, 112555. [CrossRef] [PubMed]
88. Wallace, D.R.; Taalab, Y.M.; Heinze, S.; Tariba Lovaković, B.; Pizent, A.; Renieri, E.; Tsatsakis, A.; Farooqi, A.A.; Javorac, D.; Andjelkovic, M. Toxic-metal-induced alteration in miRNA expression profile as a proposed mechanism for disease development. *Cells* **2020**, *9*, 901. [CrossRef] [PubMed]
89. Soni, D.K.; Biswas, R. Role of non-coding RNAs in post-transcriptional regulation of lung diseases. *Front. Genet.* **2021**, *12*, 767348. [CrossRef]
90. Riolo, G.; Cantara, S.; Marzocchi, C.; Ricci, C. miRNA targets: From prediction tools to experimental validation. *Methods Protoc.* **2020**, *4*, 1. [CrossRef] [PubMed]
91. Indrieri, A.; Carrella, S.; Carotenuto, P.; Banfi, S.; Franco, B. The pervasive role of the miR-181 family in development, neurodegeneration, and cancer. *Int. J. Mol. Sci.* **2020**, *21*, 2092. [CrossRef]
92. Kalayinia, S.; Arjmand, F.; Maleki, M.; Malakootian, M.; Singh, C.P. MicroRNAs: Roles in cardiovascular development and disease. *Cardiovasc. Pathol.* **2021**, *50*, 107296. [CrossRef]
93. Prasad, A.; Sharma, N.; Prasad, M. Noncoding but coding: Pri-miRNA into the action. *Trends Plant Sci.* **2021**, *26*, 204–206. [CrossRef]
94. Jin, W.; Wang, J.; Liu, C.-P.; Wang, H.-W.; Xu, R.-M. Structural basis for pri-miRNA recognition by Drosha. *Mol. Cell* **2020**, *78*, 423–433. [CrossRef] [PubMed]
95. Hill, M.; Tran, N. miRNA interplay: Mechanisms and consequences in cancer. *Dis. Model. Mech.* **2021**, *14*, dmm047662. [CrossRef] [PubMed]
96. Klicka, K.; Grzywa, T.M.; Mielniczuk, A.; Klinke, A.; Włodarski, P.K. The role of miR-200 family in the regulation of hallmarks of cancer. *Front. Oncol.* **2022**, *12*, 965231. [CrossRef] [PubMed]
97. Paul, S.; Bravo Vázquez, L.A.; Pérez Uribe, S.; Roxana Reyes-Pérez, P.; Sharma, A. Current status of microRNA-based therapeutic approaches in neurodegenerative disorders. *Cells* **2020**, *9*, 1698. [CrossRef] [PubMed]
98. Çakmak, H.A.; Demir, M. MicroRNA and cardiovascular diseases. *Balkan Med. J.* **2020**, *37*, 60. [CrossRef] [PubMed]
99. Pan, S.; Wu, W.; Ren, F.; Li, L.; Li, Y.; Li, W.; Wang, A.; Liu, D.; Dong, Y. MiR-346-5p promotes colorectal cancer cell proliferation in vitro and in vivo by targeting FBXL2 and activating the β-catenin signaling pathway. *Life Sci.* **2020**, *244*, 117300. [CrossRef] [PubMed]
100. Guo, T.; Wang, J.; Cheng, G.; Huang, H. miR-590-5p may regulate colorectal cancer cell viability and migration by targeting PDCD4. *Exp. Ther. Med.* **2020**, *20*, 55. [CrossRef]
101. He, Z.; Dang, J.; Song, A.; Cui, X.; Ma, Z.; Zhang, Y. The involvement of miR-150/β-catenin axis in colorectal cancer progression. *Biomed. Pharmacother.* **2020**, *121*, 109495. [CrossRef] [PubMed]
102. Song, G.-L.; Xiao, M.; Wan, X.-Y.; Deng, J.; Ling, J.-D.; Tian, Y.-G.; Li, M.; Yin, J.; Zheng, R.-Y.; Tang, Y. MiR-130a-3p suppresses colorectal cancer growth by targeting Wnt Family Member 1 (WNT1). *Bioengineered* **2021**, *12*, 8407–8418. [CrossRef]
103. Gao, Z.; Jiang, J.; Hou, L.; Zhang, B. Dysregulation of MiR-144-5p/RNF187 axis contributes to the progression of colorectal cancer. *J. Transl. Med.* **2022**, *10*, 65–75. [CrossRef]
104. Du, G.; Yu, X.; Chen, Y.; Cai, W. MiR-1-3p Suppresses Colorectal Cancer Cell Proliferation and Metastasis by Inhibiting YWHAZ-Mediated Epithelial–Mesenchymal Transition. *Front. Oncol.* **2021**, *11*, 634596. [CrossRef] [PubMed]
105. Wang, C.; Peng, R.; Zeng, M.; Zhang, Z.; Liu, S.; Jiang, D.; Lu, Y.; Zou, F. An autoregulatory feedback loop of miR-21/VMP1 is responsible for the abnormal expression of miR-21 in colorectal cancer cells. *Cell Death Dis.* **2020**, *11*, 1067. [CrossRef] [PubMed]

106. Jiang, R.; Chen, X.; Ge, S.; Wang, Q.; Liu, Y.; Chen, H.; Xu, J.; Wu, J. MiR-21-5p induces pyroptosis in colorectal cancer via TGFBI. *Front. Oncol.* **2021**, *10*, 610545. [CrossRef] [PubMed]
107. Lin, X.; Han, L.; Gu, C.; Lai, Y.; Lai, Q.; Li, Q.; He, C.; Meng, Y.; Pan, L.; Liu, S. MiR-452-5p promotes colorectal cancer progression by regulating an ERK/MAPK positive feedback loop. *Aging* **2021**, *13*, 7608. [CrossRef] [PubMed]
108. Pan, M.; Chen, Q.; Lu, Y.; Wei, F.; Chen, C.; Tang, G.; Huang, H. MiR-106b-5p regulates the migration and invasion of colorectal cancer cells by targeting FAT4. *Biosci. Rep.* **2020**, *40*, BSR20200098. [CrossRef] [PubMed]
109. Wang, H.; Yan, B.; Zhang, P.; Liu, S.; Li, Q.; Yang, J.; Yang, F.; Chen, E. MiR-496 promotes migration and epithelial-mesenchymal transition by targeting RASSF6 in colorectal cancer. *J. Cell. Physiol.* **2020**, *235*, 1469–1479. [CrossRef] [PubMed]
110. Zhang, X.; Li, T.; Han, Y.-N.; Ge, M.; Wang, P.; Sun, L.; Liu, H.; Cao, T.; Nie, Y.; Fan, D. miR-125b promotes colorectal cancer migration and invasion by dual-targeting CFTR and CGN. *Cancers* **2021**, *13*, 5710. [CrossRef]
111. Arabsorkhi, Z.; Gharib, E.; Yaghmoorian Khojini, J.; Farhadieh, M.E.; Nazemalhosseini-Mojarad, E.; Zali, M.R. miR-298 plays a pivotal role in colon cancer invasiveness by targeting PTEN. *J. Cell. Physiol.* **2020**, *235*, 4335–4350. [CrossRef]
112. Chen, X.; Wang, A.-l.; Liu, Y.-y.; Zhao, C.-x.; Zhou, X.; Liu, H.-l.; Lin, M.-b. MiR-429 involves in the pathogenesis of colorectal cancer via directly targeting LATS2. *Oxid. Med. Cell. Longev.* **2020**, *2020*, 5316276. [CrossRef]
113. Li, S.; Hou, X.; Wu, C.; Han, L.; Li, Q.; Wang, J.; Luo, S. MiR-645 promotes invasiveness, metastasis and tumor growth in colorectal cancer by targeting EFNA5. *Biomed. Pharmacother.* **2020**, *125*, 109889. [CrossRef]
114. Sun, N.; Zhang, L.; Zhang, C.; Yuan, Y. miR-144-3p inhibits cell proliferation of colorectal cancer cells by targeting BCL6 via inhibition of Wnt/β-catenin signaling. *Cell. Mol. Biol. Lett.* **2020**, *25*, 19. [CrossRef] [PubMed]
115. Chen, J.; Chen, S.; Zhuo, L.; Zhu, Y.; Zheng, H. Regulation of cancer stem cell properties, angiogenesis, and vasculogenic mimicry by miR-450a-5p/SOX2 axis in colorectal cancer. *Cell Death Dis.* **2020**, *11*, 173. [CrossRef] [PubMed]
116. Machackova, T.; Vychytilova-Faltejskova, P.; Souckova, K.; Trachtova, K.; Brchnelova, D.; Svoboda, M.; Kiss, I.; Prochazka, V.; Kala, Z.; Slaby, O. MiR-215-5p reduces liver metastasis in an experimental model of colorectal cancer through regulation of ECM-receptor interactions and focal adhesion. *Cancers* **2020**, *12*, 3518. [CrossRef] [PubMed]
117. Wang, G.; Cao, X.; Lai, S.; Luo, X.; Feng, Y.; Wu, J.; Ning, Q.; Xia, X.; Wang, J.; Gong, J. Altered p53 regulation of miR-148b and p55PIK contributes to tumor progression in colorectal cancer. *Oncogene* **2015**, *34*, 912–921. [CrossRef] [PubMed]
118. Ma, Q.; Wang, X.; Li, Z.; Li, B.; Ma, F.; Peng, L.; Zhang, Y.; Xu, A.; Jiang, B. microRNA-16 represses colorectal cancer cell growth in vitro by regulating the p53/survivin signaling pathway. *Oncol. Rep.* **2013**, *29*, 1652–1658. [CrossRef] [PubMed]
119. Yasudome, R.; Seki, N.; Asai, S.; Goto, Y.; Kita, Y.; Hozaka, Y.; Wada, M.; Tanabe, K.; Idichi, T.; Mori, S. Molecular pathogenesis of colorectal cancer: Impact of oncogenic targets regulated by tumor suppressive miR-139-3p. *Int. J. Mol. Sci.* **2022**, *23*, 11616. [CrossRef] [PubMed]
120. Li, Y.; Yang, J.; Wang, H.; Qiao, W.; Guo, Y.; Zhang, S.; Guo, Y. FNDC3B, targeted by miR-125a-5p and miR-217, promotes the proliferation and invasion of colorectal cancer cells via PI3K/mTOR signaling. *Onco Targets Ther.* **2020**, *13*, 3501–3510. [CrossRef]
121. Zhang, H.; Shen, Z.; Zhou, Y.; Zhang, Z.; Wang, Q.; Zhang, M.; Wang, S.; Wang, B. Downregulation of miR-654-3p in colorectal cancer indicates poor prognosis and promotes cell proliferation and invasion by targeting SRC. *Front. Genet.* **2020**, *11*, 577948. [CrossRef]
122. Yang, Y.; Lin, Z.; Han, Z.; Wu, Z.; Hua, J.; Zhong, R.; Zhao, R.; Ran, H.; Qu, K.; Huang, H. miR-539 activates the SAPK/JNK signaling pathway to promote ferropotosis in colorectal cancer by directly targeting TIPE. *Cell Death Discov.* **2021**, *7*, 272. [CrossRef]
123. Chang, S.; Sun, G.; Zhang, D.; Li, Q.; Qian, H. MiR-3622a-3p acts as a tumor suppressor in colorectal cancer by reducing stemness features and EMT through targeting spalt-like transcription factor 4. *Cell Death Dis.* **2020**, *11*, 592. [CrossRef]
124. Liu, H.-T.; Ma, R.-R.; Lv, B.-B.; Zhang, H.; Shi, D.-B.; Guo, X.-Y.; Zhang, G.-H.; Gao, P. LncRNA-HNF1A-AS1 functions as a competing endogenous RNA to activate PI3K/AKT signalling pathway by sponging miR-30b-3p in gastric cancer. *Br. J. Cancer* **2020**, *122*, 1825–1836. [CrossRef]
125. Ala, U. Competing endogenous RNAs, non-coding RNAs and diseases: An intertwined story. *Cells* **2020**, *9*, 1574. [CrossRef] [PubMed]
126. Xu, G.; Xu, W.-Y.; Xiao, Y.; Jin, B.; Du, S.-D.; Mao, Y.-l.; Zhang, Z.-T. The emerging roles of non-coding competing endogenous RNA in hepatocellular carcinoma. *Cancer Cell Int.* **2020**, *20*, 496. [CrossRef] [PubMed]
127. Mitra, A.; Pfeifer, K.; Park, K.-S. Circular RNAs and competing endogenous RNA (ceRNA) networks. *Transl. Cancer Res.* **2018**, *7* (Suppl. 5), S624. [CrossRef] [PubMed]
128. Li, D.; Zhang, J.; Li, J. Role of miRNA sponges in hepatocellular carcinoma. *Clin. Chim. Acta.* **2020**, *500*, 10–19. [CrossRef] [PubMed]
129. Li, K.; Yao, T.; Zhang, Y.; Li, W.; Wang, Z. NEAT1 as a competing endogenous RNA in tumorigenesis of various cancers: Role, mechanism and therapeutic potential. *Int. J. Biol. Sci.* **2021**, *17*, 3428. [CrossRef] [PubMed]
130. Qi, X.; Lin, Y.; Liu, X.; Chen, J.; Shen, B. Biomarker discovery for the carcinogenic heterogeneity between colon and rectal cancers based on lncRNA-associated ceRNA network analysis. *Front. Oncol.* **2020**, *10*, 535985. [CrossRef] [PubMed]
131. Fang, C.; Wang, L.; Gong, C.; Wu, W.; Yao, C.; Zhu, S. Long non-coding RNAs: How to regulate the metastasis of non–small-cell lung cancer. *J. Cell. Mol. Med.* **2020**, *24*, 3282–3291. [CrossRef]
132. Yang, X.; Xie, Z.; Lei, X.; Gan, R. Long non-coding RNA GAS5 in human cancer. *Oncol. Lett.* **2020**, *20*, 2587–2594. [CrossRef]
133. Pisignano, G.; Ladomery, M. Post-transcriptional regulation through long non-coding rnas (lncrnas). *Non-Coding RNA* **2021**, *7*, 29. [CrossRef]

134. Mattick, J.S.; Amaral, P.P.; Carninci, P.; Carpenter, S.; Chang, H.Y.; Chen, L.-L.; Chen, R.; Dean, C.; Dinger, M.E.; Fitzgerald, K.A. Long non-coding RNAs: Definitions, functions, challenges and recommendations. *Nat. Rev. Mol. Cell Biol.* **2023**, *24*, 430–447. [CrossRef] [PubMed]
135. Neve, B.; Jonckheere, N.; Vincent, A.; Van Seuningen, I. Long non-coding RNAs: The tentacles of chromatin remodeler complexes. *Cell. Mol. Life Sci.* **2021**, *78*, 1139–1161. [CrossRef] [PubMed]
136. Gil, N.; Ulitsky, I. Regulation of gene expression by cis-acting long non-coding RNAs. *Nat. Rev. Genet* **2020**, *21*, 102–117. [CrossRef] [PubMed]
137. Liu, Y.; Ding, W.; Yu, W.; Zhang, Y.; Ao, X.; Wang, J. Long non-coding RNAs: Biogenesis, functions, and clinical significance in gastric cancer. *Mol. Ther.-Oncolytics* **2021**, *23*, 458–476. [CrossRef] [PubMed]
138. Shi, Y.; Liu, J.-B.; Deng, J.; Zou, D.-Z.; Wu, J.-J.; Cao, Y.-H.; Yin, J.; Ma, Y.-S.; Da, F.; Li, W. The role of ceRNA-mediated diagnosis and therapy in hepatocellular carcinoma. *Hereditas* **2021**, *158*, 44. [CrossRef] [PubMed]
139. Quinn, J.J.; Chang, H.Y. Unique features of long non-coding RNA biogenesis and function. *Nat. Rev. Genet* **2016**, *17*, 47–62. [CrossRef] [PubMed]
140. Tsagakis, I.; Douka, K.; Birds, I.; Aspden, J.L. Long non-coding RNAs in development and disease: Conservation to mechanisms. *J. Pathol. Inform.* **2020**, *250*, 480–495. [CrossRef] [PubMed]
141. Sun, B.; Liu, C.; Li, H.; Zhang, L.; Luo, G.; Liang, S.; Lü, M. Research progress on the interactions between long non-coding RNAs and microRNAs in human cancer. *Oncol. Lett.* **2020**, *19*, 595–605. [CrossRef] [PubMed]
142. Atef, M.; Amer, A.; Hafez, Y.; Elsebaey, M.; Saber, S.; Abd El-Khalik, S. Long non-coding RNA EGFR-AS1 in colorectal cancer: Potential role in tumorigenesis and survival via miRNA-133b sponge and EGFR/STAT3 axis regulation. *J. Biomed. Sci.* **2021**, *78*, 122–129. [CrossRef]
143. Shang, A.; Wang, W.; Gu, C.; Chen, W.; Lu, W.; Sun, Z.; Li, D. Long non-coding RNA CCAT1 promotes colorectal cancer progression by regulating miR-181a-5p expression. *Aging* **2020**, *12*, 8301. [CrossRef]
144. Bi, C.; Cui, H.; Fan, H.; Li, L. LncRNA LINC01116 promotes the development of colorectal cancer by targeting miR-9-5p/STMN1. *Onco Targets Ther.* **2020**, *13*, 10547–10558. [CrossRef] [PubMed]
145. Li, Q.; Yue, W.; Li, M.; Jiang, Z.; Hou, Z.; Liu, W.; Ma, N.; Gan, W.; Li, Y.; Zhou, T. Downregulating long Non-coding RNAs CTBP1-AS2 inhibits colorectal cancer development by modulating the miR-93-5p/TGF-β/SMAD2/3 pathway. *Front. Oncol.* **2021**, *11*, 626620. [CrossRef] [PubMed]
146. Ke, D.; Wang, Q.; Ke, S.; Zou, L.; Wang, Q. Long-non coding RNA SNHG16 supports colon cancer cell growth by modulating miR-302a-3p/AKT axis. *Pathol. Oncol. Res.* **2020**, *26*, 1605–1613. [CrossRef] [PubMed]
147. Yu, Z.; Wang, Y.; Deng, J.; Liu, D.; Zhang, L.; Shao, H.; Wang, Z.; Zhu, W.; Zhao, C.; Ke, Q. Long non-coding RNA COL4A2-AS1 facilitates cell proliferation and glycolysis of colorectal cancer cells via miR-20b-5p/hypoxia inducible factor 1 alpha subunit axis. *Bioengineered* **2021**, *12*, 6251–6263. [CrossRef] [PubMed]
148. Han, H.; Li, H.; Zhou, J. Long non-coding RNA MIR503HG inhibits the proliferation, migration and invasion of colon cancer cells via miR-107/Par4 axis. *Exp. Cell Res.* **2020**, *395*, 112205. [CrossRef] [PubMed]
149. Dong, X.; Yang, Z.; Yang, H.; Li, D.; Qiu, X. Long non-coding RNA MIR4435-2HG promotes colorectal cancer proliferation and metastasis through miR-206/YAP1 axis. *Front. Oncol.* **2020**, *10*, 160. [CrossRef] [PubMed]
150. Liu, H.; Li, A.; Sun, Z.; Zhang, J.; Xu, H. Long non-coding RNA NEAT1 promotes colorectal cancer progression by regulating miR-205-5p/VEGFA axis. *Hum. Cell* **2020**, *33*, 386–396. [CrossRef] [PubMed]
151. Luo, Y.; Chen, J.-J.; Lv, Q.; Qin, J.; Huang, Y.-Z.; Yu, M.-H.; Zhong, M. Long non-coding RNA NEAT1 promotes colorectal cancer progression by competitively binding miR-34a with SIRT1 and enhancing the Wnt/β-catenin signaling pathway. *Cancer Lett.* **2019**, *440*, 11–22. [CrossRef] [PubMed]
152. Shengnan, J.; Dafei, X.; Hua, J.; Sunfu, F.; Xiaowei, W.; Liang, X. Long non-coding RNA HOTAIR as a competitive endogenous RNA to sponge miR-206 to promote colorectal cancer progression by activating CCL2. *J. Cancer* **2020**, *11*, 4431. [CrossRef]
153. Li, S.; Wu, T.; Zhang, D.; Sun, X.; Zhang, X. The long non-coding RNA HCG18 promotes the growth and invasion of colorectal cancer cells through sponging miR-1271 and upregulating MTDH/Wnt/β-catenin. *Clin. Exp. Pharmacol. Physiol.* **2020**, *47*, 703–712. [CrossRef]
154. Li, X.; Chen, W.; Jia, J.; You, Z.; Hu, C.; Zhuang, Y.; Lin, Z.; Liu, Y.; Yang, C.; Xu, R. The long non-coding RNA-RoR promotes the tumorigenesis of human colorectal cancer by targeting miR-6833-3p through SMC4. *Onco Targets Ther.* **2020**, *13*, 2573–2581. [CrossRef] [PubMed]
155. He, C.; Fu, Y.; Chen, Y.; Li, X. Long non-coding RNA SNHG8 promotes autophagy as a ceRNA to upregulate ATG7 by sponging microRNA-588 in colorectal cancer. *Oncol. Lett.* **2021**, *22*, 577. [CrossRef] [PubMed]
156. Ruan, Z.; Deng, H.; Liang, M.; Xu, Z.; Lai, M.; Ren, H.; Deng, X.; Su, X. Downregulation of long non-coding RNA MAFG-AS1 represses tumorigenesis of colorectal cancer cells through the microRNA-149-3p-dependent inhibition of HOXB8. *Cancer Cell Int.* **2020**, *20*, 511. [CrossRef] [PubMed]
157. Lai, F.; Deng, W.; Fu, C.; Wu, P.; Cao, M.; Tan, S. Long non-coding RNA SNHG6 increases JAK2 expression by targeting the miR-181 family to promote colorectal cancer cell proliferation. *J. Gene Med.* **2020**, *22*, e3262. [CrossRef] [PubMed]
158. Zheng, W.; Li, H.; Zhang, H.; Zhang, C.; Zhu, Z.; Liang, H.; Zhou, Y. Long noncoding RNA RHPN1-AS1 promotes colorectal cancer progression via targeting miR-7-5p/OGT axis. *Cancer Cell Int.* **2020**, *20*, 54. [CrossRef] [PubMed]

159. Zheng, Y.; Nie, P.; Xu, S. Long noncoding RNA CASC21 exerts an oncogenic role in colorectal cancer through regulating miR-7-5p/YAP1 axis. *Biomed. Pharmacother.* **2020**, *121*, 109628. [CrossRef]
160. Yan, F.; Ma, Y.; Liu, L.; Li, L.; Deng, J.; Sun, J. Long noncoding RNA HOXD-AS1 promotes the proliferation, migration, and invasion of colorectal cancer via the miR-526b-3p/CCND1 axis. *J. Surg. Res.* **2020**, *255*, 525–535. [CrossRef]
161. Zhu, X.; Bu, F.; Tan, T.; Luo, Q.; Zhu, J.; Lin, K.; Huang, J.; Luo, C.; Zhu, Z. Long noncoding RNA RP11-757G1. 5 sponges miR-139-5p and upregulates YAP1 thereby promoting the proliferation and liver, spleen metastasis of colorectal cancer. *J. Exp. Clin. Cancer Res.* **2020**, *39*, 207. [CrossRef]
162. Huang, F.K.; Zheng, C.Y.; Huang, L.K.; Lin, C.Q.; Zhou, J.F.; Wang, J.X. Long non-coding RNA MCF2L-AS1 promotes the aggressiveness of colorectal cancer by sponging miR-874-3p and thereby up-regulating CCNE1. *J. Gene Med.* **2021**, *23*, e3285. [CrossRef]
163. Lu, W.; Huang, Z.; Wang, J.; Liu, H. Long non-coding RNA DANCR accelerates colorectal cancer progression via regulating the miR-185-5p/HMGA2 axis. *J. Biochem.* **2022**, *171*, 389–398. [CrossRef]
164. Xu, G.; Wang, H.; Yuan, D.; Yao, J.; Meng, L.; Li, K.; Zhang, Y.; Dang, C.; Zhu, K. RUNX1-activated upregulation of lncRNA RNCR3 promotes cell proliferation, invasion, and suppresses apoptosis in colorectal cancer via miR-1301-3p/AKT1 axis in vitro and in vivo. *Clin. Transl. Oncol.* **2020**, *22*, 1762–1777. [CrossRef] [PubMed]
165. Li, C.; Pan, B.; Liu, X.; Qin, J.; Wang, X.; He, B.; Pan, Y.; Sun, H.; Xu, T.; Xu, X. Long intergenic non-coding RNA LINC00485 exerts tumor-suppressive activity by regulating miR-581/EDEM1 axis in colorectal cancer. *Aging* **2021**, *13*, 3866. [CrossRef] [PubMed]
166. Liu, G.; Zhao, H.; Song, Q.; Li, G.; Lin, S.; Xiong, S. Long non-coding RNA DPP10-AS1 exerts anti-tumor effects on colon cancer via the upregulation of ADCY1 by regulating microRNA-127-3p. *Aging* **2021**, *13*, 9748. [CrossRef] [PubMed]
167. Zhu, K.; Wang, Y.; Liu, L.; Li, S.; Yu, W. Long non-coding RNA MBNL1-AS1 regulates proliferation, migration, and invasion of cancer stem cells in colon cancer by interacting with MYL9 via sponging microRNA-412-3p. *Clin. Res. Hepatol. Gastroenterol.* **2020**, *44*, 101–114. [CrossRef] [PubMed]
168. Yin, S.L.; Xiao, F.; Liu, Y.F.; Chen, H.; Guo, G.C. Long non-coding RNA FENDRR restrains the aggressiveness of CRC via regulating miR-18a-5p/ING4 axis. *J. Cell. Biochem.* **2020**, *121*, 3973–3985. [CrossRef] [PubMed]
169. Dai, W.; Zeng, W.; Lee, D. lncRNA MCM3AP-AS1 inhibits the progression of colorectal cancer via the miR-19a-3p/FOXF2 axis. *J. Gene Med.* **2021**, *23*, e3306. [CrossRef] [PubMed]
170. Chu, Q.; Zheng, W.; Su, H.; Zhang, L.; Chang, R.; Gao, W.; Xu, T. A highly conserved circular RNA, circRasGEF1B, enhances antiviral immunity by regulating the miR-21-3p/MITA pathway in lower vertebrates. *Virol. J.* **2021**, *95*, e02145-20. [CrossRef] [PubMed]
171. Guo, W.; Zhang, B.; Sun, C.; Duan, H.Q.; Liu, W.X.; Mu, K.; Zhao, L.; Li, H.R.; Dong, Z.Y.; Cui, Q. Circular RNA derived from TIMP2 functions as a competitive endogenous RNA and regulates intervertebral disc degeneration by targeting miR-185-5p and matrix metalloproteinase 2. *Int. J. Mol. Med.* **2020**, *46*, 621–632. [CrossRef] [PubMed]
172. Tang, X.; Ren, H.; Guo, M.; Qian, J.; Yang, Y.; Gu, C. Review on circular RNAs and new insights into their roles in cancer. *Comput. Struct. Biotechnol. J.* **2021**, *19*, 910–928. [CrossRef]
173. Zhou, W.-Y.; Cai, Z.-R.; Liu, J.; Wang, D.-S.; Ju, H.-Q.; Xu, R.-H. Circular RNA: Metabolism, functions and interactions with proteins. *Mol. Cancer* **2020**, *19*, 172. [CrossRef]
174. Huang, Y.; Zhu, Q. Mechanisms regulating abnormal circular RNA biogenesis in cancer. *Cancers* **2021**, *13*, 4185. [CrossRef] [PubMed]
175. Pitolli, C.; Marini, A.; Sette, C.; Pagliarini, V. Non-canonical splicing and its implications in brain physiology and cancer. *Int. J. Mol. Sci.* **2022**, *23*, 2811. [CrossRef] [PubMed]
176. Xu, B.; Meng, Y.; Jin, Y. RNA structures in alternative splicing and back-splicing. *Wiley Interdiscip. Rev. RNA* **2021**, *12*, e1626. [CrossRef] [PubMed]
177. Van Der Steen, N.; Lyu, Y.; Hitzler, A.K.; Becker, A.C.; Seiler, J.; Diederichs, S. The circular RNA landscape of non-small cell lung cancer cells. *Cancers* **2020**, *12*, 1091. [CrossRef] [PubMed]
178. Zhao, X.; Zhong, Y.; Wang, X.; Shen, J.; An, W. Advances in circular RNA and its applications. *Int. J. Med. Sci.* **2022**, *19*, 975. [CrossRef]
179. Ma, Y.; Zhang, X.; Wang, Y.-Z.; Tian, H.; Xu, S. Research progress of circular RNAs in lung cancer. *Cancer Biol. Ther.* **2019**, *20*, 123–129. [CrossRef] [PubMed]
180. Chen, L.-L. The expanding regulatory mechanisms and cellular functions of circular RNAs. *Nat. Rev. Mol. Cell Biol.* **2020**, *21*, 475–490. [CrossRef] [PubMed]
181. Robic, A.; Kühn, C. Beyond back splicing, a still poorly explored world: Non-canonical circular RNAs. *Genes* **2020**, *11*, 1111. [CrossRef]
182. Huang, A.; Zheng, H.; Wu, Z.; Chen, M.; Huang, Y. Circular RNA-protein interactions: Functions, mechanisms, and identification. *Theranostics* **2020**, *10*, 3503. [CrossRef]
183. Yang, Y.; Yujiao, W.; Fang, W.; Linhui, Y.; Ziqi, G.; Zhichen, W.; Zirui, W.; Shengwang, W. The roles of miRNA, lncRNA and circRNA in the development of osteoporosis. *Biol. Res.* **2020**, *53*, 40. [CrossRef]
184. Li, T.; Wang, W.C.; McAlister, V.; Zhou, Q.; Zheng, X. Circular RNA in colorectal cancer. *J. Cell. Mol. Med.* **2021**, *25*, 3667–3679. [CrossRef] [PubMed]

185. Bhuyan, R.; Bagchi, A. Prediction of the differentially expressed circRNAs to decipher their roles in the onset of human colorectal cancers. *Gene* **2020**, *762*, 145035. [CrossRef] [PubMed]
186. Tang, Q.; Hann, S.S. Biological roles and mechanisms of circular RNA in human cancers. *Onco Targets Ther.* **2020**, *13*, 2067–2092. [CrossRef] [PubMed]
187. Zhang, X.; Xu, Y.; Yamaguchi, K.; Hu, J.; Zhang, L.; Wang, J.; Tian, J.; Chen, W. Circular RNA circVAPA knockdown suppresses colorectal cancer cell growth process by regulating miR-125a/CREB5 axis. *Cancer Cell Int.* **2020**, *20*, 103. [CrossRef] [PubMed]
188. Wang, J.; Luo, J.; Liu, G.; Li, X. Circular RNA hsa_circ_0008285 inhibits colorectal cancer cell proliferation and migration via the miR-382-5p/PTEN axis. *Biochem. Biophys. Res. Commun.* **2020**, *527*, 503–510. [CrossRef] [PubMed]
189. Geng, Y.; Zheng, X.; Hu, W.; Wang, Q.; Xu, Y.; He, W.; Wu, C.; Zhu, D.; Wu, C.; Jiang, J. Has_circ_0009361 acts as the sponge of miR-582 to suppress colorectal cancer progression by regulating APC2 expression. *Clin. Sci.* **2019**, *133*, 1197–1213. [CrossRef] [PubMed]
190. Chen, P.; Yao, Y.; Yang, N.; Gong, L.; Kong, Y.; Wu, A. Circular RNA circCTNNA1 promotes colorectal cancer progression by sponging miR-149-5p and regulating FOXM1 expression. *Cell Death Dis.* **2020**, *11*, 557. [CrossRef] [PubMed]
191. Hao, S.; Qu, R.; Hu, C.; Wang, M.; Li, Y. A circular RNA derived from golgi glycoprotein 1 mRNA regulates KRAS expression and promotes colorectal cancer progression by targeting microRNA-622. *Onco Targets Ther.* **2020**, *13*, 12637–12648. [CrossRef] [PubMed]
192. Yan, D.; Liu, W.; Liu, Y.; Zhu, X. Circular RNA circ_0065378 upregulates tumor suppressor candidate 1 by competitively binding with miR-4701-5p to alleviate colorectal cancer progression. *J. Gastroenterol. Hepatol.* **2022**, *37*, 1107–1118. [CrossRef]
193. Wang, J.; Zhang, Y.; Song, H.; Yin, H.; Jiang, T.; Xu, Y.; Liu, L.; Wang, H.; Gao, H.; Wang, R. The circular RNA circSPARC enhances the migration and proliferation of colorectal cancer by regulating the JAK/STAT pathway. *Mol. Cancer* **2021**, *20*, 81. [CrossRef]
194. Yin, W.; Xu, J.; Li, C.; Dai, X.; Wu, T.; Wen, J. Circular RNA circ_0007142 facilitates colorectal cancer progression by modulating CDC25A expression via miR-122-5p. *Onco Targets Ther.* **2020**, *13*, 3689–3701. [CrossRef] [PubMed]
195. Wen, T.; Wu, H.; Zhang, L.; Li, K.; Xiao, X.; Zhang, L.; Zhang, Y. Circular RNA circ_0007142 regulates cell proliferation, apoptosis, migration and invasion via miR-455-5p/SGK1 axis in colorectal cancer. *Anti-Cancer Drugs* **2021**, *32*, 22–33. [CrossRef] [PubMed]
196. Xie, L.; Pan, Z. Circular RNA circ_0000467 regulates colorectal cancer development via miR-382-5p/EN2 axis. *Bioengineered* **2021**, *12*, 886–897. [CrossRef] [PubMed]
197. Ren, C.; Zhang, Z.; Wang, S.; Zhu, W.; Zheng, P.; Wang, W. Circular RNA hsa_circ_0001178 facilitates the invasion and metastasis of colorectal cancer through upregulating ZEB1 via sponging multiple miRNAs. *Biol. Chem.* **2020**, *401*, 487–496. [CrossRef] [PubMed]
198. Xiao, H.; Liu, M. Circular RNA hsa_circ_0053277 promotes the development of colorectal cancer by upregulating matrix metallopeptidase 14 via miR-2467-3p sequestration. *J. Cell. Physiol.* **2020**, *235*, 2881–2890. [CrossRef] [PubMed]
199. Wang, X.; Ren, Y.; Ma, S.; Wang, S. Circular RNA 0060745, a novel circRNA, promotes colorectal cancer cell proliferation and metastasis through miR-4736 sponging. *Onco Targets Ther.* **2020**, *13*, 1941–1951. [CrossRef] [PubMed]
200. Li, C.; Zhou, H. Circular RNA hsa_circRNA_102209 promotes the growth and metastasis of colorectal cancer through miR-761-mediated Ras and Rab interactor 1 signaling. *Cancer Med.* **2020**, *9*, 6710–6725. [CrossRef] [PubMed]
201. Liu, K.; Mou, Y.; Shi, X.; Liu, T.; Chen, Z.; Zuo, X. Circular RNA 100146 promotes colorectal cancer progression by the MicroRNA 149/HMGA2 Axis. *Mol. Cell. Biol.* **2021**, *41*, e00445-20. [CrossRef] [PubMed]
202. Yang, L.; Sun, H.; Liu, X.; Chen, J.; Tian, Z.; Xu, J.; Xiang, B.; Qin, B. Circular RNA hsa_circ_0004277 contributes to malignant phenotype of colorectal cancer by sponging miR-512-5p to upregulate the expression of PTMA. *J. Cell. Physiol.* 2020; *ahead of print.*
203. Lu, C.; Fu, L.; Qian, X.; Dou, L.; Cang, S. Knockdown of circular RNA circ-FARSA restricts colorectal cancer cell growth through regulation of miR-330-5p/LASP1 axis. *Arch. Biochem. Biophys.* **2020**, *689*, 108434. [CrossRef]
204. Hao, Q.; Zhang, Z. hsa_circRNA_000166 facilitated cell growth and limited apoptosis through targeting miR-326/LASP1 axis in colorectal cancer. *Gastroent. Res. Pract.* **2020**, *2020*, 8834359. [CrossRef]
205. Yang, B.; Du, K.; Yang, C.; Xiang, L.; Xu, Y.; Cao, C.; Zhang, J.; Liu, W. CircPRMT5 circular RNA promotes proliferation of colorectal cancer through sponging miR-377 to induce E2F3 expression. *J. Cell. Mol. Med.* **2020**, *24*, 3431–3437. [CrossRef] [PubMed]
206. Chen, L.-Y.; Wang, L.; Ren, Y.-X.; Pang, Z.; Liu, Y.; Sun, X.-D.; Tu, J.; Zhi, Z.; Qin, Y.; Sun, L.-N. The circular RNA circ-ERBIN promotes growth and metastasis of colorectal cancer by miR-125a-5p and miR-138-5p/4EBP-1 mediated cap-independent HIF-1α translation. *Mol. Cancer* **2020**, *19*, 164. [CrossRef]
207. Zhao, J.P.; Chen, L.L. Circular RNA MAT2B induces colorectal cancer proliferation via sponging miR-610, resulting in an increased E2F1 expression. *Cancer Manag. Res.* **2020**, *12*, 7107–7116. [CrossRef] [PubMed]
208. Zhang, B.; Yang, S.; Wang, J. Circ_0084615 is an oncogenic circular RNA in colorectal cancer and promotes DNMT3A expression via repressing miR-599. *Ppathol. Res. Pract.* **2021**, *224*, 153494. [CrossRef]
209. Zhang, Z.-J.; Zhang, Y.-H.; Qin, X.-J.; Wang, Y.-X.; Fu, J. Circular RNA circDENND4C facilitates proliferation, migration and glycolysis of colorectal cancer cells through miR-760/GLUT1 axis. *Eur. Rev. Med. Pharmacol. Sci.* **2020**, *24*, 2387–2400.
210. Chen, F.; Guo, L.; Di, J.; Li, M.; Dong, D.; Pei, D. Circular RNA ubiquitin-associated protein 2 enhances autophagy and promotes colorectal cancer progression and metastasis via miR-582-5p/FOXO1 signaling. *J. Genet. Genom.* **2021**, *48*, 1091–1103. [CrossRef]

211. Fang, G.; Chen, T.; Mao, R.; Huang, X.; Ji, L. Circular RNA circ_0089153 acts as a competing endogenous RNA to regulate colorectal cancer development by the miR-198/SUMO-specific peptidase 1 (SENP1) axis. *Bioengineered* **2021**, *12*, 5664–5678. [CrossRef] [PubMed]
212. Wang, X.; Tao, G.; Huang, D.; Liang, S.; Zheng, D. Circular RNA NOX4 promotes the development of colorectal cancer via the microRNA-485-5p/CKS1B axis. *Oncol. Rep.* **2020**, *44*, 2009–2020. [CrossRef] [PubMed]
213. He, J.; Chu, Z.; Lai, W.; Lan, Q.; Zeng, Y.; Lu, D.; Jin, S.; Xu, H.; Su, P.; Yin, D. Circular RNA circHERC4 as a novel oncogenic driver to promote tumor metastasis via the miR-556-5p/CTBP2/E-cadherin axis in colorectal cancer. *J. Hematol. Oncol.* **2021**, *14*, 194. [CrossRef]
214. Liu, X.; Qin, Y.; Tang, X.; Wang, Y.; Bian, C.; Zhong, J. Circular RNA circ_0000372 contributes to the proliferation, migration and invasion of colorectal cancer by elevating IL6 expression via sponging miR-495. *Anti-Cancer Drugs* **2021**, *32*, 296–305. [CrossRef]
215. Wang, R.; Wang, J.; Chen, Y.; Chen, Y.; Xi, Q.; Sun, L.; Zhang, X.; Zhang, G.; Ding, X.; Shi, T. Circular RNA circLDLR facilitates cancer progression by altering the miR-30a-3p/SOAT1 axis in colorectal cancer. *Cell Death Discov.* **2022**, *8*, 314. [CrossRef] [PubMed]
216. Yang, B.-L.; Liu, G.-Q.; Li, P.; Li, X.-H. Circular RNA CUL2 regulates the development of colorectal cancer by modulating apoptosis and autophagy via miR-208a-3p/PPP6C. *Aging* **2022**, *14*, 497. [CrossRef] [PubMed]
217. Miao, X.; Xi, Z.; Zhang, Y.; Li, Z.; Huang, L.; Xin, T.; Shen, R.; Wang, T. Circ-SMARCA5 suppresses colorectal cancer progression via downregulating miR-39-3p and upregulating ARID4B. *Dig. Liver Dis.* **2020**, *52*, 1494–1502. [CrossRef] [PubMed]
218. Wen, C.; Feng, X.; Yuan, H.; Gong, Y.; Wang, G. Circ_0003266 sponges miR-503-5p to suppress colorectal cancer progression via regulating PDCD4 expression. *BMC Cancer* **2021**, *21*, 284. [CrossRef] [PubMed]

International Journal of *Molecular Sciences*

MDPI

Article

Synthesis, Characterization, Cytotoxicity, Cellular Imaging, Molecular Docking, and ADMET Studies of Piperazine-Linked 1,8-Naphthalimide-Arylsulfonyl Derivatives

Ashanul Haque [1,2], Khalaf M. Alenezi [1,2], Ahmed Al-Otaibi [1,2], Abdulmohsen Khalaf Dhahi Alsukaibi [1,2], Ataur Rahman [3], Ming-Fa Hsieh [4,*], Mei-Wen Tseng [4] and Wai-Yeung Wong [5,*]

1 Department of Chemistry, College of Science, University of Ha'il, Ha'il 81451, Saudi Arabia; a.haque@uoh.edu.sa (A.H.); k.alenezi@uoh.edu.sa (K.M.A.); ahmed.alotaibi@uoh.edu.sa (A.A.-O.); a.alsukaibi@uoh.edu.sa (A.K.D.A.)
2 Medical and Diagnostic Research Centre, University of Ha'il, Ha'il 55473, Saudi Arabia
3 Jamia Senior Secondary School, Jamia Millia Islamia, New Delhi 110025, India; arahman4@jmi.ac.in
4 Department of Biomedical Engineering, Chung Yuan Christian University, 200 Chung Pei Road, Chung Li District, Taoyuan City 32023, Taiwan; g11275004@cycu.edu.tw
5 Department of Applied Biology and Chemical Technology, The Hong Kong Polytechnic University, Hung Hom, Kowloon, Hong Kong, China
* Correspondence: mfhsieh@cycu.edu.tw (M.-F.H.); wai-yeung.wong@polyu.edu.hk (W.-Y.W.)

Abstract: To reduce the mortality and morbidity associated with cancer, new cancer theranostics are in high demand and are an emerging area of research. To achieve this goal, we report the synthesis and characterization of piperazine-linked 1,8-naphthalimide-arylsulfonyl derivatives (SA1–SA7). These compounds were synthesized in good yields following a two-step protocol and characterized using multiple analytical techniques. In vitro cytotoxicity and fluorescent cellular imaging of the compounds were assessed against non-cancerous fibroblast (3T3) and breast cancer (4T1) cell lines. Although the former study indicated the safe nature of the compounds (viability = 82–95% at 1 µg/mL), imaging studies revealed that the designed probes had good membrane permeability and could disperse in the whole cell cytoplasm. In silico studies, including molecular docking, molecular dynamics (MD) simulation, and ADME/Tox results, indicated that the compounds had the ability to target CAIX-expressing cancers. These findings suggest that piperazine-linked 1,8-naphthalimide-arylsulfonyl derivatives are potential candidates for cancer theranostics and a valuable backbone for future research.

Keywords: 1,8-naphthalimide; arylsulfonyl; cellular imaging; characterization; docking; synthesis

check for updates

Citation: Haque, A.; Alenezi, K.M.; Al-Otaibi, A.; Alsukaibi, A.K.D.; Rahman, A.; Hsieh, M.-F.; Tseng, M.-W.; Wong, W.-Y. Synthesis, Characterization, Cytotoxicity, Cellular Imaging, Molecular Docking, and ADMET Studies of Piperazine-Linked 1,8-Naphthalimide-Arylsulfonyl Derivatives. *Int. J. Mol. Sci.* **2024**, *25*, 1069. https://doi.org/10.3390/ijms25021069

Academic Editor: Antonio Carrieri

Received: 4 December 2023
Revised: 6 January 2024
Accepted: 8 January 2024
Published: 15 January 2024

1. Introduction

Accounting for nearly 8 million deaths and 14 million new cases annually, cancer remains the second leading cause of death worldwide [1]. Cancer therapy relies almost entirely on long-established technologies: surgery, radiotherapy, and chemotherapy, supplemented by more recent targeted therapies and emerging stem-cell and gene therapies [2]. Even though new potential treatments are regularly reported, the main therapeutic challenge remained largely unsolved for the past century: the complete eradication of rapidly growing malignant cells. The reasons behind our current failure to solve this challenge are manifold, including late detection, metastasis, lack of selective "magic bullet" drugs, varying responses to treatments, and the almost inevitable development of drug resistance [3]. The factor that may best improve cancer patient outcomes is earlier detection [3]. The earlier a tumor is detected and the more localized the tumor is, the better the prognosis of any therapeutic regimen [4].

1,8-Naphthalimides (I, Figure 1) are a class of π-conjugated planar molecules with wide-ranging applications [5,6]. Multiple functionalization sites, easy derivatization, and functionality-dependent chemico-biological properties are some notable features offered by this class of compounds [7]. It has been reported that 1,8-naphthalimide derivatives with two carbon-separated amines show excellent anti-tumor activity [5]. Amonafide and mitonafide (II, Figure 1) are well-known examples that impart activity via DNA intercalation and have entered clinical trials [8]. Based on this notion, various research has been carried out. For example, Kamal and coworkers [9] showed that a compound (III, Figure 1) of the type 1,8-naphthalimide-piperazine-aminobenzothiazole is a potent inhibitor of colon and lung cancers, with activity in the micromolar range. These compounds intercalate between the DNA strands and inhibit topoisomerase-II [8]. In addition, 1,8-naphthalimide derivatives have been extensively studied for the imaging and tracking of organelles such as the endoplasmic reticulum, lipid droplets, the plasma membrane, the nucleus, the Golgi apparatus, etc. [7]. Intrigued by these features and in the quest for new theranostics, we synthesized seven new 1,8-naphthalimide-arylsulfonyl derivatives (IV, Figure 1) from 1,8-naphthalic anhydride. To enhance the anticancer activity of the compounds, ethyl-separated piperazine was attached to the *N*-side of the napthalimide, followed by aryl sulfonyl group insertion. The compounds were characterized by standard analytical techniques and evaluated for in vitro toxicity against non-cancerous fibroblast (3T3) and breast cancer (4T1) cell lines. Since alkyl piperazine and aryl piperazine coumarin hybrids have been found to selectively target cancer-associated CAIX [10], which is also expressed in 4T1 cells, in silico studies (molecular docking and MD simulations) were carried out.

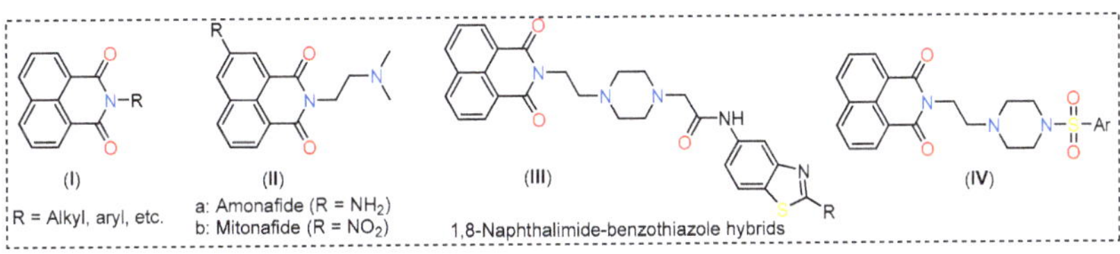

Figure 1. Some examples of 1,8-naphthalimide derivatives.

2. Results and Discussion

2.1. Synthesis and Characterization

In the present work, seven new piperazine-linked 1,8-naphthalimide-arylsulfonyl derivatives (SA1-SA7) were synthesized, with good yields (70–82%). All the synthesized compounds obtained were light yellow solids and stable at room temperature. All compounds were structurally characterized using ^{1}H-NMR, ^{13}C-NMR spectroscopy, and mass spectrometry (Figures S1–S7, Supplementary File), and the data are provided in the experimental section (Section 3.1). ^{1}H-NMR and ^{13}C-NMR showed the signals expected for aromatic, N-substituted piperazine and linker (ethyl) protons and carbons, respectively. All compounds displayed molecular ion peaks for $[M + H]^+$ and/or $[M + Na]^+$ peaks, supporting the structure of the final compounds.

2.2. Optical Studies

The ultraviolet–visible (UV-Vis) absorption and fluorescence emission spectra were measured in dilute DCM and are shown in Figure 2a, and the data are collected in Table 1. Compounds SA1–SA7 displayed high- (236–240 nm) and low-energy (~333 and 350 nm) bands in the UV region, typical for 1,8-naphthalimide derivatives [11]. The well-overlapped peaks indicate that different arylsulfonyl groups did not affect the $\pi \rightarrow \pi^*$ absorption of the 1,8-napthalimide core and that there was a lack of electronic communication between two aromatic systems separated by a saturated (ethylene) linker [11–13]. Compared to DCM, compounds were red-shifted slightly (2–3 nm) in a more polar solvent DMSO (Figure S8, Supplementary File). Like absorption, the emission profile of the compounds was similar (Figure 2b). When excited at 340 nm, monomeric emission based on the 1,8-napthalimide core was noted at around ~401–402 nm (Table 1), which can be ascribed to being assigned to a $^1\pi$–π^* emitting state [11,14]. Furthermore, a low-energy band at approximately 450 nm was observed in the case of SA2 and SA5. This band could be attributed to the emission from an aggregate in the excited state that involves aromatic interactions. However, the absence of this band in the other compounds requires further investigation. The quantum yields (Φ) of the compounds were found to range between 0.030 and 0.071.

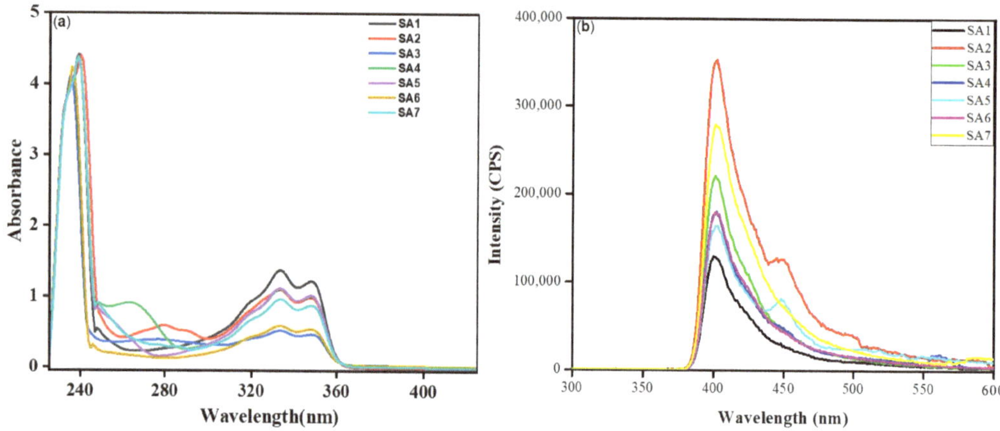

Figure 2. (**a**) Absorption (45 μM) and (**b**) emission (λ_{ex} = 340 nm, 100 μg/mL) spectra of SA1–SA7 in DCM at RT.

Table 1. Photophysical data of compounds SA1–SA7.

Sample	Absorption ($\lambda_{max.}$ nm)		Emission ($\lambda_{em.}$ nm) [1]	Stokes Shift (nm) [1]	Quantum Yield (Φ)
	DCM	DMSO			
SA1	238, 333, **348**	259, 336, **351**	401	50	0.030
SA2	240, 333, **348**	261, 327, **336**	401	65	0.071
SA3	236, 333, **349**	261, 336, **350**	401	51	0.048
SA4	238, 333, **348**	260, 336, **352**	401	49	0.042
SA5	238, 334, **348**	260, 336, **352**	402	50	0.035
SA6	236, 333, **349**	260, 336, **352**	401	49	0.033
SA7	239, 333, **348**	260, 336, **351**	401	50	0.041

The wavelength of the lowest energy is in boldface. [1] In DCM.

2.3. Biological Studies

2.3.1. Cytotoxicity

For any biological application, it is crucial that molecules be non-toxic and have high cyto-compatibility. Therefore, before evaluating the cellular imaging application of SA1–SA7, the cytotoxicity was assessed against the normal fibroblast (3T3) and breast cancer (4T1) cell lines. The results of the study are provided in Figures 3 and 4. When tested against normal cells, compounds SA1–SA7 showed minor toxicity (viability = 82–95% at 1 μg/mL, Figure 3). Such high cellular viability is advantageous for cell/tissue staining (depending on the specific imaging requirement) and others. Overall, the low concentration (1 μg/mL) of SA samples against non-cancerous cells confirmed that the compounds could be used for cellular imaging.

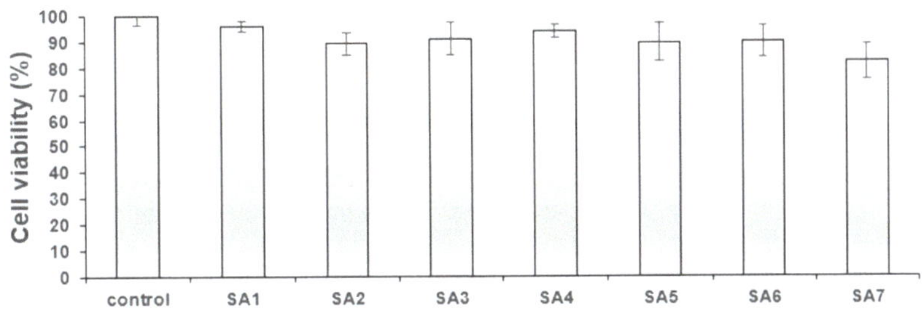

Figure 3. % Cell viability of SA1–SA7 against the 3T3 fibroblast cell line at 1 μg/mL.

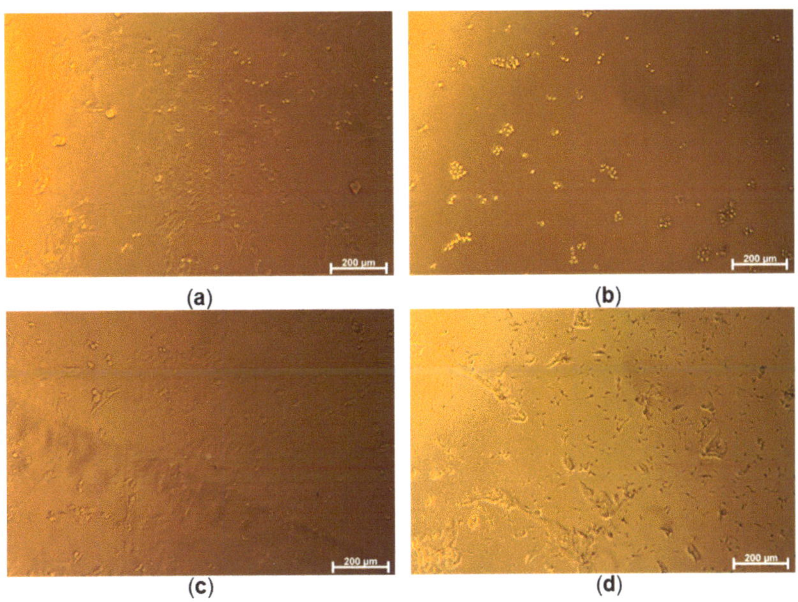

Figure 4. Optical microscope images of 4T1 cells cultured without samples for control groups ((**a**): control, (**b**): PC, (**c**): NC) and the cells co-cultured with a concentration of 20.22 μM of SA1 sample (**d**).

Following this, the anticancer potential of SA1–SA7 was assessed against the mouse breast cancer 4T1 cell line, which shares a close resemblance with human breast cancer in terms of growth and metastasis. The bright-field cellular images of the 4T1 cells (Figure 4) show that the cells were properly cultured to respond to three control conditions, e.g., well spread on culture well (Cont), under strong toxic stress (PC), and minor toxic stress (NC). It was noted that cells treated with the compounds led to toxic effects depending on the concentrations of the drugs (Figure 4). Even though precipitation of drugs occurs at higher concentrations (20.22 μM of SA1, Figure 4), the culture medium saturated with SA1 is believed to endow cytotoxicity. Figure S9 (Supplementary File) shows the dose-viability dependence of SA1–SA7. The SA samples exerted cancer cell-killing potential differently. Among all samples, SA5 was found to be the most potent (cytotoxic) compound against 4T1 breast cancer cells (viability < 80% at 0.7 μM).

2.3.2. Fluorescence Imaging

The fluorescence imaging ability of SA1–SA7 was determined on 3T3 fibroblast cells in a concentration range showing high cell viability, i.e., 1 μg/mL. As noted in the previous section, SA samples were minor/non-toxic to non-cancerous cells at the concentration of 1 μg/mL. Figure 5a–c displays the control images of both bright-field and fluorescent images. The cells co-cultured without SA samples had the typical shape of fibroblasts (Figure 5a), whereas the well-distributed cell nuclei stained with DAPI were seen yet the cellular fluorescence of the SA sample was absent (Figures 5b and 5c, respectively). In contrast to control images, the cells co-cultured with SA1–SA7 were observed under a bright field (Figure 5d), via nuclei staining (Figure 5e), and with fluorescent SA4 (Figure 5f), respectively.

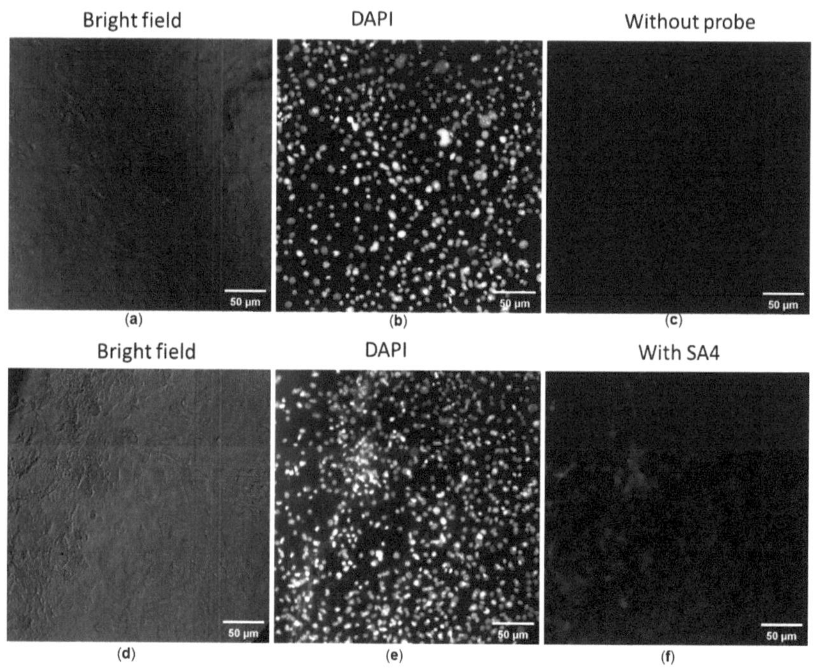

Figure 5. Fluorescence images of 3T3 fibroblast cells without (**a**–**c**) and with (**d**–**f**) probe SA4. Cells were stained with 4′,6-diamidino-2-phenylindole (DAPI) for the imaging of cell nuclei (**b**,**e**), blank cells (**c**) and SA4-treated cells (**f**). Fluorescent filter cube for cell nuclei: Ex = 340–380 nm, Em = 435–485 nm. Another filter cube (Ex = 465–495 nm, Em = 512–558 nm) was used to observe the distribution of SA samples.

We noted a substantial uptake of the compounds when incubated with NIH/3T3 cells for 30 min. Compounds SA1–SA7 readily entered cells and yielded green fluorescence bioimages. At the same time, DAPI, a nuclei-staining dye, showed clear fluorescence images in the blue channel (Figure 6). An overlaid image indicated that the probes entered the cell, and fluorescence signals were localized in the perinuclear area of the cytosol, indicating that the dyes were distributed in the whole cell cytoplasm. Spindle-like cellular morphology could be seen, and varying cellular fluorescent intensity emitted was observed from different SA samples, even though the concentration of SA samples added to the cell culture medium was the same. The present study found that SA4 yielded the strongest fluorescent intensity. It is to be noted that multiple factors determine the fluorescence intensity, including the intracellular concentration of the probe, the thickness of the cultured cells, the camera constants of the optical microscope, etc. Overall, 1,8-naphthalimide-arylsulfonyl derivatives displayed both cytotoxic and fluorescent properties and hold great potential for cancer theranostics.

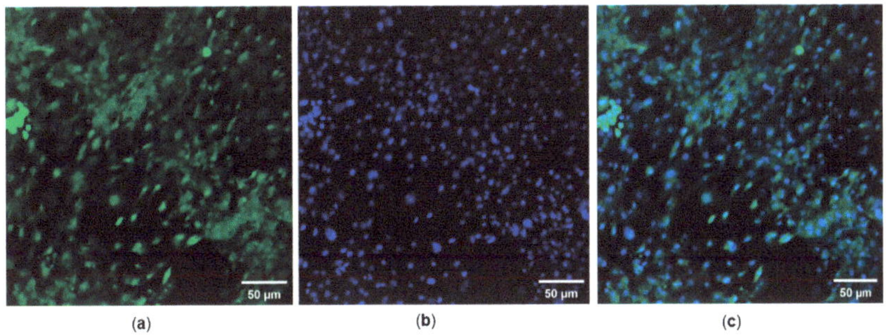

Figure 6. (**a**) FITC- and (**b**) DAPI-staining images of NIH/3T3 cells treated with 1 μg/mL SA7. (**c**) Merged DAPI and FITC fluorescence images. Magnification objective 200×.

2.4. Computational Studies

2.4.1. Molecular Docking

Carbonic anhydrases (CAs) have emerged as a promising druggable target among several targets [15]. CAs are ubiquitous zinc metalloenzymes present in animals, fungi, bacteria, algae, and the cytoplasm of green plants and are responsible for many biological processes, such as pH homeostasis, ion transport, respiration, etc. [16]. Several isoforms of CAs exist in humans and perform different functions [16]. Of the different types of CAs, CAIX is overexpressed in various cancerous cells (e.g., breast, colorectum, etc.). In particular, CAIX is absent in normal cells but is overexpressed in cancerous cells, including 4T1 mammary tumors [17], making it a promising target for designing and developing new agents [18]. Sulfonamides are classical inhibitors of CAs and have a high affinity towards CA enzymes, and several probes bearing sulfonamide cores are known of [19]. Moreover, several non-classical small heterocyclic inhibitors lacking sulfonamide functionality have also been developed. For example, Supuran and coworkers found an efficient inhibition of CA isoforms, especially CAIX, by rhodanine-N-carboxylate derivatives [20]. Before that study, the same group found that coumarin linked to thiazolidinone via a pyrazole linker and coumarin-linked 1,2,3-triazoles selectively target CAIX. Coumarin-alkyl piperazine and aryl piperazine hybrids for the inhibition of CAIX were also assessed [10]. Using experimental and in silico tools, Tiwari and coworkers [21] demonstrated that triazolo-pyrimidine urea derivatives exhibit excellent binding affinity towards CAIX. Therefore, we were prompted to evaluate the affinity of 1,8-naphthalimide-arylsulfonyl derivatives towards the CAIX protein.

Molecule docking of SA1-SA7 and CAIX protein (PDB: 5FL4) was carried out using the AutoDock4 tool. The docking study revealed that CAIX in complex with compound SA7 yielded the best binding affinity (−8.61 kcal/mol), followed by SA2 (−8.39 kcal/mol) > SA4 (−8.04 kcal/mol) > SA5 (−7.95 kcal/mol)~SA3 (−7.92 kcal/mol) > SA6 (−7.65 kcal/mol) > SA1 (−7.39 kcal/mol). Among other interactions, compounds SA1–SA7 formed two to five H-bonds with the receptor, involving residues Arg6, Trp9, Val130, Asn66, Arg64, His68, Gln71, Gly71, Leu91, Gln92, Ala128, Thr200, Thr201, and Pro202 (Figure 7a–d, Table 2 and Figure S10, Supplementary File). The involvement of these residues in interaction has already been reported in both classical [22] and non-classical [21] inhibitors of CAs. To compare the results, docking was also carried out for the established CAIX inhibitor SLC-0111 (4-(3-(4-fluorophenyl)ureido)benzenesulfonamide). This compound showed a binding affinity of −8.39 kcal/mol (Figure S11).

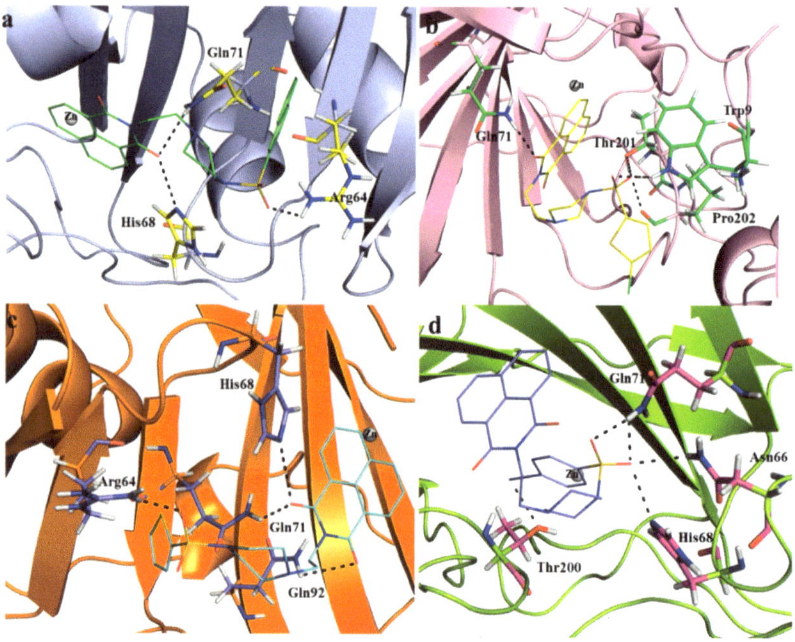

Figure 7. Docking results showing the binding residues of the CAIX protein with compounds SA2, SA4, SA5, and SA7 (labelled (**a–d**), respectively). H-bonds are shown with black dotted lines.

Table 2. Docking and molecular absorption, distribution, metabolism, and excretion (ADME) properties of the studied compounds.

Code	Docking Results		Molecular and ADME Properties										
	Binding Affinity (kcal/mol)	No. of Amino Acids Involved in H-Bonding	CNS	MW	Dipole Moment	SASA	FOSA	FISA	PISA	WPSA	Volume	HBD	HBA
SA1	−7.39	4	−2	494.52	8.90	730.17	162.83	186.15	381.02	0.155	1374.89	0	10.5
SA2	−8.39	3	1	499.58	3.00	785.48	169.23	118.06	497.58	0.599	1457.18	0	9.5
SA3	−7.92	5	−2	494.52	5.97	752.79	168.74	223.12	360.29	0.644	1389.51	0	10.5
SA4	−8.04	4	1	489.99	7.90	730.25	168.54	129.12	319.80	112.773	1337.84	0	9.5
SA5	−7.95	4	1	455.55	8.28	704.55	168.57	129.16	364.64	42.176	1292.57	0	9.5
SA6	−7.65	2	0	450.51	6.15	709.72	169.15	153.91	386.03	0.621	1305.80	0	11
SA7	−8.61	4	−2	474.53	5.76	751.89	168.66	197.45	385.18	0.6	1381.68	0	11

Molecular weight (MW); hydrogen bond donor (HBD); hydrogen bond acceptor (HBA); volume; dipole.

2.4.2. Molecular Dynamics (MD) Simulation

To gain further knowledge of the molecular interaction and the structural stability of the ligand–receptor pairs, MD simulations on a 100 ns scale were performed on the top four complexes containing ligands SA2, SA4, SA5, and SA7. Factors such as root mean square deviation (RMSD) and root mean square fluctuation (RMSF) were considered and compared (Table 3 and Figure 8). The RMSD value of the complexes varied in the order SA7 < SA2 < SA5 < SA4. The obtained results disclose a pattern wherein, within the initial 10 ns of the MD simulation, the ligand-bound protein complexes underwent equilibration, yielding stability over the entire 100 ns duration. Marginal instability was observed in the case of compound SA4 after 30 ns, marked by a slight deviation between 30 and 60 ns during the simulation. For compound SA7, the RMSD plot illustrated stabilization, with minor perturbations noted between 30 and 40 ns; however, the overall trajectory was consistent and stable (Figure 8a). These findings indicate that the protein–ligand complexes achieved substantial stability throughout the simulation.

Table 3. Average RMSD of all the shortlisted compounds for CAIX.

Code	Protein RMSD (nm)	SD	Ligand RMSD (nm)	SD	RMSF (nm)	SD	MM/PBSA (kJ/mol)
SA2	1.57	0.16	1.82	0.46	0.81	0.35	−36.78
SA4	1.71	0.26	1.40	0.25	0.82	0.45	−37.93
SA5	1.63	0.25	1.66	0.29	0.84	0.47	−36.61
SA7	1.47	0.14	1.85	0.20	0.81	0.38	−28.21

SD: standard deviation.

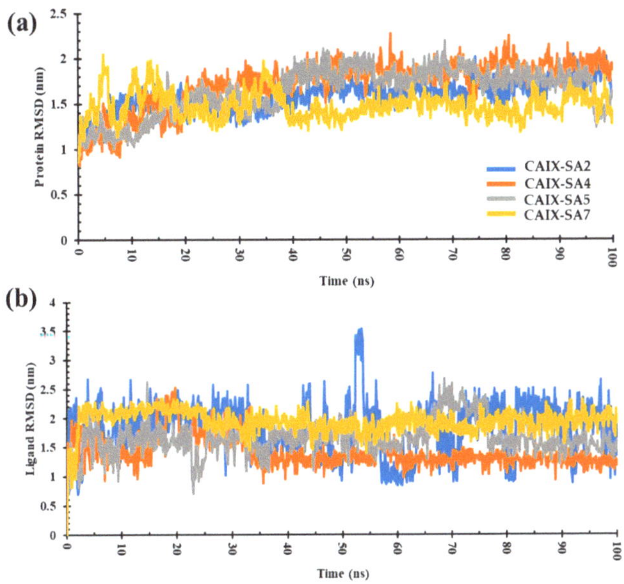

Figure 8. (**a**) RMSD plot of CAIX protein bound to ligands SA2, SA4, SA5, and SA7. (**b**) RMSD plot of ligands binding to CAIX. Illustration of the RMSD trends over a 100 ns simulation period.

The RMSD values for ligands SA2, SA4, SA5, and SA7 were calculated to be 1.82 nm, 1.40 nm, 1.66 nm, and 1.85 nm, respectively (Figure 8b). These values were obtained by tracking the conformational changes of the ligands throughout the 100 ns MD simulation. During the MD simulation, the conformational integrity of each ligand was rigorously assessed. The results unequivocally demonstrate that all ligands, namely, SA2, SA4, SA5, and SA7, maintained their structural integrity over the entire 100 ns simulation period. These results affirm the stability and robustness of the ligand structures in the complex. Negligible deviations in RMSD values further underscore the structural stability of the ligands within the protein-binding site.

The investigation of root mean square fluctuations (RMSFs) aimed to identify crucial residues involved in interactions with a ligand. The average RMSF values for compounds SA2, SA4, SA5, and SA7 were 0.81, 0.82, 0.84, and 0.81 nm, respectively. Notably, within the local domain of the CAIX protein, two substantial fluctuations were observed in the loop regions. The first notable fluctuation encompassed Asn14 and Arg19, as well as Glu79-Pro84, Glu149-Asn154, and Pro234-Leu239 residues (Figure 9). This phenomenon can be attributed to the tendency of the N and C terminals to exhibit more significant fluctuations compared to other parts of the protein. However, it is noteworthy that the compound

SA7 exhibited a stable trajectory, suggesting higher protein stability. These outcomes bear significance in rational drug design and optimization, highlighting the potential of these ligands for targeted therapeutic interventions.

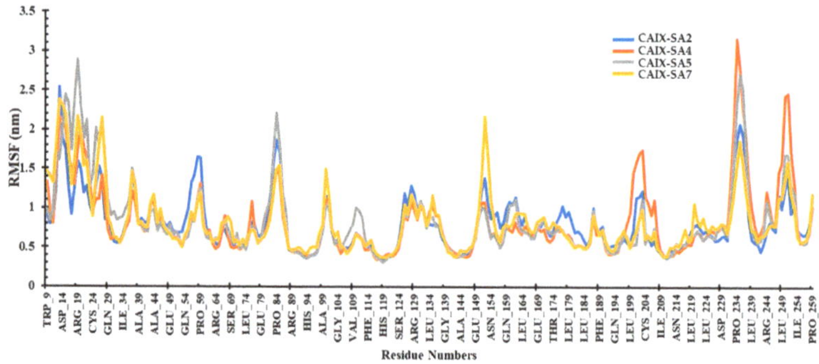

Figure 9. RMSF plot of the CAIX protein bound to ligands SA2, SA4, SA5, and SA7 during the simulation period.

2.4.3. Molecular Mechanics Poisson–Boltzmann Surface Area (MM/PBSA)

Molecular mechanics Poisson–Boltzmann surface area (MM/PBSA) is one of the most popular methods for predicting binding free energy (ΔG) [23]. This value offers a better understanding of the potential for complex formation than the binding affinity obtained from docking. The MM-PBSA analysis of contributing residues identifies crucial amino acids that can be instrumental in designing inhibitors. The ΔG value of SA2, SA4, SA5, and SA7 was computed for the last 20 ns (80–100 ns) of the trajectories. The g_mmpbsa tool was configured to extract snapshots from the simulated trajectories at 20 ps intervals, resulting in 1000 frames captured from each trajectory. The estimated values of the calculated ΔG were −36.78 kJ/mol, −37.93 kJ/mol, −36.61 kJ/mol, and −28.21 kJ/mol for SA2, SA4, SA5, and SA7, respectively.

2.4.4. Drug Likeness and Bioavailability Studies

To underpin the drug likeness and other pharmacokinetic profiles, we estimated the absorption, distribution, metabolism, and excretion and toxicity (ADMET) of the compounds using SwissADME [24], pKCSM [25], and the Qikprop module [26,27]. The results (Table 4; Table S1, Supplementary File) are discussed below.

Table 4. Representative drug property data with simple colors and shapes [28]. Polar surface area (PSA) values are indicated as horizontal bars, molecular weight (Mol. Wt.) with graphical pies and grey-scale shading; log P with green-yellow-red coloring; and rotatable bond count. Small horizontal bars and light color of the properties indicates more drug-likeness.

Comp. Code	TPSA	Mol. Wt.	logP	# RBs
SA1	133.9	494.5	1.95	6
SA2	88.1	499.6	3.63	5
SA3	133.9	494.5	2.03	6
SA4	116.3	490.0	3.41	5
SA5	116.3	455.6	2.81	5
SA6	101.0	450.5	2.02	5
SA7	111.9	474.5	2.53	5

The rules of five (Ro5) is a set of rules developed by Christopher Lipinski in 1997 and provides a reasonable estimate of the drug likeness of the compounds [29]. Although various exceptions and variants to this rule exist [30,31], analysis of the features/physiochemical parameters suggested by the rules provides essential information about the molecule that can be correlated with in vitro or in vivo activities. According to the rules, a molecule with an octanol–water partition coefficient of less than 5 ($\log P \leq 5$), a molecular weight below 500 (MW $\leq$ 500), fewer than 10 hydrogen bond acceptors (HBA < 10), and fewer than 5 hydrogen bond donors (HBD < 5) likely act as a drug (i.e., good absorption or permeation). Later, parameters such as the topological polar surface area (TPSA $\leq$ 140 Å^2) and the number of rotatable bonds (RB $\leq$ 13 RBs) have also been added to this rule, which directly/indirectly influences drug metabolism in the human body [32,33]. A quick overview of Tables 2 and 4 indicate that all compounds showed excellent drug-like features. Compared to the other compounds of the same series, SA2 and SA4 showed relatively lower hydrophilicity but were still within the limit ($\log P < 5$). It is said that a compound that violates two or more properties might not behave like a drug [32]. However, none of the compounds displayed two or more violations. In addition, the predicted total molecular solvent accessible surface area (SASA) was found to be between 730 and 785 (Table 2) and was within the range of 300–1000. Similarly, FOSA (hydrophobic solvent accessible surface area), FISA (hydrophilic solvent accessible surface area), and PISA (carbon-pie solvent accessible surface area) were also within the defined limits (0–750, 7–330, and 0–450, respectively). On the other hand, WPSA (weakly polar solvent accessible surface area) was in the range of 0.15–0.64 for all compounds except SA4 and SA5. Despite this, it was within the allowed range (0–175). Other descriptors generated using Qikprop also supported the proposition that the compounds bear good drug-like behavior (Table S1, Supplementary Information). For example, the octanol/water partition coefficient (QP logPo/w) for the compound ranged between 1.31 and 3.41 (recommended value: 2.0 to 6.5), whereas human serum albumin binding (QPlog Khsa) was found to be between −0.746 and 0.095 (recommended value: −1.5 to 1.5). The predicted apparent Caco-2 cell permeability was moderate to high (QP logHERG recommended value: >80% high and <25% poor). Similarly, the blood–brain partition coefficient was −0.39 to −1.72 (QP log BB, recommended value: −3.0 to 1.5), and human oral absorption was ~ 59–82% (%HOA recommended value: <25% = low, >80% high). Other descriptors were also found to be within the recommended range. It is worth noting that SA3, which showed relatively high viability against 3T3 and low toxicity against the breast cancer 4T1 cell line, showed one deviation in the RO3 during Qikprop analysis. This compound exhibited a low human oral absorption (%HOA) and a large polar surface area (PSA) compared to the other candidates. These findings align with an earlier study that suggested that descriptors such as lipophilicity, number of HBAs/HBDs, size, and rigidity are some of the crucial parameters of a heterocyclic core and determine its biological activity [34].

We also determined the absorption, distribution, metabolism, and excretion (ADME) properties of the compounds using the SwissADME [24] webtool. The results are displayed in radar (Figure S12a–g) and Brain Or IntestinaL EstimateD permeation (BOILED-egg, Figure S12h–n) forms. The ADME properties or bioavailability are depicted as six vertices of a pink hexagon using lipophilicity, molecular size, insaturation, fraction of sp^3 hybridized carbons, and flexibility descriptors. Molecules falling in the radar's pink region are considered molecules with optimum drug likeness. Except for compound SA2 (one offshoot in insatu.), there was no offshoot/deviation, and the saturation side was within the pink region. Overall, the greater the number of offshoots, the lesser the drug likeness. As is clear from the figures, among the studied physicochemical properties, there was only one insatu deviation in compound SA2, which was also evident in the Ro5 studies. In a BOILED-egg plot, a molecule falling in the white region exhibits the highest probability of being absorbed by human gastrointestinal absorption (HIA). On the other hand, those falling in the yellow region (yolk) have a high probability of permeating the brain [35]. Those falling in the grey region indicate low HIA and BB penetration. In our case, all molecules fell within the white region, and none showed BB permeation. It is also noteworthy that compound SA6 is a substrate of P-gp and effluxed by P-gp (as shown by the blue point).

2.4.5. Toxicological Prediction

The toxicity profiles of the compounds were predicted using the web-based platform pkCSM, and the results are provided in Table 5 [25]. Although this was theoretically calculated and needs further experimental results to draw conclusions, some findings are noteworthy. For example, among all of the compounds, only one (SA6) showed no AMES toxicity (mutagenic) and showed good anticancer activity against 4T1 cells. Similarly, only SA2 was predicted to be the inhibitor of hERG I. Although none of the compounds possessed skin sensitization, all of the compounds were predicted to be hepatotoxic. The maximum tolerated dose for human use was highest for compound SA2. As discussed before, the compounds bore stark drug-like features and most of the descriptors were favorable (Table S1, Supplementary Information). It was noted that the QP logHERG value of the compounds was below −5, which might be a concern. However, it should be noted that all of these values were calculated, and therefore, further in vivo studies are needed to confirm these observations.

Table 5. Toxicity prediction of compounds SA1–SA7.

Parameters	SA1	SA2	SA3	SA4	SA5	SA6	SA7
AMES toxicity	Yes	Yes	Yes	Yes	Yes	No	Yes
Max. tolerated dose	−0.284	0.102	−0.233	−0.135	−0.171	−0.129	−0.184
hERG I inhibitor	No	Yes	No	No	No	No	No
hERG II inhibitor	Yes	Yes	Yes	Yes	Yes	Yes	Yes
Oral rat acute tox. (LD_{50})	2.97	2.179	2.965	2.664	2.562	2.643	2.541
Chronic oral rat tox. (LOAEL)	1.934	1.37	2.026	1.309	1.382	0.435	1.331
Hepatotoxicity	Yes	Yes	Yes	Yes	Yes	Yes	Yes
Skin sensitization	No	No	No	No	No	No	No
T. Pryiformis toxicity	0.291	0.286	0.292	0.32	0.323	0.302	0.31
Minnow toxicity	1.591	0.954	1.314	0.5	0.858	−0.044	0.869

3. Materials and Methods

All reagents and solvents were obtained from Sigma Aldrich (Germany) and used as received. Melting points (m.p.) were determined using an open capillary method and were uncorrected. Thin-layer chromatography was performed on silica gel-coated aluminum sheets (Merck, Germany). ^{1}H- and ^{13}C-NMR were collected on Bruker Spectrospin DPX 300 MHz spectrometer (Bruker Analytic GmbH, Berlin, Germany). The data were processed on MestreNova version 6.0.2–5475. The chemical shift values were recorded on the δ scale (expressed in ppm) and the coupling constants (*J*) in Hertz. Tetramethylsilane (TMS) was used as an internal standard. The following abbreviations were used for reporting spectra: s = singlet, d = doublet, dd = double doublet, m = multiplet. Mass spectra of the compounds were recorded on an Agilent 6200 series TOF/6500 series Q-TOF 10.1 (48.0) (Agilent Technologies). Absorption data were collected on a Jasco V-570 spectrophotometer (JASCO Corporation, Japan). Emission spectra were obtained on Fluoromax-4 Spectrofluorometer (HORIBA, Japan) and data were processed using FluorEssence software (V3.9). Quantum yields (at room temperature) of the compound were measured relative to a reference coumarin-153 (Φ = 0.547 in ethanol) at room temperature [36]. All QY were measured within 10% error. Φ was calculated using the following equation:

$$\frac{\Phi s}{\Phi r} = \frac{(A_r)(\eta_s^2)(I_s)}{(A_s)(\eta_r^2)(I_r)}$$

where r and s stand for the reference and sample, respectively. *A* is the absorbance at the excitation wavelength, η is the index of refraction of the solvent, and *I* is the integrated luminescence intensity.

3.1. General Protocol for the Synthesis

The piperazine-linked 1,8-naphthalimide-arylsulfonyl derivatives (SA1–SA7) used in this study were synthesized following the protocol depicted in Scheme 1. Intermediate compound 2-(2-(piperazin-1-yl)ethyl)-1H-benzo[de]isoquinoline-1,3(2*H*)-dione (3, Scheme 1) was obtained by refluxing a mixture of 1,8-naphthalimide (1, 1 mmol) with 2-(piperazin-1-yl)ethan-1-amine (2, 1.1 mmol) overnight in toluene use triethylamine as a base. The final compounds (SA1–SA7) were created via a substitution reaction between substituted aryl sulfonyl chloride and (2) in DCM. The chemical composition of the compounds was determined via multiple spectroscopic techniques (vide-infra).

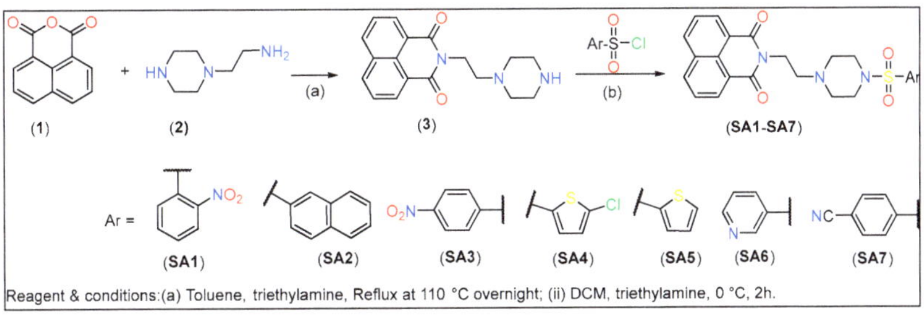

Scheme 1. Synthesis of piperazine-linked 1,8-naphthalimide-arylsulfonyl derivatives.

2-(2-(4-((2-Nitrophenyl)sulfonyl)piperazin-1-yl)ethyl)-1H-benzo[de]isoquinoline-1,3 (2H)-dione (SA1)

Yield: 0.120 g, 75%; m.p. = 170–172 °C; ^{1}H-NMR (CDCl$_3$) δ (ppm): δ 8.54 (dd, J = 7.3, 1.2 Hz, 2H), 8.19 (dd, J = 8.3, 1.2 Hz, 2H), 7.97–7.87 (m, 1H), 7.73 (dd, J = 8.2, 7.3 Hz, 2H), 7.69–7.64 (m, 2H), 7.59–7.55 (m, 1H), 4.28 (t, J = 6.8 Hz, 2H), 3.26 (t, J = 5.0 Hz, 4H), 2.71 (t, J = 6.8 Hz, 2H), 2.65 (t, J = 5.0 Hz, 4H). ^{13}C NMR (101 MHz, CDCl$_3$): δ 164.32, 148.53, 134.20, 133.78, 131.67, 131.58, 131.38, 131.08, 131.03, 128.23, 127.08, 124.14, 122.57, 55.47, 52.69, 46.10, 37.39. Calculated mass: 494.52 for $C_{24}H_{22}N_4O_6S$; observed mass (*m*/*z*) 495.13 [M+H]$^+$.

2-(2-(4-(Naphthalen-2-ylsulfonyl)piperazin-1-yl)ethyl)-1H-benzo[de]isoquinoline-1,3 (2H)-dione (SA2)

Yield: 0.140 g, 70%; m.p. = 210–212 °C (dec.); ^{1}H-NMR (CDCl$_3$) δ (ppm): δ 8.46 (dd, J = 7.3, 1.2 Hz, 2H), 8.31–8.26 (m, 1H), 8.14 (dd, J = 8.3, 1.1 Hz, 2H), 7.98–7.85 (m, 3H), 7.73–7.56 (m, 5H), 4.22 (t, J = 6.8 Hz, 2H), 3.46 (s, 2H), 3.05 (s, 4H), 2.68 (d, J = 7.3 Hz, 4H). ^{13}C NMR (101 MHz, CDCl$_3$): δ 164.29, 134.94, 134.13, 132.63, 132.25, 131.60, 131.31, 129.36, 129.24, 129.20, 128.91, 128.15, 127.99, 127.63, 127.02, 123.22, 122.50, 55.33, 52.45, 46.26, 37.38. Calculated mass: 499.58 for $C_{28}H_{25}N_3O_4S$; observed mass (*m*/*z*) 500.16 [M+H]$^+$.

2-(2-(4-((4-Nitrophenyl)sulfonyl)piperazin-1-yl)ethyl)-1H-benzo[de]isoquinoline-1,3 (2H)-dione (SA3)

Yield: 0.150 g, 77%; m.p. = 183–184 °C; ^{1}H-NMR (CDCl$_3$) δ (ppm): δ 8.50 (dd, J = 7.3, 1.2 Hz, 2H), 8.36–8.30 (m, 2H), 8.19 (dd, J = 8.3, 1.1 Hz, 2H), 7.93–7.87 (m, 2H), 7.72 (dd, J = 8.3, 7.3 Hz, 2H), 4.25 (t, J = 6.7 Hz, 2H), 3.01 (s, 4H), 2.74–2.62 (m, 6H). ^{13}C NMR (101 MHz, CDCl$_3$): δ 164.31, 150.18, 141.70, 134.22, 131.64, 131.32, 129.04, 128.19, 127.07, 124.37, 122.54, 55.29, 52.28, 46.21, 37.36. Calculated mass: 494.52 for $C_{24}H_{22}N_4O_6S$; observed mass (*m*/*z*) 495.13 [M+H]$^+$.

2-(2-(4-((5-Chlorothiophen-2-yl)sulfonyl)piperazin-1-yl)ethyl)-1H-benzo[de]isoquinoline-1,3(2H)-dione (SA4)

Yield: 0.130 g, 80%; m.p. = 188–190 °C; ^{1}H-NMR (CDCl$_3$) δ (ppm): δ 8.54 (dd, J = 7.3, 1.1 Hz, 2H), 8.20 (dd, J = 8.3, 1.1 Hz, 2H), 7.74 (dd, J = 8.3, 7.3 Hz, 2H), 7.28–7.25 (m, 1H), 6.93 (d, J = 4.0 Hz, 1H), 4.27 (t, J = 6.8 Hz, 2H), 3.04 (t, J = 4.9 Hz, 4H), 2.75–2.66 (m, 6H). ^{13}C NMR (101 MHz, CDCl$_3$): δ 164.33, 137.44, 134.22, 133.85, 132.01, 131.67, 131.37, 128.22, 127.20,

127.09, 122.56, 55.31, 55.24, 46.22, 37.43. Calculated mass: 489.99 for $C_{22}H_{20}ClN_3O_4S_2$; observed mass (*m*/*z*) 489.06 [M+H]+, 512.04 $[M+Na]^+$.

2-(2-(4-(Thiophen-2-ylsulfonyl)piperazin-1-yl)ethyl)-1H-benzo[de]isoquinoline-1,3(2H)-dione (SA5)

Yield: 0.110 g, 73%; m.p. = 183–185 °C; ^{1}H-NMR ($CDCl_3$) δ (ppm): δ 8.53 (dd, J = 7.3, 1.2 Hz, 2H), 8.19 (dd, J = 8.3, 1.1 Hz, 2H), 7.72 (dd, J = 8.3, 7.3 Hz, 2H), 7.57 (dd, J = 5.1, 1.4 Hz, 1H), 7.50 (dd, J = 3.8, 1.3 Hz, 1H), 7.10 (dd, J = 5.0, 3.8 Hz, 1H), 4.27 (t, J = 6.9 Hz, 2H), 3.05 (t, J = 4.9 Hz, 4H), 2.76–2.64 (m, 6H). ^{13}C NMR (101 MHz, $CDCl_3$): δ 164.28, 135.92, 134.17, 132.61, 132.20, 131.67, 131.35, 128.22, 127.74, 127.06, 122.58, 77.46, 77.38, 77.14, 76.83, 55.35, 52.31, 46.22, 37.4. Calculated mass: 455.55 for $C_{22}H_{21}ClN_3O_4S_2$; observed mass (*m*/*z*) 456.10 $[M+H]^+$.

2-(2-(4-(Pyridin-3-ylsulfonyl)piperazin-1-yl)ethyl)-1H-benzo[de]isoquinoline-1,3(2H)-dione (SA6)

Yield: 0.135 g, 82%; m.p. = 202–204 °C; ^{1}H-NMR ($CDCl_3$) δ (ppm): δ 8.93 (d, J = 2.3 Hz, 1H), 8.77 (dd, J = 4.9, 1.7 Hz, 1H), 8.50 (dd, J = 7.2, 1.1 Hz, 2H), 8.17 (dd, J = 8.3, 1.1 Hz, 2H), 8.00 (dt, J = 8.0, 2.0 Hz, 1H), 7.71 (dd, J = 8.2, 7.3 Hz, 2H), 7.44 (dd, J = 8.0, 4.9 Hz, 1H), 4.24 (t, J = 6.8 Hz, 2H), 3.02 (d, J = 4.9 Hz, 4H), 2.72–2.64 (m, 6H). ^{13}C NMR (101 MHz, $CDCl_3$): δ 164.28, 153.43, 148.58, 135.56, 134.17, 132.58, 131.65, 131.34, 128.20, 127.05, 123.80, 122.54, 55.35, 52.31, 46.08, 37.36. Calculated mass: 450.51 for $C_{23}H_{22}N_4O_4S$; observed mass (*m*/*z*) 451.40 $[M+H]^+$.

4-((4-(2-(1,3-Dioxo-1H-benzo[de]isoquinolin-2(3H)-yl)ethyl)piperazin-1-yl)sulfonyl) benzonitrile (SA7)

Yield: 0.140 g, 83%; m.p. = 158–160 °C; ^{1}H-NMR ($CDCl_3$) δ (ppm): 8.50 (dd, J = 7.3, 1.2 Hz, 2H), 8.19 (dd, J = 8.4, 1.2 Hz, 2H), 7.84–7.76 (m, 4H), 7.72 (dd, J = 8.2, 7.3 Hz, 2H), 4.24 (t, J = 6.7 Hz, 2H), 2.99 (d, J = 4.9 Hz, 4H), 2.67 (dt, J = 15.5, 5.8 Hz, 6H) ^{13}C NMR (101 MHz, $CDCl_3$): δ 164.28, 140.21, 134.20, 132.91, 131.65, 131.29, 128.43, 128.19, 127.07, 122.55, 117.40, 116.53, 55.30, 52.29, 46.19, 37.37. Calculated mass: 474.53 for $C_{25}H_{22}N_4O_4S$; observed mass (*m*/*z*) 475.25 $[M+H]^+$.

3.2. Biological Studies

For the anticancer studies, we established a stable cell culture for mouse breast cancer cell line 4T1 (ATCC: CRL-2539) cultured in the medium composition RPMI-1640 (Gibco; Grand Island, New York, USA), 10% fetal bovine serum (Gibco, Mexico), 4.5 g/L glucose, 1.5 g/L $NaHCO_3$, 1 mM sodium pyruvate (Gibco; Grand Island, New York, NY, USA), and 1% penicillin–streptomycin (10,000 U/mL). We observed typical cell morphology after the IV passage of cells unfrozen from liquid nitrogen. Then, the cells were used for the subsequent anticancer experiment. A non-cancerous fibroblast cell line (3T3/NIH, ATCC: CRL-1658) was employed for the fluorescence imaging assay. The cells were cultured in 90% Dulbecco's modified Eagle's medium with 4 mM L-glutamine, 2 mM sodium pyruvate, 1% non-essential amino acid, 1% penicillin/streptomycin/amphotericin, and 10% fetal bovine serum.

3.2.1. Cytotoxicity Assay

A cellular viability assay was employed for the cytotoxicity of SA samples. The lower viability indicates that the samples are toxic toward 4T1 cells. A cell viability reagent (MTT, Sigma-Aldrich, St Louis, MI, USA) containing yellow tetrazolium salt (3-(4,5-dimethylthiazol-2-yl)-2,5-diphenyltetrazolium bromide) can be converted to purple formazan salt to indicate the cell viability. The 4T1 breast cancer cells were cultured at a density of 5.0×10^3 cells per well in 96-well plates. The cells were treated with SA samples for 24 h. The control groups were abbreviated as Cont (100% culture medium), representing normal cell culture; PC (20% DMSO and 80% culture medium), representing toxic condition; and NC (0.1% DMSO and 99.9%), representing minimum toxic condition. MTT reagent was added to 96-well plates to measure the cytotoxic effect of the SA compounds on 4T1 cells.

3.2.2. Fluorescence Imaging Assay

A total of 5.0×10^4 of 3T3 cells per well in a 96-well plate were co-cultured with the 1.0 μg/mL of SA samples for 72 h. The viability of the 3T3 cells was assayed by using 10% CCK-8 reagent for 60 min to determine the toxicity of the SA samples. Afterwards, the cells were fixed with glutaraldehyde, followed by staining with 0.1 μg/mL of 4′,6-diamidino-2-phenylindole (DAPI) for the imaging of the nuclei. An inverted fluorescent optical microscope (Nikon, Eclipse Ti-S, Japan) was utilized. The fluorescent light source was Nikon Intensilight C-HGFI, and two fluorescent filter cubes (Ex = 340–380 nm, Em = 435–485 nm, corresponding to DAPI, and Ex = 465–495 nm, Em = 512–558 nm, corresponding to SA samples) were used to observe the distribution of SA samples in the cytoplasm. A scientific-grade CCD (Dhyana, model: 400 BSI, Tucsen, Fuzhou, China) was used to record the images. The magnification of the OM was 200×. The original fluorescent images were recorded in greyscale. ImageJ software (an open-source software, NIH Image, version 1.54f, Bethesda, MD, USA) was employed to merge and convert two fluorescent images into visible colors, e.g., blue for cell nuclei and green for SA1–SA7 samples.

3.3. Computational Details

Molecular docking was performed on a computer with a Windows 10 operating system and 32 GB RAM. MD simulations were conducted on a Linux operating system using a Dell workstation with 32 GB RAM and a ZOTAC NVIDIA RTX 3060 Twin Edge 12 GB GDDR6 graphics card.

3.3.1. Docking (Ligands, Receptor, and Grid Preparation)

The chemical structure of SA1–SA7 was drawn using Marvin sketch (http://www.chemaxon.com, accessed on 29 September 2023) and energy minimization was carried out by utilizing the MM2 force field. Using the Autodock tools (ADT), Gasteiger charges were added to the ligands and saved in PDBQT format. The target protein (PDB ID: 5FL4) was downloaded from the protein databank [37]. Water molecules present in the receptor were removed, and polar hydrogen atoms were added. In the target protein, zinc (Zn^{2+}) is a cofactor already present in the crystal structure of 5FL4. The partial charge was determined using AutoDock 4 before saving it in PDBQT format [38].

A grid box of size 94 × 90 × 94 Å with a 0.35 Å spacing was created to accommodate the entire protein [39]. The grid was centered at coordinates x = 6.965 Å, y = −21.786 Å, z = 57.614 Å to optimize ligand-binding orientation exploration. Default parameters for the Lamarckian genetic algorithm (LGA) were applied to generate the best molecular conformation of the molecules [40]. To assess binding poses, up to 9 conformers per ligand were considered during the docking procedure. Output files were visualized using a PyMol visualizer [41].

3.3.2. Molecular Dynamics (MD) Simulation

MD simulations at 100 ns was conducted using the Desmond module (Schrödinger suite 2020) [42]. The complex, including the explicit solvent system, was investigated by utilizing the OPLS3e force field, which was integrated into the Desmond (v12.8.117 Release 2021-22) software. The molecular system underwent solvation with crystallographic water (TIP3P) molecules [43]. Periodic boundary conditions were implemented in an orthorhombic setup, extending 10 Å beyond the solute in all directions to create a buffer region. Removal of overlapping water molecules ensued, followed by neutralizing the system by adding two Na^+ ions to achieve a concentration of 0.15 M. The Nose–Hoover thermostat and barostat ensemble (NPT) was applied to maintain the system at a constant temperature of 300 K and pressure of 1 bar, respectively. Notably, applying these techniques ensures that the simulation faithfully represents the system's behavior in a realistic environment. A hybrid energy minimization algorithm was used to prepare the system for the MD simulations. This involved an initial stage of 1000 steps using the steepest descent algorithm and subsequent steps utilizing the conjugate gradient algorithm [44]. This approach effectively optimized the system's energy landscape before commencing the MD simulations.

3.3.3. Absorption, Distribution, Metabolism, and Excretion and Toxicity (ADMET) Studies

In silico absorption, distribution, metabolism, and excretion (ADME) and toxicity studies were carried out using the web-based online platforms SwissADME [24] and pKCSM [25]. SMILE formats (as input) of the molecules were generated using Marvin 16.11.28.0, 2016, ChemAxon (http://www.chemaxon.com, accessed on 29 September 2023).

3.3.4. Determining Binding Free Energy

The molecular mechanics Poisson–Boltzmann surface area (MM/PBSA) method was employed to calculate the relative binding free energy within the protein–ligand complex [45]. MM/PBASA assesses various interaction energies, including electrostatic interactions, van der Waals interactions, polar solvation energy, and nonpolar solvation. From the final 20 ns of simulation trajectories for each protein–ligand complex, the binding free energy (ΔG) was computed.

4. Conclusions

In summary, we synthesized and characterized seven new piperazine-linked 1,8-naphthalimide-arylsulfonyl derivatives that are non-toxic towards normal cells but show activity against breast cancer cells. Despite the limited solubility of the compounds, they could enter the cytoplasm and impart cytotoxicity. Modeling studies predicted that compounds could act by inhibiting the CAIX enzyme, which is often expressed in different cancers. The results presented in this work support the proposition that piperazine-linked 1,8-naphthalimide-arylsulfonyl derivatives are potential candidates for cancer theranostics.

Supplementary Materials: The following supporting information can be downloaded at: https://www.mdpi.com/article/10.3390/ijms25021069/s1.

Author Contributions: Conceptualization, A.H. and M.-F.H.; methodology, K.M.A., A.A.-O., A.K.D.A. and M.-W.T.; software, A.H. and A.R.; validation, M.-F.H. and W.-Y.W.; formal analysis, A.H. and A.R.; writing—original draft, A.H. and M.-W.T.; writing—review and editing, M.-F.H. and W.-Y.W.; funding acquisition, A.H. All authors have read and agreed to the published version of the manuscript.

Funding: This research has been funded by Scientific Research Deanship at University of Ha'il–Saudi Arabia through project number MDR-22 006.

Institutional Review Board Statement: Not applicable.

Informed Consent Statement: Not applicable.

Data Availability Statement: Data are contained within the article.

Acknowledgments: This research has been funded by Scientific Research Deanship at University of Ha'il–Saudi Arabia through project number MDR-22 006.

Conflicts of Interest: The authors declare no conflicts of interest.

References

1. Stewart, B.; Wild, C.P. *World Cancer Report 2014*; World Health Organization: Geneva, Switzerland, 2015.
2. Ali, I.; Salim, K.; Rather, M.A.; Wani, W.A.; Haque, A. Advances in nano drugs for cancer chemotherapy. *Curr. Cancer Drug Targets* **2011**, *11*, 135–146. [CrossRef] [PubMed]
3. DeSantis, C.E.; Lin, C.C.; Mariotto, A.B.; Siegel, R.L.; Stein, K.D.; Kramer, J.L.; Alteri, R.; Robbins, A.S.; Jemal, A. Cancer treatment and survivorship statistics, 2014. *CA Cancer J. Clin.* **2014**, *64*, 252–271. [CrossRef]
4. Greenlee, R.T.; Murray, T.; Bolden, S.; Wingo, P.A. Cancer statistics, 2000. *CA Cancer J. Clin.* **2000**, *50*, 7–33. [CrossRef] [PubMed]
5. Banerjee, S.; Veale, E.B.; Phelan, C.M.; Murphy, S.A.; Tocci, G.M.; Gillespie, L.J.; Frimannsson, D.O.; Kelly, J.M.; Gunnlaugsson, T. Recent advances in the development of 1,8-naphthalimide based DNA targeting binders, anticancer and fluorescent cellular imaging agents. *Chem. Soc. Rev.* **2013**, *42*, 1601–1618. [CrossRef] [PubMed]
6. Tomczyk, M.D.; Walczak, K.Z. 1,8-Naphthalimide based DNA intercalators and anticancer agents. A systematic review from 2007 to 2017. *Eur. J. Med. Chem.* **2018**, *159*, 393–422. [CrossRef] [PubMed]

7. Yu, H.; Guo, Y.; Zhu, W.; Havener, K.; Zheng, X. Recent advances in 1,8-naphthalimide-based small-molecule fluorescent probes for organelles imaging and tracking in living cells. *Coord. Chem. Rev.* **2021**, *444*, 214019. [CrossRef]
8. Tandon, R.; Luxami, V.; Kaur, H.; Tandon, N.; Paul, K. 1,8-Naphthalimide: A potent DNA intercalator and target for cancer therapy. *Chem. Rec.* **2017**, *17*, 956–993. [CrossRef] [PubMed]
9. Rao, N.S.; Nagesh, N.; Nayak, V.L.; Sunkari, S.; Tokala, R.; Kiranmai, G.; Regur, P.; Shankaraiah, N.; Kamal, A. Design and synthesis of DNA-intercalative naphthalimide-benzothiazole/cinnamide derivatives: Cytotoxicity evaluation and topoisomerase-IIα inhibition. *MedChemComm* **2019**, *10*, 72–79.
10. Buran, K.; Bua, S.; Poli, G.; Önen Bayram, F.E.; Tuccinardi, T.; Supuran, T.C. Novel 8-substituted coumarins that selectively inhibit human carbonic anhydrase IX and XII. *Int. J. Mol. Sci.* **2019**, *20*, 1208. [CrossRef]
11. Plyusnin, V.F.; Kupryakov, A.S.; Grivin, V.P.; Shelton, A.H.; Sazanovich, I.V.; Meijer, A.J.; Weinstein, J.A.; Ward, M.D. Photophysics of 1,8-naphthalimide/Ln (III) dyads (Ln = Eu, Gd): Naphthalimide→Eu (III) energy-transfer from both singlet and triplet states. *Photochem. Photobiol. Sci.* **2013**, *12*, 1666–1679. [CrossRef]
12. O'Neil, A.T.; Chalard, A.; Malmström, J.; Kitchen, J.A. White light and colour-tunable emission from a single component europium-1,8-naphthalimide thin film. *Dalton Trans.* **2023**, *52*, 2255–2261. [CrossRef]
13. Wang, Z.; Liu, N.; Li, H.; Chen, P.; Yan, P. The Role of Blue-Emissive 1,8-Naphthalimidopyridine N-Oxide in Sensitizing EuIII Photoluminescence in Dimeric Hexafluoroacetylacetonate Complexes. *Eur. J. Inorg. Chem.* **2017**, *2017*, 2211–2219. [CrossRef]
14. Langdon-Jones, E.E.; Williams, C.F.; Hayes, A.J.; Lloyd, D.; Coles, S.J.; Horton, P.N.; Groves, L.M.; Pope, S.J. Luminescent 1,8-Naphthalimide-Derived ReI Complexes: Syntheses, Spectroscopy, X-ray Structure and Preliminary Bioimaging in Fission Yeast Cells. *Eur. J. Inorg. Chem.* **2017**, *2017*, 5279–5287. [CrossRef]
15. Haapasalo, J.; Nordfors, K.; Haapasalo, H.; Parkkila, S. The Expression of Carbonic Anhydrases II, IX and XII in Brain Tumors. *Cancers* **2020**, *12*, 1723. [CrossRef]
16. Maren, T.H. Carbonic anhydrase: Chemistry, physiology, and inhibition. *Physiol. Rev.* **1967**, *47*, 595–781. [CrossRef]
17. Lou, Y.; McDonald, P.C.; Oloumi, A.; Chia, S.; Ostlund, C.; Ahmadi, A.; Kyle, A.; auf dem Keller, U.; Leung, S.; Huntsman, D. Targeting tumor hypoxia: Suppression of breast tumor growth and metastasis by novel carbonic anhydrase IX inhibitors. *Cancer Res.* **2011**, *71*, 3364–3376. [CrossRef] [PubMed]
18. Winum, J.Y.; Rami, M.; Scozzafava, A.; Montero, J.L.; Supuran, C. Carbonic anhydrase IX: A new druggable target for the design of antitumor agents. *Med. Res. Rev.* **2008**, *28*, 445–463. [CrossRef]
19. Supuran, C.T. Carbonic anhydrases: Novel therapeutic applications for inhibitors and activators. *Nat. Rev. Drug Discov.* **2008**, *7*, 168–181. [CrossRef] [PubMed]
20. Chinchilli, K.K.; Akunuri, R.; Ghouse, S.M.; Soujanya, D.; Angeli, A.; Parupalli, R.; Arifuddin, M.; Yaddanapudi, V.M.; Supuran, C.T.; Nanduri, S. Design, synthesis, and structure–activity studies of new rhodanine derivatives as carbonic anhydrase II, IX inhibitors. *Arch. Pharm.* **2023**, *356*, e2300205. [CrossRef] [PubMed]
21. Kumari, S.; Idrees, D.; Mishra, C.B.; Prakash, A.; Ahmad, F.; Hassan, M.I.; Tiwari, M. Design and synthesis of a novel class of carbonic anhydrase-IX inhibitor 1-(3-(phenyl/4-fluorophenyl)-7-imino-3H-[1,2,3] triazolo [4,5d] pyrimidin 6 (7H) yl) urea. *J. Mol. Graph. Modell.* **2016**, *64*, 101–109. [CrossRef] [PubMed]
22. Wang, Y.; Guo, H.; Tang, G.; He, Q.; Zhang, Y.; Hu, Y.; Wang, Y.; Lin, Z. A selectivity study of benzenesulfonamide derivatives on human carbonic anhydrase II/IX by 3D-QSAR, molecular docking and molecular dynamics simulation. *Comput. Biol. Chem.* **2019**, *80*, 234–243. [CrossRef]
23. Tuccinardi, T. What is the current value of MM/PBSA and MM/GBSA methods in drug discovery? *Expert Opin. Drug Discov.* **2021**, *16*, 1233–1237. [CrossRef] [PubMed]
24. Daina, A.; Michielin, O.; Zoete, V. SwissADME: A free web tool to evaluate pharmacokinetics, drug-likeness and medicinal chemistry friendliness of small molecules. *Sci. Rep.* **2017**, *7*, 42717. [CrossRef] [PubMed]
25. Pires, D.E.; Blundell, T.L.; Ascher, D.B. pkCSM: Predicting small-molecule pharmacokinetic and toxicity properties using graph-based signatures. *J. Med. Chem.* **2015**, *58*, 4066–4072. [CrossRef] [PubMed]
26. Shahbazi, S.; Kaur, J.; Kuanar, A.; Kar, D.; Singh, S.; Sobti, R.C. Risk of Late-Onset Alzheimer's Disease by Plasma Cholesterol: Rational In Silico Drug Investigation of Pyrrole-Based HMG-CoA Reductase Inhibitors. *ASSAY Drug Dev. Technol.* **2017**, *15*, 342–351. [CrossRef]
27. Omoboyowa, D.A. Exploring molecular docking with E-pharmacophore and QSAR models to predict potent inhibitors of 14-α-demethylase protease from Moringa spp. *Pharmacol. Res. Mod. Chin. Med.* **2022**, *4*, 100147. [CrossRef]
28. Ritchie, T.J.; Ertl, P.; Lewis, R. The graphical representation of ADME-related molecule properties for medicinal chemists. *Drug Discov. Today* **2011**, *16*, 65–72. [CrossRef] [PubMed]
29. Lipinski, C.A.; Lombardo, F.; Dominy, B.W.; Feeney, P.J. Experimental and computational approaches to estimate solubility and permeability in drug discovery and development settings. *Adv. Drug Deliv. Rev.* **1997**, *23*, 3–25. [CrossRef]
30. Lipinski, C.A. Lead-and drug-like compounds: The rule-of-five revolution. *Drug Discov. Today Technol.* **2004**, *1*, 337–341. [CrossRef]
31. Lipinski, C.A. Drug-like properties and the causes of poor solubility and poor permeability. *J. Pharmacol. Toxicol. Methods* **2000**, *44*, 235–249. [CrossRef]
32. Keller, T.H.; Pichota, A.; Yin, Z. A practical view of 'druggability'. *Curr. Opin. Chem. Biol.* **2006**, *10*, 357–361. [CrossRef]
33. Veber, D.F.; Johnson, S.R.; Cheng, H.-Y.; Smith, B.R.; Ward, K.W.; Kopple, K.D. Molecular properties that influence the oral bioavailability of drug candidates. *J. Med. Chem.* **2002**, *45*, 2615–2623. [CrossRef] [PubMed]

34. Hamdan, A.; Kasabri, V.; Al-Hiari, Y.; Arabiyat, S.; AlAlawi, S.; Bustanji, Y. Dual anti-inflammatory and antiglycation propensities of a potentially novel class of functionalized fluoroquinolones. *J. Heterocycl. Chem.* **2020**, *57*, 663–675. [CrossRef]
35. Daina, A.; Zoete, V. A boiled-egg to predict gastrointestinal absorption and brain penetration of small molecules. *ChemMedChem* **2016**, *11*, 1117. [CrossRef] [PubMed]
36. Melhuish, W.H. Quantum efficiencies of fluorescence of organic substances: Effect of solvent and concentration of the fluorescent solute1. *J. Phys. Chem.* **1961**, *65*, 229–235. [CrossRef]
37. Leitans, J.; Kazaks, A.; Balode, A.; Ivanova, J.; Zalubovskis, R.; Supuran, C.T.; Tars, K. Efficient expression and crystallization system of cancer-associated carbonic anhydrase isoform IX. *J. Med. Chem.* **2015**, *58*, 9004–9009. [CrossRef] [PubMed]
38. Morris, G.M.; Huey, R.; Lindstrom, W.; Sanner, M.F.; Belew, R.K.; Goodsell, D.S.; Olson, A.J. AutoDock4 and AutoDockTools4: Automated docking with selective receptor flexibility. *J. Comput. Chem.* **2009**, *30*, 2785–2791. [CrossRef]
39. Morris, G.M.; Huey, R.; Olson, A.J. Using autodock for ligand-receptor docking. *Curr. Protoc. Bioinform.* **2008**, *24*, 8.14.11–18.14.40. [CrossRef]
40. Morris, G.M.; Goodsell, D.S.; Halliday, R.S.; Huey, R.; Hart, W.E.; Belew, R.K.; Olson, A.J. Automated docking using a Lamarckian genetic algorithm and an empirical binding free energy function. *J. Comput. Chem.* **1998**, *19*, 1639–1662. [CrossRef]
41. DeLano, W.L. Pymol: An open-source molecular graphics tool. *CCP4 Newsl. Protein Crystallogr.* **2002**, *40*, 82–92.
42. *Maestro*; Schrödinger, LLC: New York, NY, USA, 2017.
43. Mark, P.; Nilsson, L. Structure and dynamics of the TIP3P, SPC, and SPC/E water models at 298 K. *J. Phys. Chem. A* **2001**, *105*, 9954–9960. [CrossRef]
44. Hager, W.W.; Zhang, H. A new conjugate gradient method with guaranteed descent and an efficient line search. *SIAM J. Optim.* **2005**, *16*, 170–192. [CrossRef]
45. Kumari, R.; Kumar, R.; Consortium, O.S.D.D.; Lynn, A. g_mmpbsa A GROMACS tool for high-throughput MM-PBSA calculations. *J. Chem. Inf. Model.* **2014**, *54*, 1951–1962. [CrossRef] [PubMed]

International Journal of *Molecular Sciences*

MDPI

Article

POSS Engineering of Multifunctional Nanoplatforms for Chemo-Mild Photothermal Synergistic Therapy

Zhengye Gu [1,†], Xiaochuan Geng [2,†], Shanyi Guang [3,*] and Hongyao Xu [1,*]

1 State Key Laboratory for Modification of Chemical Fibers and Polymer Materials, College of Materials Science, Engineering & Research Center for Analysis and Measurement, Donghua University, Shanghai 201620, China; guzhengye@126.com

2 Department of Radiology, Renji Hospital, Shanghai Jiao Tong University School of Medicine, Shanghai 200127, China; gengxiaochuan1949@126.com

3 College of Chemistry and Chemical Engineering, Donghua University, Shanghai 201620, China

* Correspondence: gsy@dhu.edu.cn (S.G.); hongyaoxu@dhu.edu.cn (H.X.)

† These authors contributed equally to this work.

Abstract: Chemo-mild photothermal synergistic therapy can effectively inhibit tumor growth under mild hyperthermia, minimizing damage to nearby healthy tissues and skin while ensuring therapeutic efficacy. In this paper, we develop a multifunctional study based on polyhedral oligomeric sesquisiloxane (POSS) that exhibits a synergistic therapeutic effect through mild photothermal and chemotherapy treatments (POSS-SQ-DOX). The nanoplatform utilizes SQ-N as a photothermal agent (PTA) for mild photothermal, while doxorubicin (DOX) serves as the chemotherapeutic drug for chemotherapy. By incorporating POSS into the nanoplatform, we successfully prevent the aggregation of SQ-N in aqueous solutions, thus maintaining its excellent photothermal properties both in vitro and in vivo. Furthermore, the introduction of polyethylene glycol (PEG) significantly enhances cell permeability, which contributes to the remarkable therapeutic effect of POSS-SQ-DOX NPs. Our studies on the photothermal properties of POSS-SQ-DOX NPs demonstrate their high photothermal conversion efficiency (62.3%) and stability, confirming their suitability for use in mild photothermal therapy. A combination index value (CI = 0.72) verified the presence of a synergistic effect between these two treatments, indicating that POSS-SQ-DOX NPs exhibited significantly higher cell mortality (74.7%) and tumor inhibition rate (72.7%) compared to single chemotherapy and mild photothermal therapy. This observation highlights the synergistic therapeutic potential of POSS-SQ-DOX NPs. Furthermore, in vitro and in vivo toxicity tests suggest that the absence of cytotoxicity and excellent biocompatibility of POSS-SQ-DOX NPs provide a guarantee for clinical applications. Therefore, utilizing near-infrared light-triggering POSS-SQ-DOX NPs can serve as chemo-mild photothermal PTA, while functionalized POSS-SQ-DOX NPs hold great promise as a novel nanoplatform that may drive significant advancements in the field of chemo-mild photothermal therapy.

Keywords: chemo-mild photothermal; synergistic effect; POSS-based nanoplatform; squaraine; high inhibition rate

Citation: Gu, Z.; Geng, X.; Guang, S.; Xu, H. POSS Engineering of Multifunctional Nanoplatforms for Chemo-Mild Photothermal Synergistic Therapy. *Int. J. Mol. Sci.* **2024**, *25*, 1012. https://doi.org/10.3390/ijms25021012

Academic Editor: Kang-Nan Wang

Received: 18 December 2023
Revised: 7 January 2024
Accepted: 11 January 2024
Published: 13 January 2024

1. Introduction

As a highly promising candidate, photothermal therapy (PTT) has garnered significant attention in recent years. PTT utilizes light and photothermal agents (PTAs).

To eradicate cancer cells through photo-thermal conversion, offering higher efficiency, minimal invasiveness, and fewer side effects [1–12]. However, it typically relies on local hyperthermia (above 50 °C) [13–16] for complete tumor ablation, which can result in damage to the surrounding healthy tissues, inflammation, and other undesirable biological effects during cancer treatments [17–21]. Therefore, achieving cancer cell death at mild temperatures is crucial for clinical practice [21,22]; thus, the concept of mild photothermal therapy (MPTT) has emerged. Nevertheless, in most cases, MPTT alone exhibits an inferior anticancer ability due to the presence of heat shock proteins (HSPs), which may repair cell apoptosis induced by a lower heating temperature such as 45 °C [23]. Hence, adjuvant strategies, including PTAs for the inhibition of HSPs [24], autophagy modulation [25–27], organelle targeting [28–30], and gas sensitization [31–33] are necessary to enhance the therapeutic effect of MPTT. In addition to these aforementioned strategies, with the assistance of nanotechnology, PTAs in MPTT and other agents like chemodrugs, radiosensitizers, and photosensitizers can form a nanoplatform with multimodal anti-tumor therapeutic effects [34–37]. On one hand, the tumor cell-killing performance of other treatments in combination with MPTT is improved. On the other hand, other treatments can sensitize tumor cells to be more vulnerable to mild heat, thus enhancing the tumor eradication effect of MPTT [21].

As one of the main strategies for cancer treatment, chemotherapy plays an important role in clinical cancer treatment. However, the non-specific distribution and resistance of chemotherapeutic drugs significantly limit the clinical applications of chemotherapy. The combination of MPTT with chemotherapy can solve these problems easily. First, MPTT may enable spatiotemporally controlled drug delivery by adjusting the position, power density, and working time of NIR light. Second, its mild temperature can make cancer cells more accessible and vulnerable to chemotherapeutic drugs. Third, the heat generated by PTAs can stimulate intratumoral penetration and the accumulation of drugs due to the increased tumor blood flow and vascular permeability [21]. In turn, enhanced chemotherapy efficacy can compensate for insufficient heat damage due to the mild temperature. Therefore, a more excellent anti-tumor effect can be achieved by combining MPTT with chemotherapy.

Squaraine dyes (SQs) are a well-studied class of zwitterionic dyes with stable quinoid structures and high planarity [38]. Typically, they consist of an electron-accepting squaric acid and two electron donors in a D-A-D pattern. Due to their advantageous structure, these dyes exhibit exceptional optical properties, including elevated photoconductivity, well-defined absorption in the visible and near-infrared regions, a high molar absorption coefficient, and impressive photothermal conversion efficiency (PCE) [39,40]. As a result, SQs find wide applications in various biological fields, such as biosensing [41], bioimaging [42,43], and phototherapy [44,45]. However, on account of the strong dipole–dipole and π–π stacking interactions between molecules, SQs tend to aggregate in aqueous solutions, which can negatively impact their photothermal performance and photostability [46]. Therefore, there is a challenge when using SQs for biological applications, especially in PTT, as follows: preventing the easy formation of dye aggregates while preserving their photophysical characteristics.

The emergence of polyhedral oligomeric sesquisiloxane (POSS) can effectively address this issue. A typical POSS molecule possesses a three-dimensional cage-like structure consisting of a cubic silica core and eight surrounding organic corner groups. These corner groups can be readily modified into various functional groups, facilitating the convenient preparation of multifunctional polymers with POSS cores [47–49]. Moreover, due to the nanoscale size of POSS, diverse types of functional POSS-based nanoplatforms can be constructed. Additionally, Wang et al. reported that conjugating porphyrin to POSS resulted in increased photodynamic efficiency and reduced aggregation [50]. Consequently, incorporating POSS can mitigate the aggregation caused by π–π stacking interactions between molecules. Furthermore, based on our previous research conducted in our laboratory [51,52], we discovered that introducing PEG could effectively enhance water solubility and achieve amphipathy for POSS-based nanoparticles, thereby significantly improving cell permeability and the biocompatibility of these nanoplatforms while simultaneously enhancing their potential applications in biological imaging [53], drug delivery [54,55], and tissue regeneration [56].

Hence, based on the aforementioned introduction, a multifunctional POSS-based nanoplatform (POSS-SQ-DOX) was developed for mild chemo-photothermal synergistic tumor therapy. Firstly, POSS-PEG was synthesized via click chemistry to enhance the cell permeability and biocompatibility of the nanoplatform. Subsequently, SQ-N and

DOX were incorporated into POSS using the same method. In this nanoplatform, SQ-N exhibited excellent photothermal properties and high PCE and served as a PTA in MPTT. Additionally, DOX acted as a chemotherapeutic drug for chemotherapy purposes. As illustrated in Scheme 1, upon in situ injection, POSS-SQ-DOX NPs accumulated within tumor cells where the mild thermal effect induced by near-infrared light at 808 nm worked synergistically with DOX to achieve an enhanced anti-tumor effect. The experimental results demonstrated that tumor growth was significantly inhibited in mice treated with POSS-SQ-DOX NPs compared to those receiving single DOX chemotherapy or MPTT alone. Furthermore, this treatment approach proved safe while ensuring highly effective anti-tumor outcomes. Overall, this study explored the synergistic anti-tumor effect of DOX chemotherapy and photothermal therapy working under mild temperatures, highlighting the great potential for utilizing the chemo-mild photothermal synergistic therapy system based on DOX and SQ-modified POSS-based nanoparticles for tumor treatments, bringing a novel strategy to combat tumors.

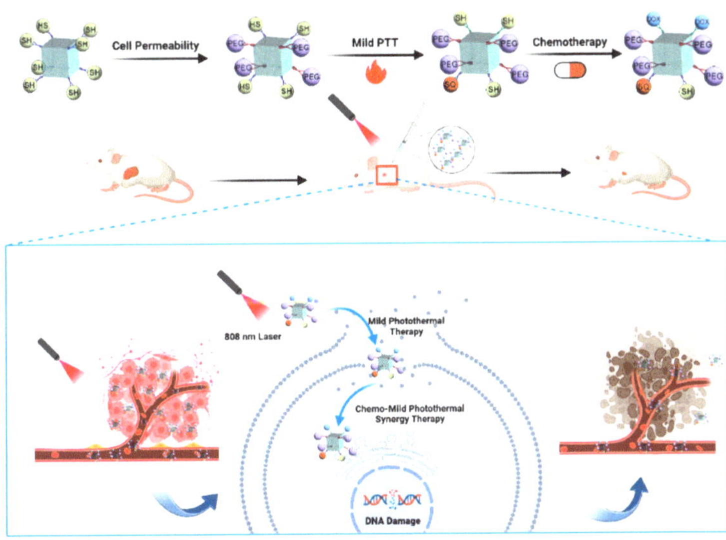

Scheme 1. Schematic illustration of POSS-SQ-DOX NPs' preparation and the chemo-mild photothermal therapeutic mechanism in vivo.

2. Results and Discussion

2.1. Preparation and Characterization of POSS-SQ-DOX NPs

PEG, SQ-N, and DOX were all chemically attached to POSS nanoparticles through click chemistry, resulting in the formation of POSS-SQ-DOX NPs (Schemes S1–S3). The structures of these compounds were confirmed using ^{1}H NMR and FITR (Figures S1–S4). The structure of POSS-SQ-DOX NPs is shown in Figure 1a. Additionally, transmission electron microscope (TEM) and dynamic laser light scattering (DLS) techniques were employed to verify the nanostructures of POSS-SQ-DOX NPs. TEM analysis revealed that the size of POSS-SQ-DOX NPs was approximately 25 nm, while DLS measurements showed a size of 24 nm, which are both larger than that of pure POSS-SH [57] (Figure 1b,c). Furthermore, the particle size stability of POSS-SQ-DOX NPs was monitored over a two-week period (Figure 1d), demonstrating minimal changes in nanoparticle size and indicating excellent stability for at least two weeks. Additionally, UV-Vis absorption spectra and fluorescence spectra were used to investigate the optical properties of POSS-SQ-DOX NPs. As depicted in Figure 1e, distinct UV absorption peaks near 818 nm and a fluorescence emission peak at 949 nm were observed for POSS-SQ-DOX NPs. The presence of near-infrared absorption characteristics provided a foundation for utilizing POSS-SQ-DOX NPs as PTAs in mild photothermal therapy. Notably, compared to SQ-N alone, its incorporation into POSS nanoparticles significantly enhanced both absorption and emission intensities for SQ-N molecules within the hybrid system due to the increased intermolecular distances between SQ-N units and weakened π–π stacking interactions (Figure S5). Furthermore, no significant changes in either the position or intensity of the absorption peak were observed during storage at room temperature for two weeks, indicating that POSS-SQ-DOX NPs displayed optical stability under suitable conditions (Figure 1f). Moreover, contact angle measurements confirmed the enhanced hydrophilicity of POSS-SQ-DOX NPs. In comparison to POSS-SH, DOX, and SQ-N, POSS-SQ-DOX NPs exhibited superior hydrophilicity due to the PEG modification (Figure S6). In summary, we successfully synthesized NIR-absorbing and highly stable POSS-SQ-DOX NPs, making them a suitable and robust system for the following research.

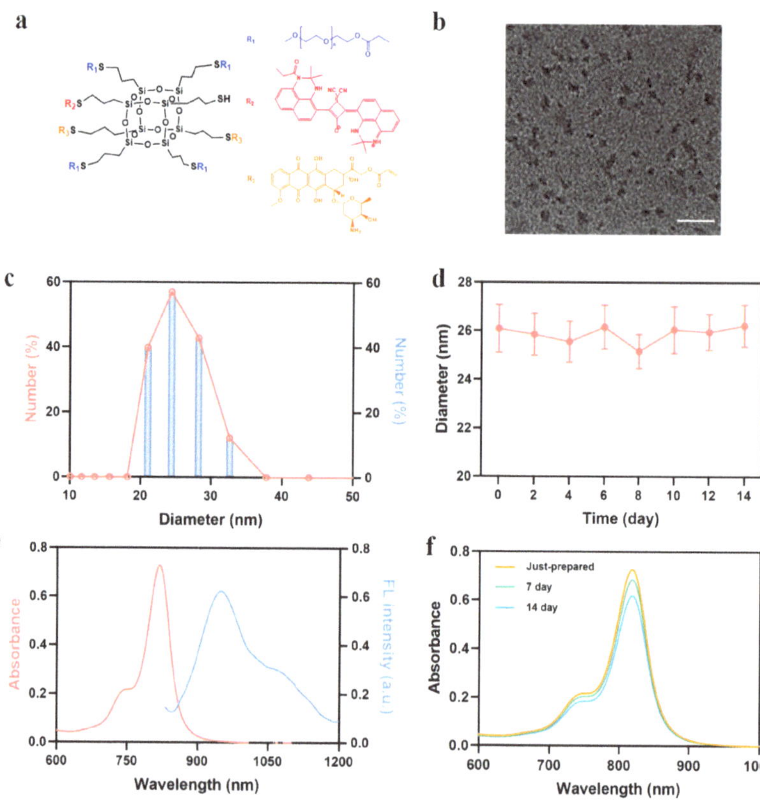

Figure 1. (**a**) The structure of POSS-SQ-DOX NPs; (**b**) TEM image of POSS-SQ-DOX NPs; (**c**) DLS data of POSS-SQ-DOX NPs; (**d**) The results of POSS-SQ-DOX NP size monitoring within two weeks; (**e**) UV-vis absorption and fluorescence spectra of POSS-SQ-DOX NPs (200 μg/mL) in an aqueous solution; (**f**) Results of UV-vis absorption spectra of POSS-SQ-DOX NPs within two weeks. The scale bar is 100 μm.

2.2. Photothermal Performance of POSS-SQ-DOX NPs

In order to evaluate the photothermal performance of POSS-SQ-DOX NPs, an 808 nm laser was employed for irradiation, and the resulting temperature changes were recorded. As shown in Figure 2a, it was observed that the temperature increments of POSS-SQ-DOX NPs exhibited a positive correlation with their concentrations under the irradiation of an 808 nm laser (0.5 W/cm^2). Specifically, when the concentration of POSS-SQ-DOX NPs reached 200 μg/mL, the temperature increased from 29.3 °C to 81.1 °C within 210 s (ΔT = 51.8 °C). In addition, at a concentration of 50 μg/mL, the temperature elevation demonstrated an upward trend as the laser power densities increased (Figure 2b). These results collectively indicate that POSS-SQ-DOX NPs possess efficient light-to-thermal energy conversion capabilities and enable precise control over heat generation. Additionally, a high PCE value (η) of up to 62.3% was calculated for POSS-SQ-DOX NPs [58,59] (Figure 2c,d). Moreover, thermal images provided a more intuitive visualization of the temperature variations associated with POSS-SQ-DOX NPs (Figure 2e). In addition, significant chemosensitization effects occur when the tumor site temperature reaches 39.0–43.0 °C because, when the temperature is increased from 37.0 to 43.0 °C, the permeability of the cell membrane is significantly enhanced, which promotes the uptake of nano-drugs by cells [5]. Thus, a laser power of 0.5 W/cm^2 and a concentration of 50 μg/mL were selected for the following experiments. Furthermore, photothermal conversion stability referred to their ability to withstand multiple excitation cycles without irreversible loss, which was an important parameter for characterizing organic dye molecules [60]. The stability of photothermal conversion was evaluated through heating–cooling cycle experiments. As shown in Figure 2f, the POSS-SQ-DOX NPs solution was irradiated with an 808 nm laser for 210 s and subsequently allowed to cool naturally to room temperature while recording the temperatures. It was evident that the temperature of POSS-SQ-DOX NPs exhibited negligible changes during five heating–cooling cycles, suggesting excellent stability and reproducibility of POSS-SQ-DOX NPs. Hence, all these experiments indicate that POSS-SQ-DOX NPs could serve as a highly efficient PTA in PTT due to their high PCE and good photochemical stability.

2.3. Evaluation Synergies Effect In Vitro

The cytotoxicity of different formulations (free DOX, free SQ-N, and different ratios of DOX and SQ-N) in HeLa cells was evaluated using Cell Counting Kit-8 (CCK-8) assays. Both single DOX and SQ-N +L exhibited cytotoxic effects on HeLa cells. However, the efficacy was suboptimal with an IC_{50} value of 19.94 μg/mL for DOX alone and 17.45 μg/mL for SQ-N +L (Figure 3a, Table 1). To determine the optimal ratio of DOX and SQ-N in nanoparticles, we assessed the impact of varying DOX contents on cell viability at a constant total concentration (Figure 3b). Furthermore, combination index (CI) values were calculated to confirm the synergistic effect between different ratios and CI < 1 indicated synergism. Our results demonstrate clear synergistic effects within the range of DOX:SQ-N +L = 1:3–3:1, with the strongest synergistic effect observed at a dose ratio of DOX:SQ-N +L was 2:1 (*n*/*n*), resulting in an IC_{50} value as low as 9.32 μg/mL for DOX and 4.33 μg/mL for SQ-N when prepared as POSS-SQ-DOX NPs. In summary, by utilizing a ration of DOX:SQ-N +L = 2:1 during preparation, not only could we ensure a synergistic therapeutic effect but also reduce chemotherapy drugs dosage while alleviating potential side effects.

Table 1. IC_{50} and CI values of DOX and SQ-N +L at different dose ratios to HeLa cells under 808 nm NIR laser irradiation.

Treatment	IC_{50} (μg/mL)		CI
	DOX	SQ-N +L	
Free DOX	19.94		
Free SQ-N +L		17.45	
DOX:SQ-N +L (1:3)	4.15	12.01	0.99
DOX:SQ-N +L (1:2)	5.43	10.68	0.88
DOX:SQ-N +L (1:1)	7.62	8.07	0.84
DOX:SQ-N +L (2:1)	9.32	4.33	0.72
DOX:SQ-N +L (3:1)	11.70	3.75	0.80

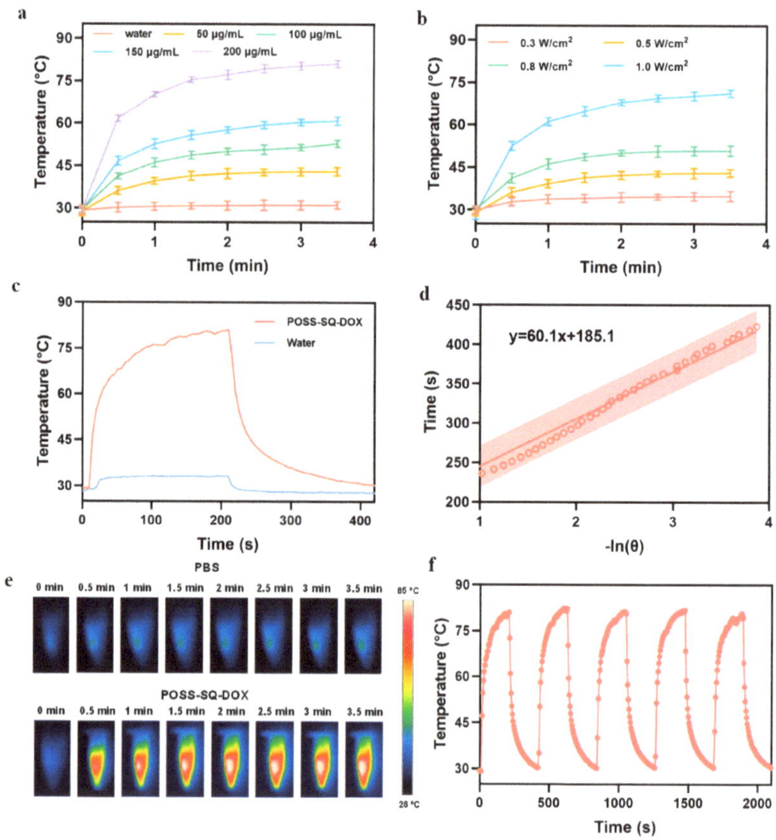

Figure 2. (**a**) Photothermal conversion of POSS-SQ-DOX NPs at different concentrations under 808 nm laser irradiation (0.5 W/cm^2); (**b**) Photothermal conversion of POSS-SQ-DOX NPs at different levels of exposure intensity (50 μg/mL); (**c**) Temperature rise curves of POSS-SQ-DOX NPs and water at 808 nm laser irradiation on/off; (**d**) Linear time data of POSS-SQ-DOX NPs versus $-\ln(\theta)$ obtained from the cooling section; (**e**) IR thermal images of POSS-SQ-DOX NP (200 μg/mL) and PBS under 808 nm laser irradiation (0.5 W/cm^2) at different exposure times; (**f**) Photothermal stability study of POSS-SQ-DOX NPs during five heating–cooling cycles.

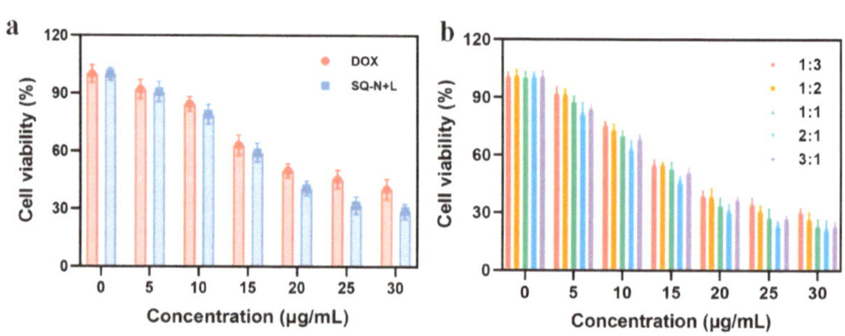

Figure 3. Cell viability of HeLa cells incubated with (**a**) free DOX and SQ-N +L at different concentrations and (**b**) different ratios of free DOX and SQ-N +L.

2.4. In Vitro Tumor Therapy

It is essential to assess the biocompatibility of POSS-SQ-DOX NPs prior to conducting further in vitro and in vivo anti-tumor therapy experiments. Moreover, all experiments were performed under 808 nm laser irradiation with a power density of 0.5 W/cm^2. The toxicity of POSS-SQ-DOX NPs was evaluated using the CCK-8 method. Figure 4a shows that even at a concentration of 200 μg/mL, the cell viability remained above 90% for POSS-SQ NPs, indicating their safety for both normal L929 cells and tumor HeLa cells. However, when incubated with POSS-SQ-DOX NPs or a free DOX treatment for 24 h, the viability of HeLa cells was only slightly above 40%, suggesting that the toxicity of POSS-SQ-DOX NPs primarily stemmed from the presence of the DOX drug (Figure 4b). Furthermore, as depicted in Figure 4c,d, it could be visually confirmed that chemo-mild photothermal synergistic therapy exhibited better efficacy compared to single chemotherapy or PTT alone. Additionally, the IC_{50} value for DOX was determined to be 10.94 μg/mL for POSS-SQ-DOX NPs, which is lower than the 19.94 μg/mL of free DOX. These results collectively indicate that chemo-mild photothermal synergistic therapy exerts a stronger inhibitory effect on HeLa cell proliferation with reduced amounts of chemotherapy drugs compared to single chemotherapy or PTT. This further supports the notion that chemo-mild photothermal synergistic therapy could achieve enhanced anti-tumor therapeutic effects.

Additionally, live-dead cell-staining experiments were carried out to further confirm the synergistic anti-tumor effect of chemo-mild photothermal synergistic therapy (Figure 4e). In these experiments, living cells were stained with Calcein AM (green), and dead cells were stained with PI (red). Compared to the DOX group, the POSS-SQ-DOX NPs group exhibited enhanced cytotoxicity against HeLa cells due to the superior cell permeability of the nanoparticles. Furthermore, the PI-positive rate was calculated for each group. As shown in Figure 4f, the PI-positive rate of the POSS-SQ-DOX +L group was 1.9-fold higher than that of the POSS-SQ +L group and 1.7-fold higher than that of the POSS-SQ-DOX group, confirming that POSS-SQ-DOX NPs achieved effective cellular ablation under 808 nm laser irradiation. These findings collectively demonstrate how POSS-SQ-DOX NPs could effectively exert a synergistic effect in chemo-mild photothermal therapy by more efficiently inhibiting tumor cell proliferation and inducing apoptosis.

2.5. Anti-Tumor Therapy In Vivo

To assess the feasibility of chemo-mild photothermal therapy in tumors, we established a HeLa tumor-bearing mouse model and utilized the following five formulations for our studies: PBS, DOX, POSS-SQ +L, POSS-SQ-DOX, and POSS-SQ-DOX +L. In subsequent experiments, the concentrations of DOX, POSS-SQ NPs, and POSS-SQ-DOX NPs were 12 μg/mL, 18 μg/mL, and 50 μg/mL, respectively. All experiments were carried out under irradiation from an 808 nm laser (0.5 W/cm^2). Notably, significant thermal signals were observed in tumor areas following the injection of POSS-SQ NPs and POSS-SQ-DOX NPs, while there was no clear single detection with PBS or DOX alone (Figure 5a). Real-time temperature changes at the tumor site were also recorded to further validate these thermal imaging results (Figure 5b). Tumor volume change data were documented, as shown in Figure 5c–g. While single chemotherapy (DOX group) and mild photothermal therapy (POSS-SQ +L group) exhibited certain inhibitory effects on tumor growth, they still resulted in an increased tumor volume with only moderate therapeutic efficacy. However, the inhibitory effect on tumors was slightly better in the POSS-SQ-DOX group compared to the DOX group due to the enhanced cell permeability provided by the use of POSS-SQ-DOX NPs. Moreover, the smallest tumor volume was observed in the group of POSS-SQ-DOX +L, indicating that this combination had a more pronounced efficacy in suppressing tumor growth. Furthermore, in sharp contrast with other experimental groups, both the tumor weight and the tumor inhibition ratio (TIR) were meticulously recorded and calculated (Figure 5h,i). The TIR value of the POSS-SQ-DOX +L group was found to be 72.7%, which exhibited a remarkable increase of 2.35 times compared to DOX alone, which is 1.46 times higher than POSS-SQ +L and 1.53 times higher than POSS-SQ-DOX. Additionally, the changes in body weight of tumor-bearing mice in each group were recorded, and it was found that the weight change trend of different groups of tumor-bearing mice was similar. Furthermore, on day 12, the body weight of the tumor-bearing mice was similar between the experimental and control groups, which demonstrated the excellent biosafety profile of POSS-SQ-DOX NPs (Figure 5j). Based on these comprehensive findings, it can be inferred that chemo-mild photothermal therapy had superior tumor-suppressive effects when compared to single chemotherapy or photothermal therapy alone without any associated toxicity.

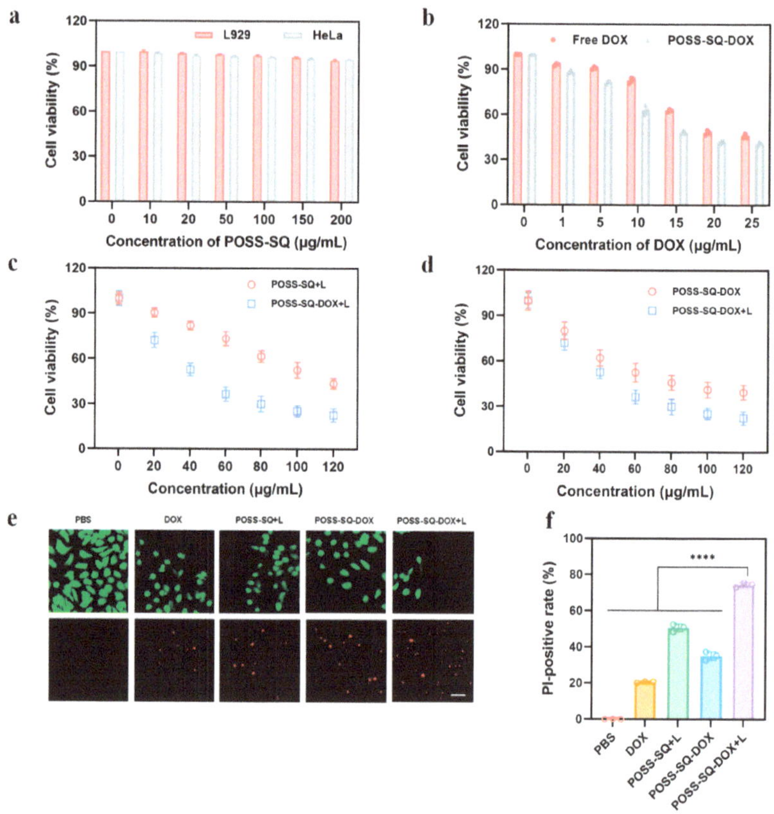

Figure 4. (**a**) Cell viability of L929 cells and HeLa cells incubated with POSS-SQ NPs at different concentrations; (**b**) Cell viability of HeLa cells incubated with free DOX and POSS-SQ-DOX NPs at different concentrations; cell viability of HeLa cells incubated (**c**) with or without DOX and (**d**) with or without 808 nm laser irradiation at different concentrations; (**e**) Confocal fluorescence imaging of HeLa cells co-stained with calcein AM (green color, staining live cells) and propidium iodide (red color, staining dead cells) after incubation with PBS, DOX (12 μg/mL), POSS-SQ +L (18 μg/mL), POSS-SQ-DOX and POSS-SQ-DOX +L (50 μg/mL). The scale bar is 100 μm; (**f**) Quantitative analysis of dead HeLa cells following different treatments. (Data are presented as the mean ± SD, $n = 3$, **** $p < 0.0001$).

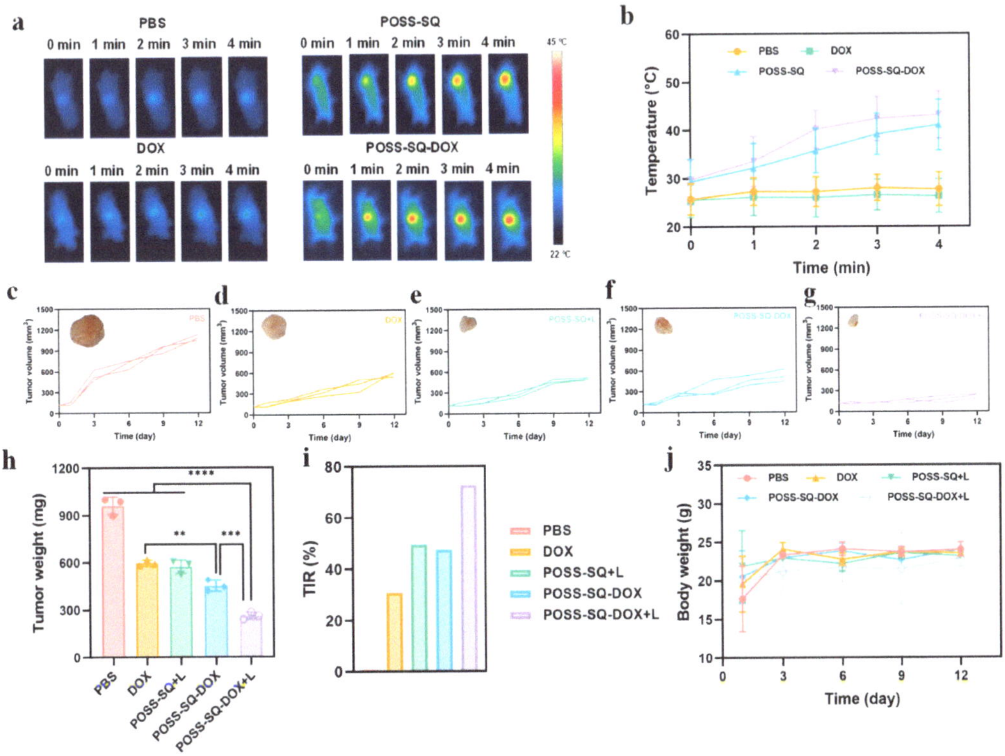

Figure 5. (**a**) IR thermal images of tumor-bearing mice upon 808 nm laser irradiation (0.5 W/cm^2) for different durations; (**b**) The changes in temperatures in tumor sites in every treatment group; (**c–g**) Tumor growth curves and the representative tumor pictures of mice in every group; (**h**) The tumor weight and (**i**) tumor inhibition rate of mice in different treatment groups; (j) Body weight curve of tumor-bearing mice in every group. (Data are presented as the mean ± SD, $n = 3$, ** $p < 0.01$, *** $p < 0.001$, and **** $p < 0.0001$).

The therapeutic effects of all groups were further evaluated through H&E, Ki-67, and TUNEL staining (Figure 6a–c). According to the results of H&E staining, tumor cells in the PBS were in good condition. However, the DOX, POSS-SQ +L, and POSS-SQ-DOX groups showed varying degrees of tumor cell lysis, indicating that a single drug or mild photothermal therapy had inhibitory effects on the tumor. In contrast, extensive damage and destruction with significant cell lysis were observed in the POSS-SQ-DOX +L group. Moreover, compared to other groups, treatment with POSS-SQ-DOX NPs with laser irradiation resulted in a significant reduction in Ki67-positive cells and an increase in TUNEL-positive cells within tumors. This may be because the cell membrane integrity was severely disrupted under the synergistic treatment of chemotherapy and MPTT, which led to cell membrane lysis and nuclear erosion and ultimately resulted in the apoptosis of tumor cells (Figure S7). These findings confirm that chemo-mild photothermal synergistic therapy is superior to chemotherapy or photothermal alone. Consistently with previous studies, these results highlight the excellent anti-tumor effects of POSS-SQ-DOX NPs by inhibiting cell proliferation and promoting cell apoptosis. Overall, these results demonstrate that this system has an excellent synergistic effect and achieves an excellent anti-tumor therapeutic outcome.

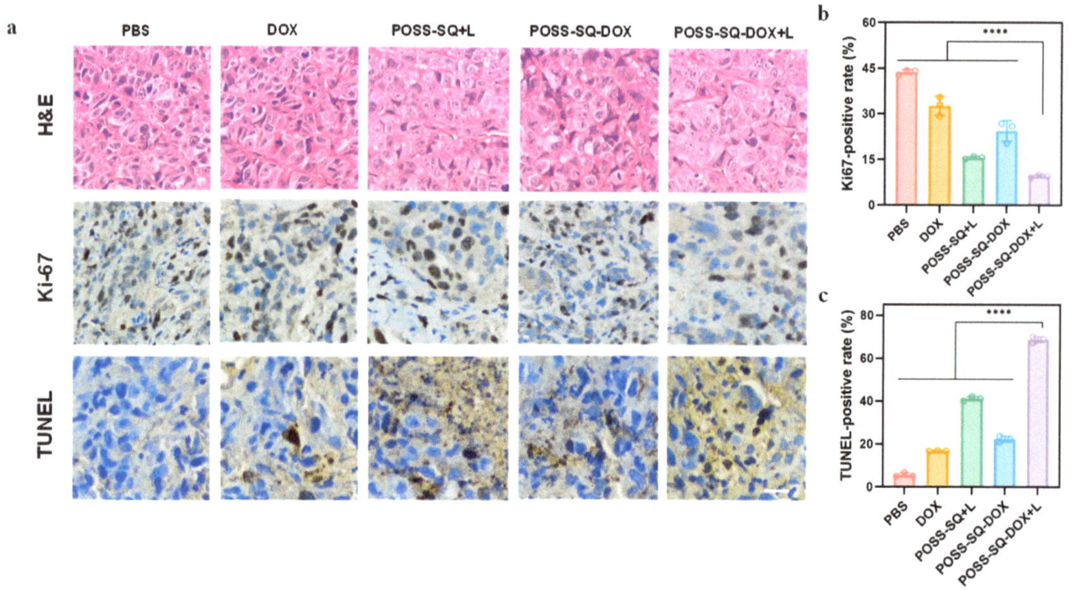

Figure 6. (**a**) H&E staining and Ki-67 and TUNEL histopathological analyses of tumor tissues from tumor-bearing mice in different treatment groups; (**b**) Ki-67-positive rates and (**c**) TUNEL-positive rates of different groups (Data are presented as the mean ± SD, $n = 3$, **** $p < 0.0001$). The scale bar is 100 µm.

2.6. In Vivo Safety Evaluation

To further evaluate the in vivo biosafety of POSS-SQ-DOX NPs, histological analyses and hemanalysis were conducted. HeLa tumor-bearing mice were injected with POSS-SQ-DOX NPs at the tumor site and subsequently exposed to 808 nm laser irradiation at a power density of 0.5 W/cm^{-2} for 5 min. After 12 days post-irradiation, the major organs, including the hearts, livers, spleens, lungs, and kidneys, were dissected from the mice and stained with H&E (Figure 7a). The results revealed no discernible organ damage in the POSS-SQ-DOX +L-treated mice compared to those treated with PBS alone, indicating its excellent biocompatibility and minor side effects for anti-tumor therapy. To further evaluate the potential toxicity of POSS-SQ-DOX NPs in vivo, two groups of mice were sacrificed for a blood routine assay after 12 days of treatment. As shown in Figure 7b–g, there was no significant difference in the parameters of blood routines between the PBS group and the POSS-SQ-DOX +L group, and all data were within the normal range. Normal white blood cell counts and lymphocyte counts indicated that there was no infection or inflammation, and the normal red blood cell counts and platelet counts indicated that there was no severe tissue damage. Altogether, the POSS-SQ-DOX NPs displayed excellent biocompatibility and biosafety and could be applied in chemo-mild PTT against cancers.

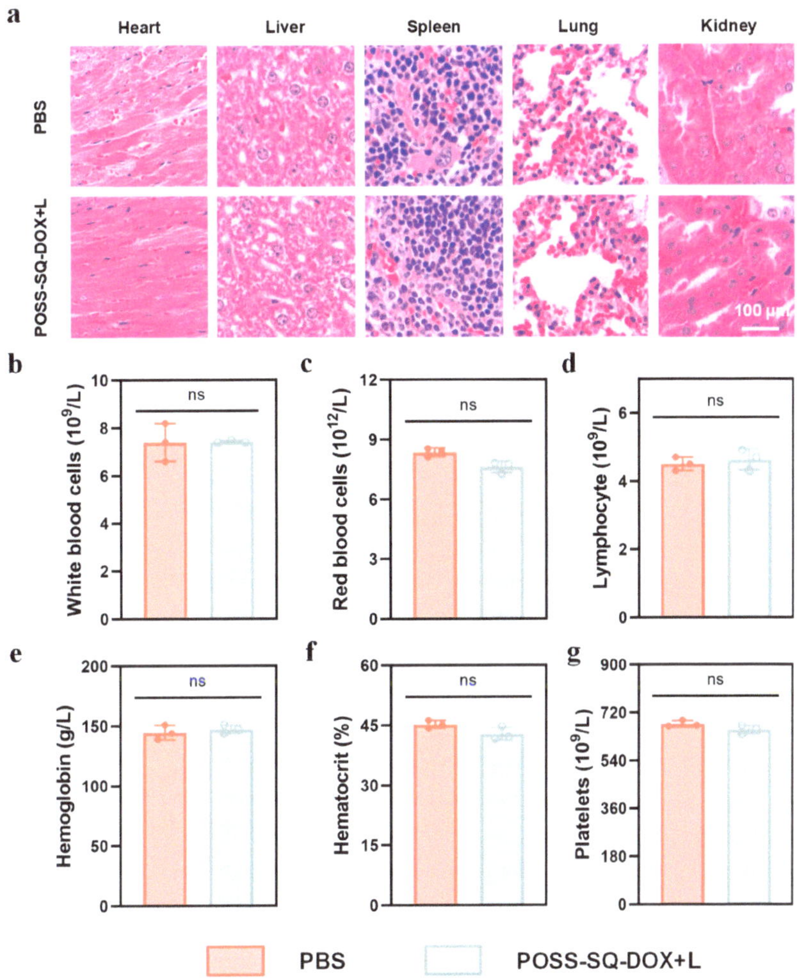

Figure 7. (**a**) H&E staining images of major organs (heart, liver, spleen, lung, and kidney) harvested from PBS-treated healthy mice and POSS-SQ-DOX +L-treated HeLa tumor-bearing mice; Routine blood assays of healthy mice and POSS-SQ-DOX +L-treated mice at the end of experiments (**b**) White blood cells; (**c**) Red blood cells; (**d**) Lymphocytes; (**e**) Hemoglobin; (**f**) Hematocrit; (**g**) Platelets. (Data are presented as the mean ± SD, $n = 3$ ^{ns} $p > 0.05$).

3. Materials and Methods

3.1. General Methods

3.1.1. Materials

1,8-naphthalenediamine, acetone, bismuth trichloride ($BiCl_3$), squaraine (SQ), 2,2-dimethoxy-2-phenylacetophenone (DMAP), potassium carbonate, n-butanol, doxorubicin, toluene, triethylamine, ethanol (C_2H_5OH), acryloyl chloride, methylene chloride (CH_2Cl_2), triethylamine, dimethyl sulfoxide (DMSO) and tetrahydrofuran (THF) were supplied by Sinopharm Chemical Reagent Co., Ltd. (Beijing, China). without further purification. Calcein AM and propidium iodide were purchased from Beyotime Biotechnology (Shanghai, China). L929 cell lines and cervical cancer cell lines (HeLa) were obtained from the Cell Bank of the Chinese Academy of Sciences. Male BALB/c mice and nude mice at the age of 6 weeks were purchased from Shanghai Silaike Experimental Animal Co., Ltd. (Shanghai, China).

3.1.2. Synthesis of DOX-AC

Doxorubicin (1.09 g, 2.0 mmol) was dissolved in 25 mL of dry dichloromethane in a three-necked round bottom flask with stirring. After dissolution, 3.0 mL of triethylamine was added to the solution, and the solution was placed in an ice–water bath with stirring. A solution of 0.35 mL (3.6 mmol) of acryloyl chloride in 12 mL of dichloromethane was placed dropwise into the flask for over 30 min. Then, the reaction lasted for 24 h at 25 °C and was monitored via the point plate. After the completion of the reaction, the solution was dried and concentrated using a rotary evaporator. The crude product was crystallized using column chromatography (silica gel, ethanol/chloroform = 1: 20), and after vacuum-drying for 8 h, the DOX-AC yielded a white solid. Yield: 69%. 1 H NMR (600 MHz, DMSO, 298 K, δ/ppm): δ10.2 (s, 1 H), 10.0 (s, 1 H), 9.3 (s, 1 H), 7.9 (m, 2 H), 7.7 (m, 2 H), 7.3 (d, J = 1.1 Hz 2 H), 7.2 (m, 3 H), 6.9 (m, 2 H), 6.8 (m, 2 H), 6.7 (d, J = 1.2 Hz, 2 H), 6.6 (m, 3 H), 6.5 (m, 3 H), 6.4 (dd, J = 8.6, 5.2 Hz, 2 H), 6.2 (dd, J = 5.2, 0.6 Hz, 2 H), 2.2 (s, 3 H). FTIR (KBr), v/cm^{-1}: 3363 cm^{-1} (-OH); 2958, 2881 cm^{-1} (-CH_2-), 1728 cm^{-1} (C=O), 1618 cm^{-1} (C=C), 1452 cm^{-1} (NH), 1121 cm^{-1} (C-O-C).

3.1.3. Synthesis of POSS-SQ NPs

In total, 828.1 mg (0.50 mmol) of POSS-PEG and 14.0 mg (0.036 mmol) of DMPA were fully dissolved in 20 mL of THF in a three-necked flask filled with nitrogen (N_2). Subsequently, 10 mL of anhydrous THF containing 288.0 mg (0.50 mmol) of SQ-AC was added to the flask. Then, the mixture was irradiated with UV light for 2 h. After the reaction, the solvent was removed via rotary evaporation, and the product was obtained as a viscous liquid. Yield: 100%. 1 H NMR (600 MHz, $CDCl_3$, 298 K, δ/ppm): δ 11.1 (s, 2 H), 9.2 (s, 2 H), 8.0–8.0 (ddd, J = 0.3, 3.6 Hz, 2 H), 7.56 (m, 3 H), 7.21 (m, 2 H), 7.1 (m, 2 H), 6.7 (d, J = 1.1 Hz, 2 H), 6.6 (m, 4 H), 6.5 (m, 6 H), 6.3 (m, 5 H), 6.0 (dd, J = 0.6, 5.2 Hz, 2 H), 3.4 (q, J = 3.5 Hz, 12 H), 1.2 (t, J = 3.5 Hz, 18 H). FTIR (KBr), v/cm^{-1}: 3411 cm^{-1} (-NH), 2923, 2846 cm^{-1} (-CH_2-), 2569 cm^{-1} (-SH), 2186 cm^{-1} (C≡N), 1742 cm^{-1} (C=O), 1613 cm^{-1} (C=C), 1440 cm^{-1} (Ar-NH), 1149 cm^{-1} (Si-O).

3.1.4. Synthesis of POSS-SQ-DOX NPs

For the synthesis of POSS-SQ-DOX NPs, up to 1.11 g (0.5 mmol) of POSS-SQ and 0.028 g (0.072 mmol) of DMPA were fully dissolved in 20 mL of THF in a three-necked flask filled with nitrogen (N_2). Subsequently, 10 mL of anhydrous THF containing 0.598 g (1.0 mmol) of DOX-AC was added into the flask. Then, the mixture was irradiated with UV light for 2 h. After the reaction, the solvent was removed via rotary evaporation, and the product was obtained as a viscous liquid. Yield: 100%.1H NMR (600 MHz, $CDCl_3$, 298 K, δ/ppm): δ 10.2 (s, 2 H), 10.1 (s, 3 H), 9.3 (s, 2 H), 8.2 (s, 6 H), 8.0 (m, 10 H), 7.7 (qt, J = 3.5, 6.7 Hz, 8 H), 7.5 (m, 10 H), 7.4 (td, J = 2.4, 3.7 Hz, 10 H), 7.2 (m, 10 H), 6.9 (m, 10 H), 6.7 (d, J = 1.2 Hz, 3 H), 6.6 (m, 12 H), 6.5 (m, 12 H), 4.3 (m, 18 H). FTIR (KBr), v/cm^{-1}: 3453 cm^{-1} (-OH), 2971, 2869 cm^{-1} (-CH_2-), 2538 cm^{-1} (-SH), 2367 cm^{-1} (C≡N), 1715 cm^{-1} (C=O), 1630 cm^{-1} (C=C), 1488 cm^{-1} (Ar-NH), 1216 cm^{-1} (C-O-C), 1132 cm^{-1} (Si-O). ESI-MS: m/z: 4670.57 [M+H]+, found 4671.57.

3.2. Characterization of POSS-SQ-DOX NPs

^{1}H NMR spectra were recorded with a Bruker AVANCE III-HD 600 MHz NMR spectrometer (Bruker, Billericca, MA, USA). Fourier transform infrared (FTIR) spectra were obtained using a Thermo Nicolet 8700 spectrometer (Thermo Scientific, Waltham, MA, USA). The morphologies of POSS-SQ-DOX NPs were observed using transmission electron microscopy (TEM, Thermo Scientific, Waltham, MA, USA) with an accelerating voltage of 120 kV. The size distribution of POSS-SQ-DOX NPs was assessed via dynamic light scattering (DLS, Brookhaven Instruments Co., New York, NY, USA). Ultraviolet–visible

(UV–vis) absorption spectra were measured with a Lambda 950 UV–vis spectrometer (PerkinElmer, Waltham, MA, USA).

3.3. Stability of POSS-SQ-DOX NPs

To evaluate the nanoparticles' stability, POSS-SQ-DOX NPs were placed at 37 °C in PBS, and the sizes were detected at different times using DLS.

3.4. Optical Performance of POSS-SQ-DOX NPs

The optical performance of POSS-SQ-DOX NPs was evaluated using UV-Vis and fluorescence spectra. The UV scanning spectra were in the range of 600–1000 nm of the POSS-SQ-DOX NP solution (200 μg/mL), and the excitation wavelength was 850 nm of fluorescence spectra. Furthermore, the optical stability of POSS-SQ-DOX NPs was verified by measuring the solution after the dilution process, with storage for 7 days and storage for 14 days at 37 °C.

3.5. Cytotoxicity and Combination Index Evaluation

CCK-8 assays were carried out to evaluate the in vitro cytotoxicity and killing effect of SQ-N, DOX, POSS-SQ NPs, and POSS-SQ-DOX NPs on the L929 cell line and HeLa cell line (accession number: SCSP-504). After seeding cells in a 96-well plate (1×10^4 cells/well) for 18 h in a humidified 5% CO_2 atmosphere at 37 °C, the culture medium was replaced by a serum-free medium containing a series different concentrations of SQ-N, DOX, POSS-SQ NPs and POSS-SQ-DOX NPs. After incubation for another 8 h, half of the parallel wells were exposed to 808 nm laser irradiation (0.5 W/cm^2) for 5 min, while the other half was kept in the dark. After a further 16 h incubation, 2-(2-methoxy-4-nitrophenyl)-3-(4-nitrophenyl)-5-(2, 4-benzene disulfonic acid) 2h tetrazolium monosodium salt (WST-8) was added into the wells and mixture were incubated for 4 h. After this, 100 μL of dimethyl sulfoxide (DMSO) was added into each well to dissolve the formazan crystal. Eventually, the absorbance was measured at a wavelength of 450 nm using a microplate reader. The cell viability (%) was calculated using the following Equation (1):

$$V\% = \left([A]_{experimental} - [A]_{black}\right) / ([A]_{control} - [A]_{black}) \times 100\% \tag{1}$$

The concept of the combination index (CI) was introduced to evaluate the synergistic, additive, and antagonistic effects of two different drugs at different dosages according to the Chou–Talalay method, and the formula for CI is shown in Equation (2):

$$CI_{50} = \frac{(D)_{DOX}}{(D_{50})_{DOX}} + \frac{(D)_{SQ\text{-}N+L}}{(D_{50})_{SQ\text{-}N+L}} \tag{2}$$

where $(D_{50})_{DOX}$ and $(D_{50})_{SQ\text{-}N+L}$ express the IC_{50} values of DOX alone and SQ-N+L alone, respectively. $(D)_{DOX}$ and $(D)_{SQ\text{-}N+L}$ express the concentrations of DOX and SQ-N+L in the combination system at the IC_{50} values. CI > 1 means antagonism, CI = 1 means additive, and CI < 1 means synergism.

3.6. Live/Dead Cell Double-Staining Assay

The in vitro anti-tumor efficacy of DOX, POSS-SQ NPs, and POSS-SQ-DOX NPs was evaluated using Calcein-AM/PI (propidium iodide) staining and a CKX 53 fluorescence microscope (Olympus, Tokyo, Japan). First, HeLa cells were incubated with PBS, DOX (12 μg/mL in DMEM, 1 mL), POSS-SQ NPs (18 μg/mL in DMEM, 1 mL), and POSS-SQ-DOX NPs (50 μg/mL in DMEM, 1 mL) in confocal dishes for 4 h. Then, the cells were kept in the dark or irradiated with an 808 nm laser (0.5 W/cm^{-2}) for 2 min. And then after 6 h of continuous incubation, the cells were co-stained with calcein AM/propidium iodide to mark viable cells and dead cells, respectively. Cells were imaged with the CKX53 fluorescence microscope via the green/red fluorescence channel.

3.7. In Vivo Synergistic Anti-Tumor Effects

Animal experiments were carried out following strict compliance with the provisions of the "Guidelines for the Use and Management of Experimental Animals" (IACUC). In order to evaluate the synergistic anti-tumor effect of chemotherapy and photothermal therapy, we selected mice with a tumor volume of about 120 mm^3 and randomly divided them into five groups ($n = 3$). The experimental groups were "PBS", "DOX", "POSS-SQ +L", "POSS-SQ-DOX", and "POSS-SQ-DOX +L". In total, 200 μL of the PBS, DOX (12 μg/mL), POSS-SQ NPs (18 μg/mL), and POSS-SQ-DOX NPs (50 μg/mL) solution were injected into mice via intratumoral injection and the groups of "POSS-SQ +L", and "POSS-SQ-DOX +L" were treated with 808 nm laser irradiation for 2 min immediately (tumor site, 0.5 W/cm^2), and then synchronously imaged every 10 s with a T540 infrared thermal camera (FLIR, Wilsonville, Oregon, USA). After that, all groups were treated with the same operation again every three days, and the tumor volume and body weight of the tumor-bearing mice in each group were measured and recorded for a total of 12 consecutive days. All animal studies were approved by the e Institutional Animal Care and Use Committee guidelines of Shanghai Jiao Tong University.

3.8. Histological Studies

All the mice in the three groups were euthanized on the 12th day after therapy. The tumors and main normal organs (heart, liver, spleen, lungs, and kidney) were resected, sliced, and stained. Cell states in the tumor sites and other main normal organs were analyzed using hematoxylin–eosin (H&E) staining; the tissues were immersed in 4% formalin solution, embedded in paraffin after dehydration, and sliced at a 4 μm thickness. For cell proliferation detection, serial tissue sections with a thickness of 4 μm were cut and then stained with an antibody against Ki67 (Servicebio, Wuhan, China) for 1 h. For cell apoptosis detection, the tumor sections were stained with terminal transferase-mediated dUTP nick end labeling (TUNEL) per the manufacturer's instructions (Promega, Madison, Wisconsin, USA). All slides were digitized with a high-resolution whole-slide scanner (Pannoramic DESK, Kongerizoge Street, Budapest, Hungary).

3.9. Blood Routine Assay

Healthy BALB/c nude mice were randomly assigned to two groups, labeled as "PBS" and "POSS-SQ-DOX +L" ($n = 3$ per group). After 12 days, the blood was sampled and detected for the blood routine examination.

3.10. Ethical Statement

The animal study protocol was approved by the Ethics Committee for Animal Experiments Affairs in Shanghai Renji Hospital, China (ra-2019-091).

3.11. Statistical Analysis

All statistical analyses were performed using GraphPad Prism 9.0 (GraphPad software, San Diego, California, USA). Statistical significance was determined using one-way analysis of variance (ANOVA) for multiple groups. All data are presented as the mean ± SD. $p < 0.05$ was considered statistically significant.

4. Conclusions

In summary, POSS-based multifunctional nanoparticles (POSS-SQ-DOX NPs) were rationally designed and successfully applied for chemo-mild photothermal synergistic anti-tumor therapy. POSS-SQ-DOX NPs were strong in stability, good in biocompatibility, and had a high photothermal conversion efficiency, which allowed POSS-SQ-DOX NPs to be used as a PTA and drug transport platform in chemo-mild photothermal therapy. Meanwhile, the calculation of the combination index (0.72, <1.0) indicated the existence of a synergistic anti-tumor effect. Furthermore, due to the better cell permeability and the presence of synergistic therapy, POSS-SQ-DOX NPs could enter into tumor cells more easily, which was conducive to the nanoparticles for more effective mild photothermal therapy. Additionally, chemotherapy also solved the problem of the poor therapeutic effect caused by low temperatures during mild photothermal therapy. Therefore, both in vitro and in vivo experiments showed the effect of chemo-mild photothermal synergistic therapy better than single chemotherapy and photothermal therapy. In conclusion, the developed POSS-SQ-DOX NPs, as a comprehensive nanotherapeutic agent, have great potential to induce chemo-mild photothermal synergistic therapy, improve anti-tumor efficacy and reduce side effects, which is expected to be further applied in clinical tumor treatments.

Supplementary Materials: The supporting information can be downloaded at: https://www.mdpi.com/article/10.3390/ijms25021012/s1.

Author Contributions: Conceptualization, Z.G.; Data curation, H.X.; Formal analysis, X.G. and H.X.; Funding acquisition, X.G., S.G. and H.X.; Investigation, Z.G.; Methodology, H.X.; Project administration, H.X.; Resources, H.X.; Supervision, S.G.; Validation, Z.G. and X.G.; Writing—original draft, Z.G.; Writing—review and editing, H.X. All authors have read and agreed to the published version of the manuscript.

Funding: This research was founded by the National Natural Science Foundation of China (Grant Nos. 52372282 to Shanyi Guang) and the Clinical Research Plan of Shanghai Shenkang Development Center (Nos. SHDC2020CR2008A and SHDC 12021103 to Xiaochuan Geng).

Institutional Review Board Statement: The study was conducted in accordance with the Declaration of Helsinki, and approved by the Ethics Committee for Animal Experiments Affairs in Shanghai Renji Hospital, China (ra-2019-091).

Informed Consent Statement: Not applicable.

Data Availability Statement: Data is contained within the article and Supplementary Materials.

Conflicts of Interest: The authors declare no conflict of interest.

References

1. Yin, W.; Yan, L.; Yu, J.; Tian, G.; Zhou, L.; Yong, Y.; Li, J.; Gu, Z.; Zhao, Y. High-Throughput Synthesis of Single-Layer MoS_2 Nanosheets as a Near-Infrared Photothermal-Triggered Drug Delivery for Effective Cancer Therapy. *ACS Nano* **2014**, *8*, 6922–6933. [CrossRef] [PubMed]
2. Yang, G.; Liu, J.; Wu, Y.; Feng, L.; Liu, Z. Near-infrared-light responsive nanoscale drug delivery systems for cancer treatment. *Coord. Chem. Rev.* **2016**, *320*, 100–117. [CrossRef]
3. Zhao, R.; Han, X.; Li, Y.; Wang, H.; Ji, T.; Zhao, Y.; Nie, G. Photothermal Effect Enhanced Cascade-Targeting Strategy for Improved Pancreatic Cancer Therapy by Gold Nanoshell@Mesoporous Silica Nanorod. *ACS Nano* **2017**, *11*, 8103–8113. [CrossRef] [PubMed]
4. Gao, P.; Wei, R.; Cui, B.; Liu, X.; Chen, Y.; Pan, W.; Li, N.; Tang, B. Ultrathin functionalized covalent organic framework nanosheets for tumor-targeted photodynamic therapy. *Chem. Commun.* **2021**, *57*, 6082–6085. [CrossRef]
5. Yang, T.; Wang, H.; Zhou, Q.; Zhang, J.; Yu, Y.; Sun, T. Mild Chemo-Photothermal Synergistic Therapy for Tumors Based on Gold-Nanoparticles Coupled with Metformin. *ACS Appl. Nano Mater.* **2023**, *6*, 5729–5736. [CrossRef]
6. Siddique, S.; Chow, J.C.L. Application of Nanomaterials in Biomedical Imaging and Cancer Therapy. *Nanomaterials* **2020**, *10*, 1700. [CrossRef]
7. Bian, W.Q.; Wang, Y.K.; Pan, Z.X.; Chen, N.P.; Li, X.J.; Wong, W.-L.; Liu, X.J.; He, Y.; Zhang, K.; Lu, Y.-J. Review of Functionalized Nanomaterials for Photothermal Therapy of Cancers. *ACS Appl. Nano Mater.* **2021**, *4*, 11353–11385. [CrossRef]
8. Ma, H.; Xue, M.Q. Recent Advances in the Photothermal Applications of Two-Dimensional Nanomaterials: Photothermal Therapy and Beyond. *J. Mater. Chem. A* **2021**, *9*, 17569–17591. [CrossRef]
9. Siddique, S.; Chow, J.C.L. Recent Advances in Functionalized Nanoparticles in Cancer Theranostics. *Nanomaterials* **2022**, *12*, 2826. [CrossRef]
10. Nasseri, B.; Alizadeh, E.; Bani, F.; Davaran, S.; Akbarzadeh, A.; Rabiee, N.; Bahadori, A.; Ziaei, M.; Bagherzadeh, M.; Saeb, M.R.; et al. Nanomaterials for Photothermal and Photodynamic Cancer Therapy. *Appl. Phys. Rev.* **2022**, *9*, 11317. [CrossRef]
11. Wang, Z.T.; Wang, M.L.; Wang, X.X.; Hao, Z.K.; Han, S.B.; Wang, T.; Zhang, H.Y. Photothermal-Based Nanomaterials and Photothermal-Sensing: An Overview. *Biosens. Bioelectron.* **2023**, *220*, 114883. [CrossRef] [PubMed]
12. Duan, S.F.; Hu, Y.L.; Zhao, Y.; Tang, K.Y.; Zhang, Z.J.; Liu, Z.L.; Wang, Y.; Guo, H.Y.; Miao, Y.C.; Du, H.D.; et al. Nanomaterials for Photothermal Cancer Therapy. *RSC Adv.* **2023**, *13*, 14443–14460. [CrossRef] [PubMed]
13. Jaque, D.; Martínez Maestro, L.; del Rosal, B.; Haro-Gonzalez, P.; Benayas, A.; Plaza, J.L.; Martín Rodríguez, E.; García Solé, J. Nanoparticles for photothermal therapies. *Nanoscale* **2014**, *6*, 9494–9530. [CrossRef] [PubMed]

14. Li, T.; Li, C.; Ruan, Z.; Xu, P.; Yang, X.; Yuan, P.; Wang, Q.; Yan, L. Polypeptide-Conjugated Second Near-Infrared Organic Fluorophore for Image-Guided Photothermal Therapy. *ACS Nano* **2019**, *13*, 3691–3702. [CrossRef] [PubMed]
15. Huang, L.; Li, Y.; Du, Y.; Zhang, Y.; Wang, X.; Ding, Y.; Yang, X.; Meng, F.; Tu, J.; Luo, L.; et al. Mild photothermal therapy potentiates anti-PD-L1 treatment for immunologically cold tumors via an all-in-one and all-in-control strategy. *Nat. Commun.* **2019**, *10*, 4871. [CrossRef] [PubMed]
16. Shao, L.; Li, Y.; Huang, F.; Wang, X.; Lu, J.; Jia, F.; Pan, Z.; Cui, X.; Ge, G.; Deng, X.; et al. Complementary autophagy inhibition and glucose metabolism with rattle-structured polydopamine@mesoporous silica nanoparticles for augmented low-temperature photothermal therapy and in vivo photoacoustic imaging. *Theranostics* **2020**, *10*, 7273–7286. [CrossRef] [PubMed]
17. Yang, Y.; Zhu, W.; Dong, Z.; Chao, Y.; Xu, L.; Chen, M.; Liu, Z. 1D Coordination Polymer Nanofibers for Low-Temperature Photothermal Therapy. *Adv. Mater.* **2017**, *29*, 1703588–1703599. [CrossRef]
18. Dong, Q.; Wang, X.; Hu, X.; Xiao, L.; Zhang, L.; Song, L.; Xu, M.; Zou, Y.; Chen, L.; Chen, Z.; et al. Simultaneous Application of Photothermal Therapy and an Anti-inflammatory Prodrug using Pyrene-Aspirin-Loaded Gold Nanorod Graphitic Nanocapsules. *Angew. Chem. Int. Edit.* **2018**, *57*, 177–181. [CrossRef]
19. Pan, W.; Zhang, X.; Gao, P.; Li, N.; Tang, B. An anti-inflammatory nanoagent for tumor-targeted photothermal therapy. *Chem. Commun.* **2019**, *55*, 9645–9648. [CrossRef]
20. Li, B.; Hao, G.; Sun, B.; Gu, Z.; Xu, Z.P. Engineering a Therapy-Induced "Immunogenic Cancer Cell Death" Amplifier to Boost Systemic Tumor Elimination. *Adv. Funct. Mater.* **2020**, *30*, 1909745–1909756. [CrossRef]
21. Yi, X.; Duan, Q.-Y.; Wu, F.-G. Low-Temperature Photothermal Therapy: Strategies and Applications. *Research* **2021**, *2021*, 9816594. [CrossRef] [PubMed]
22. Gao, G.; Sun, X.; Liang, G. Nanoagent-Promoted Mild-Temperature Photothermal Therapy for Cancer Treatment. *Adv. Funct. Mater.* **2021**, *31*, 2100738–2100751. [CrossRef]
23. Deng, X.; Guan, W.; Qing, X.; Yang, W.; Que, Y.; Tan, L.; Liang, H.; Zhang, Z.; Wang, B.; Liu, X.; et al. Ultrafast Low-Temperature Photothermal Therapy Activates Autophagy and Recovers Immunity for Efficient Antitumor Treatment. *ACS Appl. Mater. Interfaces* **2020**, *12*, 4265–4275. [CrossRef] [PubMed]
24. Liu, D.; Ma, L.; An, Y.; Li, Y.; Liu, Y.; Wang, L.; Guo, J.; Wang, J.; Zhou, J. Thermoresponsive Nanogel-Encapsulated PEDOT and HSP70 Inhibitor for Improving the Depth of the Photothermal Therapeutic Effect. *Adv. Funct. Mater.* **2016**, *26*, 4749–4759. [CrossRef]
25. Zhou, Z.; Yan, Y.; Hu, K.; Zou, Y.; Li, Y.; Ma, R.; Zhang, Q.; Cheng, Y. Autophagy inhibition enabled efficient photothermal therapy at a mild temperature. *Biomaterials* **2017**, *141*, 116–124. [CrossRef] [PubMed]
26. Chen, T.; Cen, D.; Ren, Z.; Wang, Y.; Cai, X.; Huang, J.; Di Silvio, L.; Li, X.; Han, G. Bismuth embedded silica nanoparticles loaded with autophagy suppressant to promote photothermal therapy. *Biomaterials* **2019**, *221*, 119419. [CrossRef] [PubMed]
27. Zhou, Z.; Yan, Y.; Wang, L.; Zhang, Q.; Cheng, Y. Melanin-like nanoparticles decorated with an autophagy-inducing peptide for efficient targeted photothermal therapy. *Biomaterials* **2019**, *203*, 63–72. [CrossRef]
28. Ma, Z.; Han, K.; Dai, X.; Han, H. Precisely Striking Tumors without Adjacent Normal Tissue Damage via Mitochondria-Templated Accumulation. *ACS Nano* **2018**, *12*, 6252–6262. [CrossRef]
29. Wang, H.; Chang, J.; Shi, M.; Pan, W.; Li, N.; Tang, B. A Dual-Targeted Organic Photothermal Agent for Enhanced Photothermal Therapy. *Angew. Chem. Inter. Edit.* **2019**, *58*, 1057–1061. [CrossRef]
30. Gao, P.; Pan, W.; Li, N.; Tang, B. Boosting Cancer Therapy with Organelle-Targeted Nanomaterials. *ACS Appl. Mater. Interfaces* **2019**, *11*, 26529–26558. [CrossRef]
31. Yin, H.; Guan, X.; Lin, H.; Pu, Y.; Fang, Y.; Yue, W.; Zhou, B.; Wang, Q.; Chen, Y.; Xu, H. Nanomedicine-Enabled Photonic Thermogaseous Cancer Therapy. *Adv. Sci.* **2019**, *7*, 1901954–1901965. [CrossRef]
32. Lu, Q.; Lu, T.; Xu, M.; Yang, L.; Song, Y.; Li, N. SO_2 prodrug doped nanorattles with extra-high drug payload for "collusion inside and outside" photothermal/pH triggered—Gas therapy. *Biomaterials* **2020**, *257*, 120236–120249. [CrossRef]
33. You, C.; Li, Y.; Dong, Y.; Ning, L.; Zhang, Y.; Yao, L.; Wang, F. Low-Temperature Trigger Nitric Oxide Nanogenerators for Enhanced Mild Photothermal Therapy. *ACS Biomater. Sci. Eng.* **2020**, *6*, 1535–1542. [CrossRef] [PubMed]
34. Sun, S.; Chen, J.; Jiang, K.; Tang, Z.; Wang, Y.; Li, Z.; Liu, C.; Wu, A.; Lin, H. Ce6-Modified Carbon Dots for Multimodal-Imaging-Guided and Single-NIR-Laser-Triggered Photothermal/Photodynamic Synergistic Cancer Therapy by Reduced Irradiation Power. *ACS Appl. Mater. Interfaces* **2019**, *11*, 5791–5803. [CrossRef] [PubMed]
35. She, D.; Peng, S.; Liu, L.; Huang, H.; Zheng, Y.; Lu, Y.; Geng, D.; Yin, B. Biomimic FeS_2 nanodrug with hypothermal photothermal effect by clinical approved NIR-II light for augmented chemodynamic therapy. *Chem. Eng. J.* **2020**, *400*, 125933–125943. [CrossRef]
36. Gao, G.; Jiang, Y.W.; Guo, Y.; Jia, H.R.; Cheng, X.; Deng, Y.; Yu, X.W.; Zhu, Y.X.; Guo, H.Y.; Sun, W.; et al. Enzyme-Mediated Tumor Starvation and Phototherapy Enhance Mild-Temperature Photothermal Therapy. *Adv. Funct. Mater.* **2020**, *30*, 1909391–1909403. [CrossRef]
37. Zhang, J.; Zhao, B.; Chen, S.; Wang, Y.; Zhang, Y.; Wang, Y.; Wei, D.; Zhang, L.; Rong, G.; Weng, Y.; et al. Near-Infrared Light Irradiation Induced Mild Hyperthermia Enhances Glutathione Depletion and DNA Interstrand Cross-Link Formation for Efficient Chemotherapy. *ACS Nano* **2020**, *14*, 14831–14845. [CrossRef]
38. Yao, Y.; Zhang, Y.; Zhang, J.; Yang, X.; Ding, D.; Shi, Y.; Xu, H.; Gao, X. Azulene-Containing Squaraines for Photoacoustic Imaging and Photothermal Therapy. *ACS Appl. Mater. Interfaces* **2022**, *14*, 19192–19203. [CrossRef]

39. Ilina, K.; MacCuaig, W.M.; Laramie, M.; Jeouty, J.N.; McNally, L.R.; Henary, M. Squaraine Dyes: Molecular Design for Different Applications and Remaining Challenges. *Bioconjug. Chem.* **2020**, *31*, 194–213. [CrossRef]
40. He, J.; Jo, Y.J.; Sun, X.; Qiao, W.; Ok, J.; Kim, T.i.; Li, Z.A. Squaraine Dyes for Photovoltaic and Biomedical Applications. *Adv. Funct. Mater.* **2020**, *31*, 2008201–2008233. [CrossRef]
41. Grande, V.; Doria, F.; Freccero, M.; Wurthner, F. An Aggregating Amphiphilic Squaraine: A Light-up Probe That Discriminates Parallel G-Quadruplexes. *Angew. Chem. Int. Ed. Engl.* **2017**, *56*, 7520–7524. [CrossRef] [PubMed]
42. Sreejith, S.; Ma, X.; Zhao, Y. Graphene Oxide Wrapping on Squaraine-Loaded Mesoporous Silica Nanoparticles for Bioimaging. *J. Am. Chem. Soc.* **2012**, *134*, 17346–17349. [CrossRef]
43. Zhang, D.; Zhao, Y.X.; Qiao, Z.Y.; Mayerhoffer, U.; Spenst, P.; Li, X.J.; Wurthner, F.; Wang, H. Nano-confined squaraine dye assemblies: New photoacoustic and near-infrared fluorescence dual-modular imaging probes in vivo. *Bioconjug. Chem.* **2014**, *25*, 2021–2029. [CrossRef] [PubMed]
44. Saneesh Babu, P.S.; Manu, P.M.; Dhanya, T.J.; Tapas, P.; Meera, R.N.; Surendran, A.; Aneesh, K.A.; Vadakkancheril, S.J.; Ramaiah, D.; Nair, S.A.; et al. Bis(3,5-diiodo-2,4,6-trihydroxyphenyl)squaraine photodynamic therapy disrupts redox homeostasis and induce mitochondria-mediated apoptosis in human breast cancer cells. *Sci. Rep.* **2017**, *7*, 42126. [CrossRef] [PubMed]
45. Yao, D.; Wang, Y.; Zou, R.; Bian, K.; Liu, P.; Shen, S.; Yang, W.; Zhang, B.; Wang, D. Molecular Engineered Squaraine Nanoprobe for NIR-II/Photoacoustic Imaging and Photothermal Therapy of Metastatic Breast Cancer. *ACS Appl. Mater. Interfaces* **2020**, *12*, 4276–4284. [CrossRef] [PubMed]
46. Dimitriev, O.P.; Dimitriyeva, A.P.; Tolmachev, A.I.; Kurdyukov, V.V. Solvent-induced organization of squaraine dyes in solution capillary layers and adsorbed films. *J. Phys. Chem. B* **2005**, *109*, 4561–4567. [CrossRef]
47. Yang, Y.Y.; Wang, X.; Hu, Y.; Hu, H.; Wu, D.C.; Xu, F.J. Bioreducible POSS-cored star-shaped polycation for efficient gene delivery. *ACS Appl. Mater. Interfaces* **2014**, *6*, 1044–1052. [CrossRef]
48. Pang, B.; Liu, Y.; Cao, R.; Zhang, W. Hybrid Nanoscale Vesicles of Polyhedral Oligomeric Silsesquioxane-Based Star Block Copolymers for Thermal Insulation Applications. *ACS Appl. Nano Mater.* **2022**, *5*, 7042–7050. [CrossRef]
49. Chi, H.; Wang, M.; Li, J.; Tian, H.; Chong, Y.T.; Lim, S.H.; Wang, Y.; Wang, F. Polyhedral Oligomeric Silsesquioxane as a Polarity Mediator and Reinforced Nanofiller for Fabricating Robust and Hierarchical Porous Film for Cell Bioengineering. *ACS Appl. Polym. Mater.* **2022**, *4*, 5882–5890. [CrossRef]
50. Chen, J.; Xu, Y.; Gao, Y.; Yang, D.; Wang, F.; Zhang, L.; Bao, B.; Wang, L. Nanoscale Organic-Inorganic Hybrid Photosensitizers for Highly Effective Photodynamic Cancer Therapy. *ACS Appl. Mater. Interfaces* **2018**, *10*, 248–255. [CrossRef]
51. Wei, G.; Zhang, K.; Gu, Y.; Guang, S.; Feng, J.; Xu, H. Novel multifunctional nano-hybrid polyhedral oligomeric silsesquioxane-based molecules with high cell permeability: Molecular design and application for diagnosis and treatment of tumors. *Nanoscale* **2021**, *13*, 2982–2994. [CrossRef] [PubMed]
52. Wei, G.; Gu, Y.; Lin, N.; Ning, X.; Lu, Y.; Zhao, G.; Guang, S.; Feng, J.; Xu, H. Autonomous Bionanorobots via a Cage-Shaped Silsesquioxane Vehicle for In Vivo Heavy Metal Detoxification. *ACS Appl. Mater. Interfaces* **2022**, *14*, 29238–29249. [CrossRef] [PubMed]
53. Zhou, Z.; Han, Z.; Lu, Z.R. A targeted nanoglobular contrast agent from host-guest self-assembly for MR cancer molecular imaging. *Biomaterials* **2016**, *85*, 168–179. [CrossRef] [PubMed]
54. Yang, Q.; Wu, L.; Li, L.; Zhou, Z.; Huang, Y. Subcellular co-delivery of two different site-oriented payloads for tumor therapy. *Nanoscale* **2017**, *9*, 1547–1558. [CrossRef] [PubMed]
55. Fan, L.; Wang, X.; Cao, Q.; Yang, Y.; Wu, D. POSS-based supramolecular amphiphilic zwitterionic complexes for drug delivery. *Biomater. Sci.* **2019**, *7*, 1984–1994. [CrossRef]
56. John, L. Selected developments and medical applications of organic-inorganic hybrid biomaterials based on functionalized spherosilicates. *Mater. Sci. Eng. C Mater. Biol. Appl.* **2018**, *88*, 172–181. [CrossRef] [PubMed]
57. Abudukelimu, S.; Wei, G.; Huang, J.; Zhao, G.; Wei, L.; Cui, W.; Lu, M.; Yao, W. Polyhedral oligomeric silsesquioxane (POSS)-based hybrid nanocomposite for synergistic chemo-photothermal therapy against pancreatic cancer. *Chem. Eng. J.* **2022**, *442*, 136124–136134. [CrossRef]
58. Lyu, Y.; Xie, C.; Chechetka, S.A.; Miyako, E.; Pu, K. Semiconducting Polymer Nanobioconjugates for Targeted Photothermal Activation of Neurons. *J. Am. Chem. Soc.* **2016**, *138*, 9049–9052. [CrossRef]
59. Lyu, Y.; Zeng, J.; Jiang, Y.; Zhen, X.; Wang, T.; Qiu, S.; Lou, X.; Gao, M.; Pu, K. Enhancing Both Biodegradability and Efficacy of Semiconducting Polymer Nanoparticles for Photoacoustic Imaging and Photothermal Therapy. *ACS Nano* **2018**, *12*, 1801–1810. [CrossRef]
60. Yun, K.; Guo, J.; Zhu, R.; Wang, T.; Zhang, X.; Pan, H.; Pan, W. Design of ROS-Responsive Hyaluronic Acid–Methotrexate Conjugates for Synergistic Chemo-Photothermal Therapy for Cancer. *Mol. Pharm.* **2022**, *19*, 3323–3335. [CrossRef]

International Journal of *Molecular Sciences*

MDPI

Article

Design, Synthesis, and Antitumor Activity Evaluation of Proteolysis-Targeting Chimeras as Degraders of Extracellular Signal-Regulated Kinases 1/2

Pengming Pan, Yichao He, Tongtong Geng, Zhongtang Li, Zhongjun Li and Xiangbao Meng *

State Key Laboratory of Natural and Biomimetic Drugs, School of Pharmaceutical Sciences, Peking University, Beijing 100191, China; 2011110122@pku.edu.cn (P.P.); 1910307221@pku.edu.cn (Y.H.); gtt940505@163.com (T.G.); lizhongtang@bjmu.edu.cn (Z.L.); zjli@bjmu.edu.cn (Z.L.)

* Correspondence: xbmeng@bjmu.edu.cn

Abstract: Inhibition of the extracellular signal-regulated kinases 1/2 (ERK1/2) alone or in combination with other targets has emerged as a promising treatment strategy for a variety of human tumors. In addition to the development of inhibitors, the development of ERK1/2 degraders is an alternative approach to decrease its activity. We synthesized proteolysis-targeting chimeras (PROTACs) as effective ERK1/2 degraders, among which **B1-10J** showed high degradative activity, with DC_{50} of 102 nM and cytotoxic IC_{50} of 2.2 μM against HCT116 cells. Moreover, **B1-10J** dose-dependently inhibited tumor cell migration. Xenograft experiments in nude mice demonstrated that **B1-10J** inhibited HCT116 tumor cell growth and achieved significant regression of tumors at a daily dose of 25 mg/kg.

Keywords: ERK1/2; PROTAC; antitumor; MAPK pathway

Citation: Pan, P.; He, Y.; Geng, T.; Li, Z.; Li, Z.; Meng, X. Design, Synthesis, and Antitumor Activity Evaluation of Proteolysis-Targeting Chimeras as Degraders of Extracellular Signal-Regulated Kinases 1/2. *Int. J. Mol. Sci.* **2023**, *24*, 16290. https://doi.org/10.3390/ijms242216290

Academic Editor: Tullio Florio

Received: 23 October 2023
Revised: 8 November 2023
Accepted: 9 November 2023
Published: 14 November 2023

1. Introduction

The mitogen-activated protein kinase (MAPK) pathway is one of the most essential signal transduction pathways in eukaryotic cells, capable of transducing extracellular signals and influencing a variety of fundamental cellular behaviors [1–3]. Extracellular signal-regulated kinase 1 (ERK1) and ERK2 are well-studied classic MAPKs. Human ERK1 is composed of 379 amino acid residues, while ERK2 is composed of 360 amino acid residues. ERK1 and ERK2 are highly homologous, sharing 84% of their amino acid sequence, and conduct mostly the same biological functions [4]. Through the rat sarcoma (RAS)–rapidly accelerated fibrosarcoma (RAF)–mitogen-activated protein kinase kinase (MEK)–ERK signaling pathway, ERK1/2 primarily regulates cell proliferation, survival, growth, metabolism, motility, differentiation, and development (Figure 1). Additionally, this pathway is associated with the formation of tumors. It is frequently abnormally activated in numerous human tumors, and mutations in RAS or RAF are common. Upon mutation, some initial cancer treatment strategies may fail [5–7]. As a downstream protein of RAS-RAF-MEK-ERK, ERK1/2 also plays a crucial role in the signaling network of tumor development. Compared to RAS, RAF, and MEK, ERK1/2 protein mutations occur less frequently. Increasing evidence suggests that ERK1/2 is an essential target for the development of antitumor drugs [8–11]. ERK1/2 can be targeted to treat cancer either alone or in conjunction with its upstream proteins.

Numerous ERK1/2 inhibitors are currently in development. BVD-523 (ulixertinib) is an example of orally effective, potent, and specific inhibitors of ERK1 and ERK2, with Ki values of <0.3 nM and 0.04 nM, respectively. Ulixertinib has demonstrated anticancer activity in BRAF mutant and RAS mutant cell lines in preclinical studies, as well as resistance and safety against multiple solid tumors in multiple clinical trials [12,13]. GDC-0994, an additional drug undergoing clinical trials, has an IC_{50} of 6.1 nM against ERK1 and 3.1 nM against ERK2. GDC-0994 demonstrates potent antitumor activity in HCT116 mouse xenografts and excellent PK/PD and tolerance properties [14]. Additionally, Ward et al. discovered compound 35 as a potent and selective inhibitor of ERK1/2, with an IC_{50} of <0.3 nM against ERK2. Compound 35 was investigated in a mouse antitumor study and demonstrated significant tumor regression at a daily dose of 50 mg/kg [15].

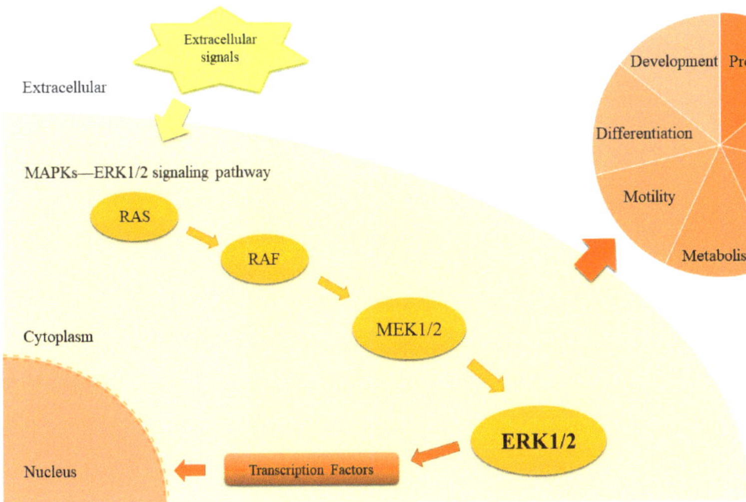

Figure 1. ERK1/2 signaling pathway.

In addition to the development of ERK1/2 inhibitors, the exploration of ERK1/2 degraders is another promising strategy to modulate ERK1/2 activity. Degraders have many advantages over inhibitors, including catalytic specificity and efficiency, minimal toxicity, and the ability to overcome target accumulation and mutation. PROTACs are the most widely used techniques for the selective degradation of proteins. A bifunctional PROTAC molecule with two covalently linked ligands recruits the target protein and E3 ubiquitin ligase to induce proteasomal degradation of the target protein via the ubiquitin–proteasome system. PROTAC has emerged as a promising targeted therapy strategy for a variety of diseases, particularly cancer [16–18]. Moreover, Li et al. reported the click-formed proteolysis-targeting chimera (CLIPATAC) technology, which effectively degraded ERK1/2 on the basis of a covalent inhibitor [19]. This result suggests that exogenous molecules can degrade ERK1/2. Therefore, we plan to design more effective ERK1/2 degraders utilizing the PROTAC technology for the discovery of new ERK1/2 modulators.

Based on previously reported ERK1/2 inhibitors, various E3 ligands and linkers were selected to synthesize PROTAC molecules. We found that ERK1/2 ligands, linkers, and E3 ligands were all critical for the degradative activities of the resulting conjugates. Finally, compound **B1-10J** was identified as the most promising compound after screening over 32 molecules with in vitro and in vivo evaluation of antitumor activity.

2. Results and Discussion

2.1. Design and Discovery of ERK1/2 Degraders

Three reported inhibitors, ulixertinib, GDC-0994, and compound 35 (Figure 2A), were selected as the ligand components of ERK1/2 [12–15] because they have high binding affinities with ERK1/2 and relatively low molecular weights. Importantly, their X-ray cocrystal structures reveal the portions of their structures that are located on the surface of the protein, which is essential for conjugation [20]. The isopropyl moiety in ulixertinib was replaced with a 3-aminopropanyl group to generate **A** (Figure 2A) for further extension. The 1-methyl-1H-pyrazol-5-amine moiety of GDC-0994 and compound 35 was changed to a 3-(4-aminobut-1-yn-1-yl)pyridin-4-amine group (Figure 2A) based on earlier probe design. Validation using molecular docking revealed that the modified molecules **A**, **B**, and **C** bind well in the ATP pocket of ERK1/2, and the exposed amino group can extend beyond the protein pocket (Figure 2A and Figure S1). An alkyl chain with 8 to 10 carbons was selected as the linker connected via an amide bond. Cereblon (CRBN) protein ligand, von Hippel–Lindau (VHL) protein ligand derivatives, and piperlongumine derivatives were chosen as ligands for E3 ligase, and adamantane, a hydrophobic tag, was also included [21–23]. The combination of the abovementioned ERK1/2 ligands, linker, and E3 ligase ligands generated 12 target molecules (Figure 2B). As depicted in Schemes 1–5, the synthesis steps followed previous reports [14,15,21,24].

A

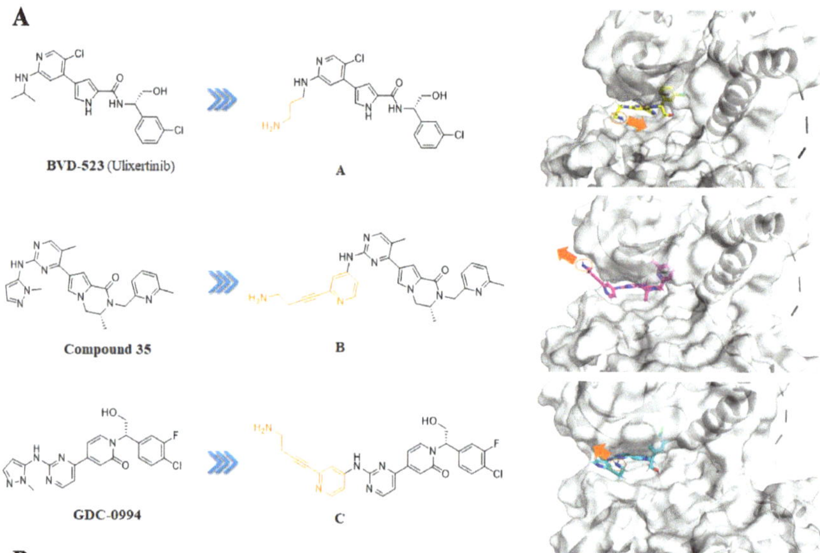

B

Compound	ERK1/2 ligand	Linker	E3 ligase Ligand
A1	A	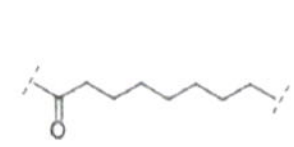	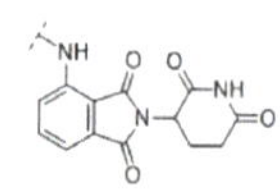
B1-8	B		
C1	C		
A2	A	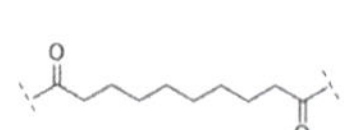	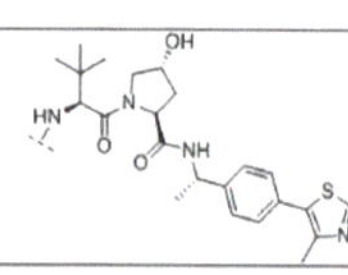
B2-10	B		
C2	C		
A3	A	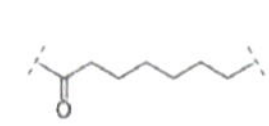	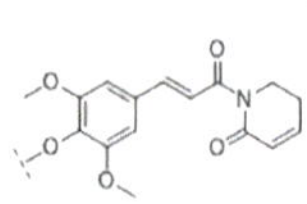
B3	B		
C3	C		
A4	A	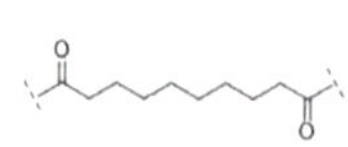	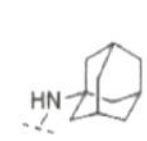
B4	B		
C4	C		

Figure 2. *Cont.*

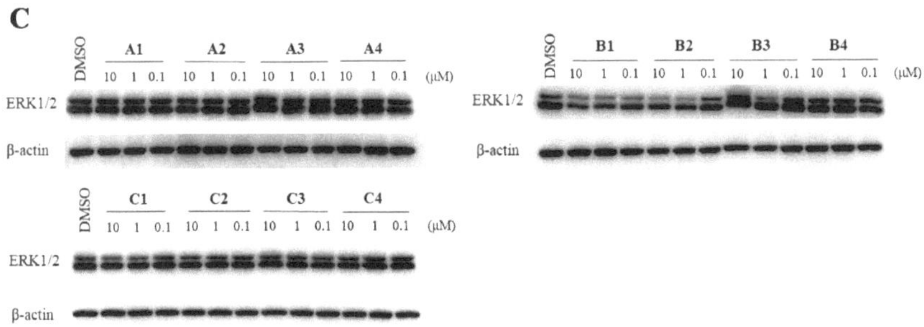

Figure 2. Design of the inhibitor-based ERK1/2 PROTACs. (**A**) Docking model of the modified inhibitors of ERK1/2 (PDB: 6GDQ). (**B**) Structures of ERK1/2 PROTACs. (**C**) A375 cells were treated with DMSO and compounds for 12 h at the indicated concentrations, followed by Western blot analysis.

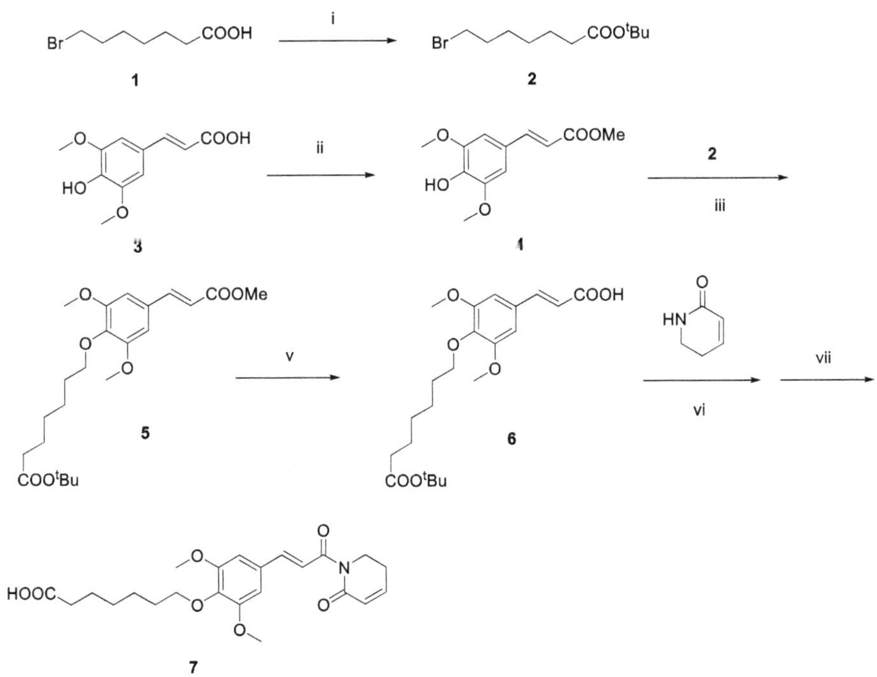

Scheme 1. Synthesis of piperlongumine derivatives.

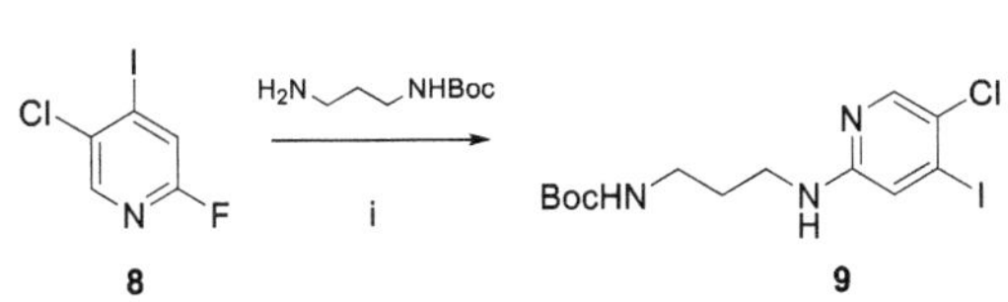

Scheme 2. *Cont.*

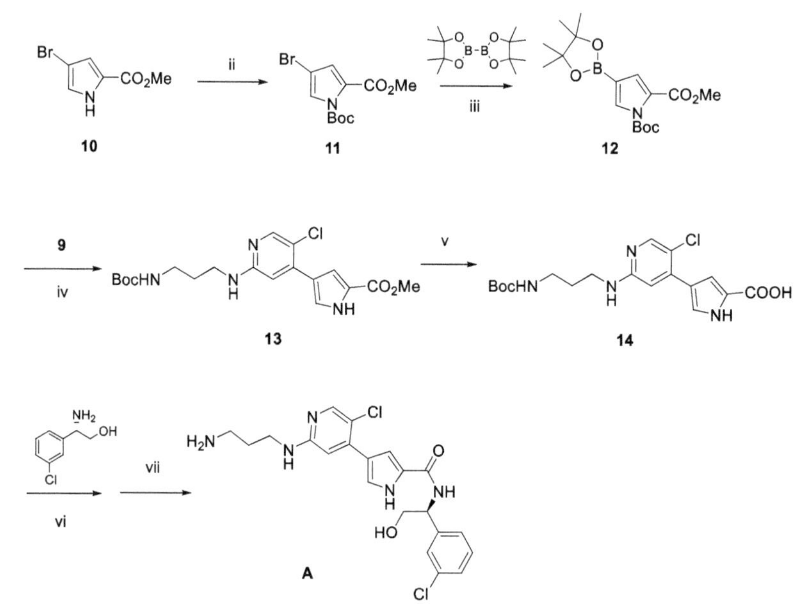

Scheme 2. Synthesis of compound A.

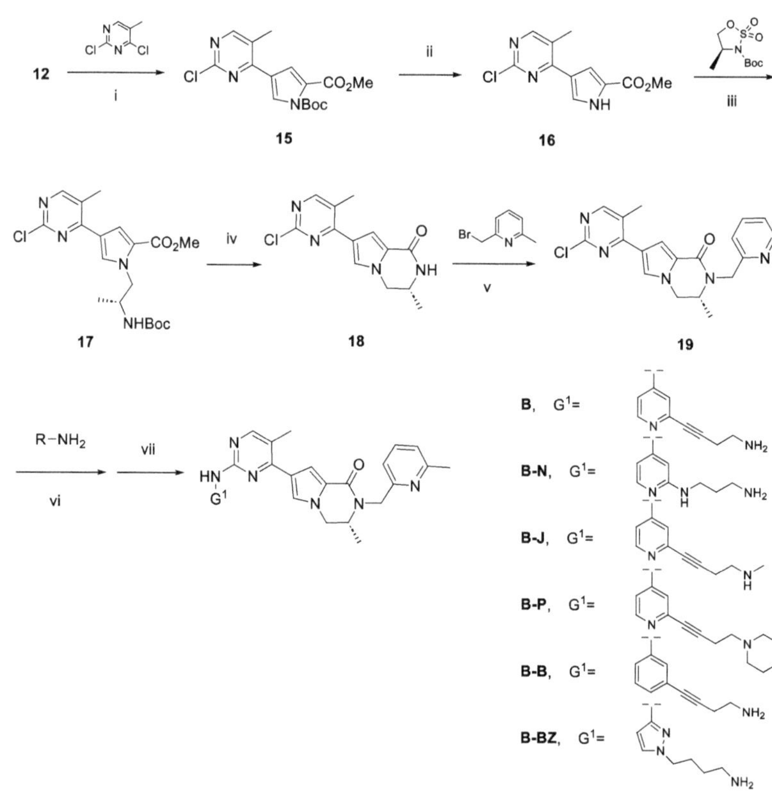

Scheme 3. Synthesis of compounds B, B-N, B-J, B-P, B-B, and B-BZ.

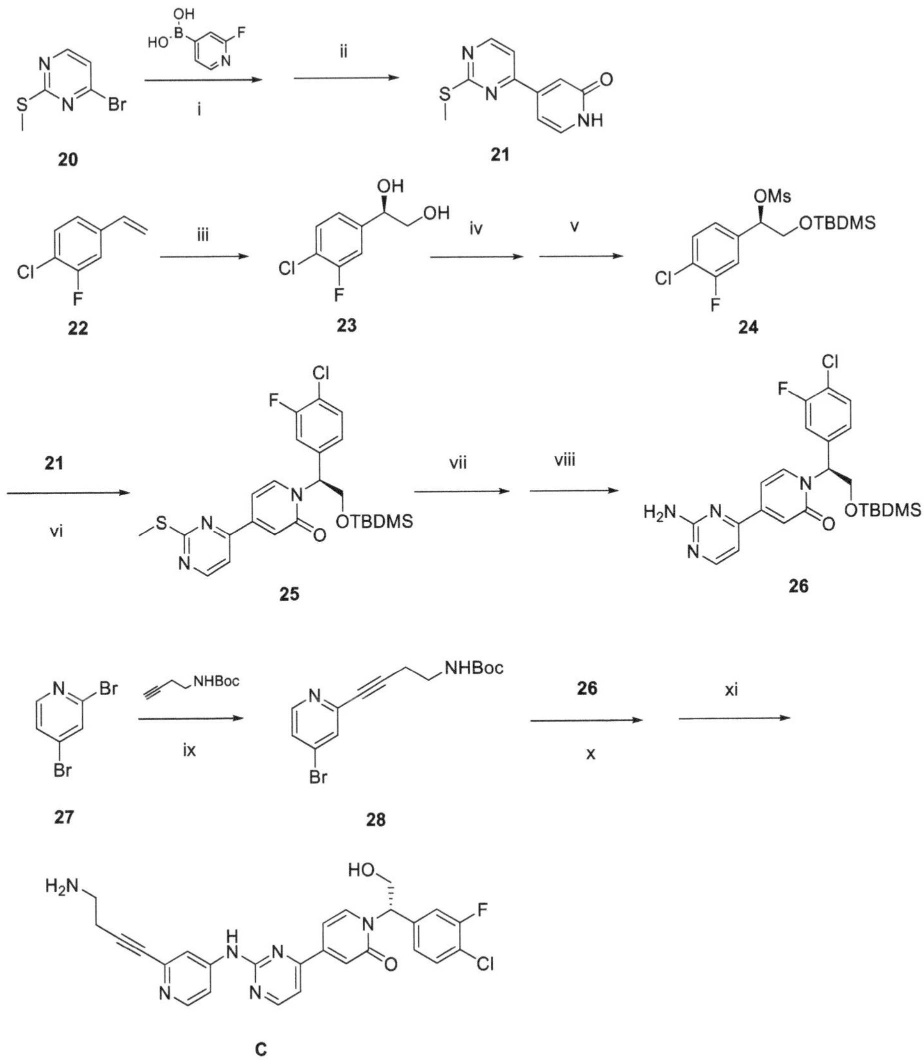

Scheme 4. Synthesis of compound C.

The synthesis of piperlongumine derivatives is described in Scheme 1. We used a method similar to the one reported in the literature to generate compound **7** [21].

The reagents and conditions were as follows: (i) $(Boc)_2O$, DMAP, *t*-BuOH, r.t., 2 h; (ii) H_2SO_4, MeOH, reflux, 3 h; (iii) K_2CO_3, DMF, 60 °C, 12 h; (iv) a. 2 M LiOH (a.q.), THF, 12 h; b. 0.5 M HCl (a.q.), 2 h; (v) a. ClCOCOCl, DMF, DCM, r.t., 2 h; b. *n*-BuLi, THF, −40 °C, 4 h; and (vi) TFA, DCM, r.t., 2 h.

The synthesis of compound **A** is described in Scheme 2. The substitution reaction of compound **8** with tert-butyl (3-aminopropyl)carbamate produced compound **9**. Compound **12** was prepared according to previous protocols [25]. Compounds **9** and **12** were submitted to Suzuki–Miyaura cross-coupling reaction, followed by hydrolysis, to generate compound **14**. Coupling compound **14** and (S)-2-amino-2-(3-chlorophenyl)ethan-1-ol using HATU in dimethylformamide (DMF) and removing the *t*-butyloxy carbonyl (Boc) groups with trifluoroacetic acid (TFA) afforded compound **A**.

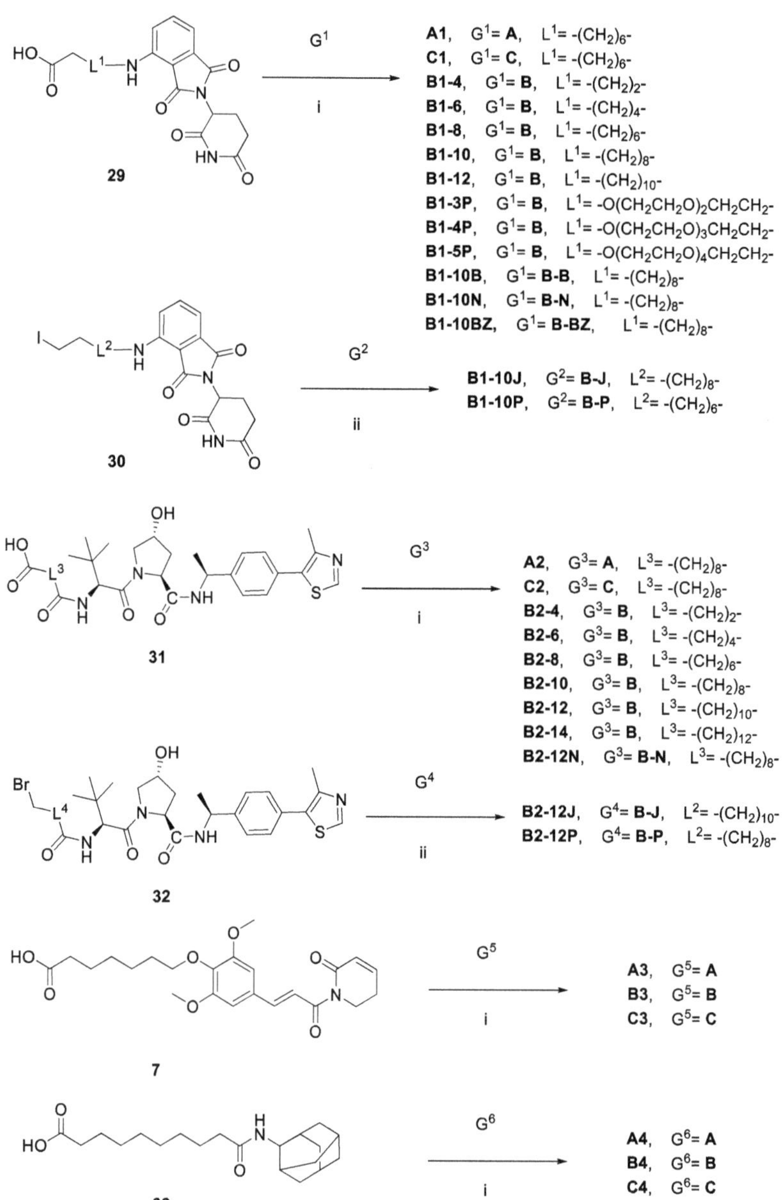

Scheme 5. Synthesis of PROTAC compounds.

The reagents and conditions were as follows: (i) DIEA, DMF, 70 °C, 12 h; (ii) $(Boc)_2O$, DMAP, DIEA, DCM, r.t., 2 h; (iii) Pd(dppf)Cl_2, KOAc, dioxane, 90 °C, 4 h; (iv) $Pd(PPh)_3$, K_2CO_3, dioxane, H_2O, 90 °C, 8h; (v) a. 2N LiOH (a.q.), THF, 12 h; b. 0.5 N HCl (a.q.), 2 h; (vi) HATU, DIEA, DCM, r.t., 12 h; and (vii) TFA, DCM, r.t., 2 h.

The synthesis of compounds **B**, **B-N**, **B-J**, **B-P**, **B-B**, and **B-BZ** is described in Scheme 3. Compound **19** was prepared according to previously reported protocols [15]. Compound **19** and corresponding amino compounds were submitted to Buchwald–Hartwig cross-coupling reaction, followed by removing the Boc groups with trifluoroacetic acid (TFA), to generate the corresponding compounds.

The reagents and conditions were as follows: (i) $Pd(PPh_3)_4$, K_3PO_4, dioxane, H_2O, 90 °C, 8 h; (ii) TFA, DCM, r.t., 2 h; (iii) K_2CO_3, 18-crown-6, dioxane, 100 °C, 12 h; (iv) a. TFA, DCM, r.t., 2 h; b. NH_4OH, MeOH, 50 °C, 2 h; (v) NaH, DMF, r.t., 4 h; (vi) BrettPhos Pd G3, Cs_2CO_3, *t*-BuOH, dioxane, 85 °C, 6 h; and (vii) TFA, DCM, r.t., 2 h.

The synthesis of compound **C** was performed according to previously reported protocols and is depicted in Scheme 4 [26].

The reagents and conditions were as follows: (i) Na_2CO_3, $Pd(dppf)Cl_2$, dioxane:H_2O (1:1), 80 °C, 1.5 h; (ii) 2 N HCl (a.q.), reflux, 2 h; (iii) AD-mix β, *t*-BuOH:H_2O (1:1), r.t. 18 h; (iv) TBDMSCl, imidazole, DCM, 0 °C, 1 h; (v) methanesulfonic anhydride, DIEA, DCM, 0 °C, 1 h; (vi) 1 M KHMDS in THF, THF:DMF (4:1), 75 °C, 20 h; (vii) *m*-CPBA, DCM, 0 °C, 2 h; (viii) NH_4OH, dioxane, r.t., 24 h; (ix) $Pd(PPh_3)_2Cl_2$, CuI, DIEA, THF, r.t., 18 h; (x) K_2CO_3, XPhos, $Pd_2(dba)_3$, MeCN, 80 °C, 18 h; and (xi) TFA, DCM, r.t., 2 h.

The synthesis of PROTAC compounds is described in Scheme 5. The substitution of iodide **30** and bromide **32**, or the formation of amides via HATU-mediated condensation reactions of acids (**7**, **29**, **31**, and **33**) with corresponding amines, was carried out to afford PROTAC compounds, respectively.

The reagents and conditions were as follows: (i) HATU, DIEA, DCM/MeOH, r.t., 8 h, and (ii) K_2CO_3, DMF, 80 °C, 6 h.

The compounds in Figure 2B were used to treat A375 cells at various concentrations. The Western blot analysis revealed that **B1** and **B2** induced downregulation of ERK1/2 protein in A375 cells, whereas other compounds did not (Figure 2C).

Therefore, compound **B** was used as a ligand for ERK1/2, and the ligands for the E3 ligase were CRBN ligand and VHL ligand, which were coupled with various lengths and types of linkers (Table 1). In total, 14 newly synthesized compounds were tested for their ability to degrade ERK1/2. These compounds were used to treat A375 cells for 48 h. Western blotting and quantitative grayscale analysis revealed that compounds with approximately 10 carbon chains had the highest activity with CRBN as the E3 ligand, while compounds with 10 to 14 carbon chains had similar degradation activity with the VHL ligand. Compounds linked with a PEG chain, to our surprise, had poor degradation activity (Figure 3A). To evaluate the ERK1/2 degradation activity of **B1-10**, **B2-10**, **B2-12**, and **B2-14**, the A375 cell line was treated with various concentrations of these compounds for 48 h, and the expression of ERK1/2 protein was quantified using Western blot with grayscale quantification. The results demonstrated that **B1-10**, **B2-12**, and **B2-14** exhibited high degradation activity, with DC_{50} values ranging from 300 nM to 400 nM and DC_{max} values exceeding 65% (Figure 3B).

For PROTAC compounds using CRBN ligands as E3 ligands, the 10-carbon chain length was selected as the linker, while for VHL ligands, the 12-carbon chain length was selected as the linker. The degradative effects of some structural modifications were investigated. Firstly, the amine group linking the ERK1/2 ligand and linker was replaced with methylamino and piperazine groups. For CRBN ligands, the methylamino- and piperazinyl-modified compounds **B1-10J** and **B1-10P** showed slightly increased DC_{50} compared to the amide counterpart **B1-10**, whereas for VHL ligands, the methylamino-modified compound **B2-12J** lost its degradative activity, and the piperazinyl-modified compound **B2-10P** showed a significantly increased DC_{50} value. Secondly, when the alkynyl group on compound **B** was replaced with flexible sp^3 N or C chains (**B1-10N**, **B1-10BZ**, and **B2-2N** in Table 2), the degradative activity disappeared. Thirdly, after the pyridine group on compound **B** was replaced with benzene and pyrazole (**B1-10B** and **B1-10BZ** in Table 2), no degradative activity was observed. These findings suggest that variations in ERK1/2 ligands can significantly affect the degradative activity of synthesized PROTACs, particularly the portion of ERK1/2 ligands located outside of the protein pocket.

Table 1. ERK1/2 PROTACs with different linker structures.

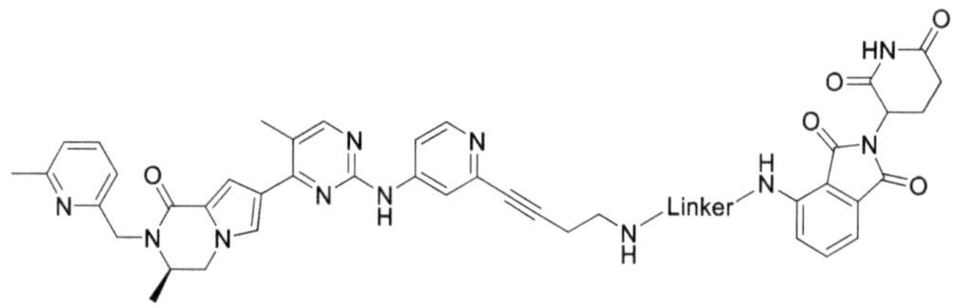

Compound	Linker
B1-4	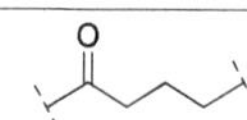
B1-6	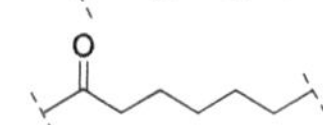
B1-10	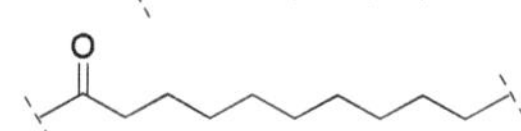
B1-12	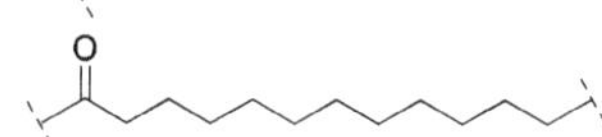
B1-3P	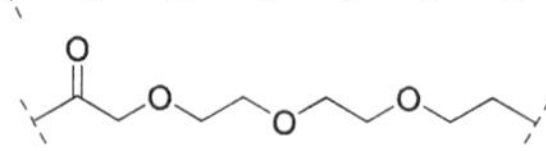
B1-4P	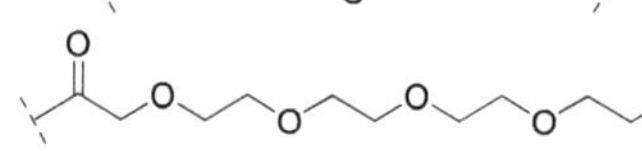
B1-5P	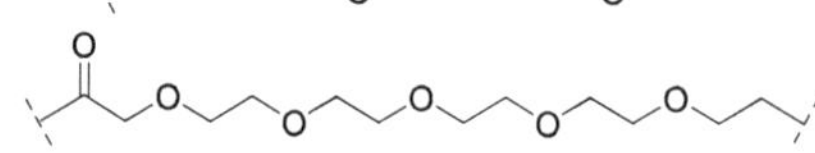

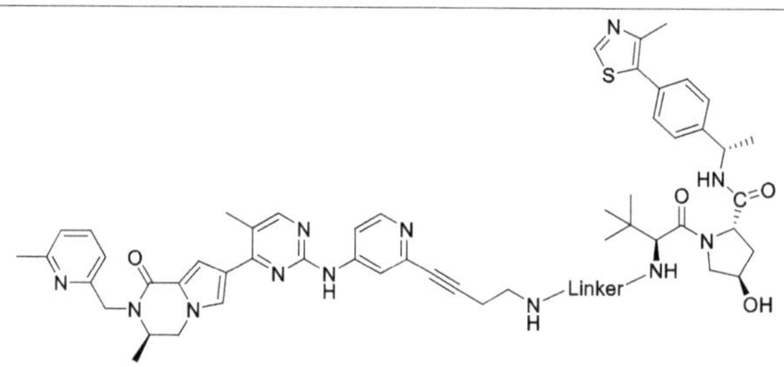

Compound	Linker
B2-4	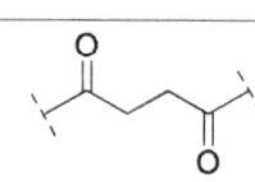
B2-6	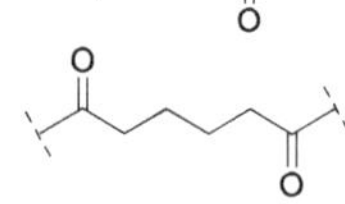
B2-8	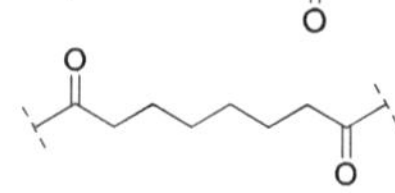
B2-12	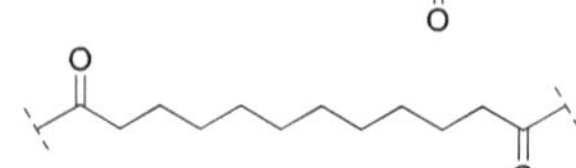
B2-14	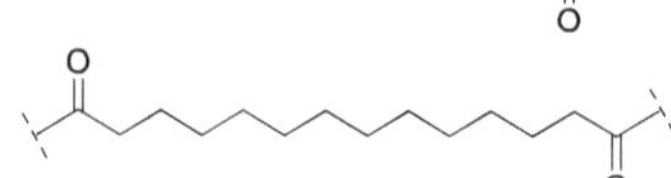

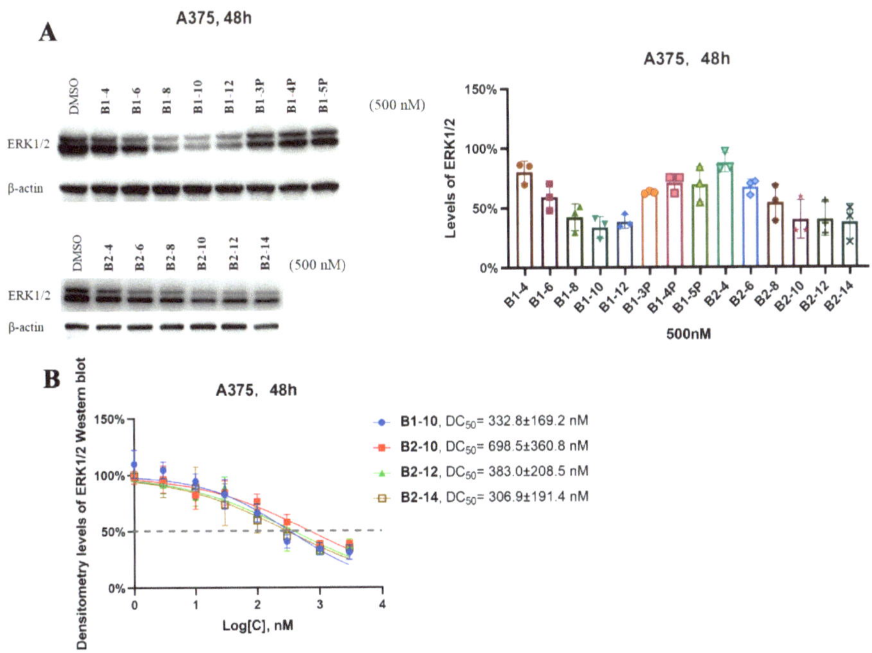

Figure 3. Determination of the degradation activity of PROTACs. (**A**) A375 cells were treated with various compounds for 48 h at a concentration of 500 nM, followed by Western blotting and grayscale quantification. (**B**) A375 cells were treated with various compounds for 48 h at the indicated concentrations, followed by Western blotting and grayscale quantification. Statistical analysis was performed using GraphPad Prism 8. The error bars represent the mean ± s.d. of n = 3 biological duplicates.

Table 2. ERK1/2 PROTACs with further structural modification.

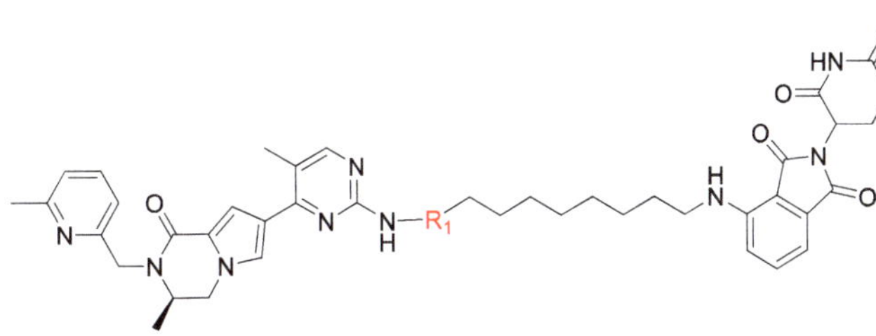

Compound	R_1	DC_{50} (nM) [a]
B1-10J	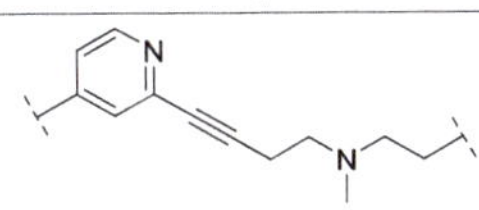	498.4 ± 369.3
B1-10P	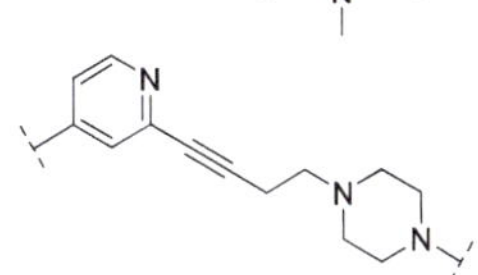	616.0 ± 389.4
B1-10N	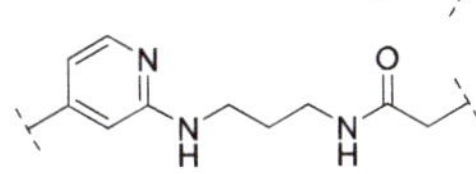	- [b]

Table 2. *Cont.*

B1-10B	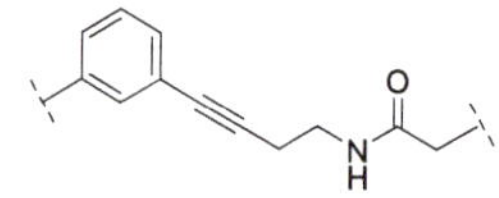	-
B1-10BZ	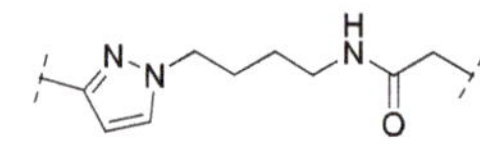	-

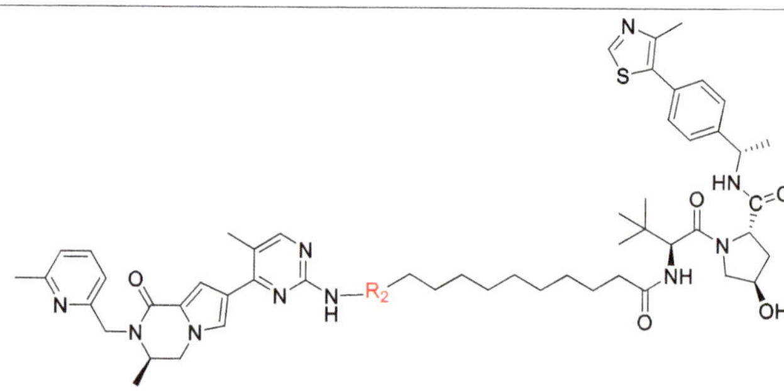

Compound	R_2	DC_{50} (nM) [a]
B2-12J	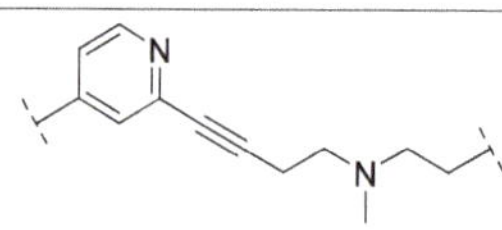	-
B2-12P	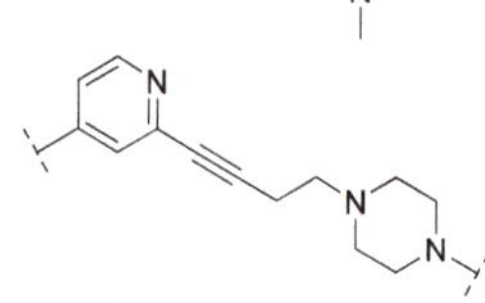	3800 ± 1824
B2-12N	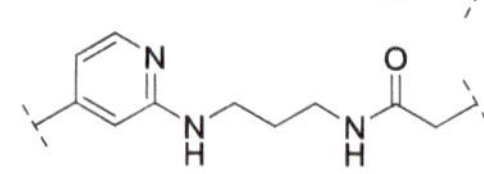	-

[a] A375 cells were treated with various compounds for 48 h at the indicated concentrations, followed by Western blotting and grayscale quantification. Statistical analysis was performed using GraphPad Prism 8. The error bars represent the mean ± s.d. of n = 3 biological duplicates. [b] - = not determined.

2.2. *Characterization of the Degradative Activity of* **B1-10J**

B1-10J was chosen as a representative compound for further characterization. It was first confirmed that **B1-10J** induced ERK1/2 degradation via the ubiquitination pathway. A375 cells were treated with or without **B1-10J**, proteasome inhibitor MG132, and E3 ligase inhibitor MLN4924 for 24 h [27,28]. The Western blot results indicated that ERK1/2 protein was only degraded by **B1-10J** alone (Figure 4A). In addition, A375 cells treated with **B1-10J** at a large range of concentrations exhibited a typical "hook" effect (Figure 4B) [29]. These findings verified that **B1-10J** induced ERK1/2 degradation through the ubiquitination pathway. Moreover, experiments were conducted to demonstrate the time-dependent degradation of ERK1/2 proteins. After A375 and HCT116 cell lines were treated with 1 μM **B1-10J**, ERK1/2 protein began to degrade at 6 h and nearly reached its highest level at 24 h. The degradation of ERK1/2 in HCT116 cells lasted up to 72 h, whereas in A375 cells, the protein level returned to normal after 48 h. Phosphorylated ERK1/2 plays a significant physiological role, and its variations were also monitored [4,30]. Interestingly, pERK1/2 exhibited an upward trend during the initial phase of the degrader treatment. pERK1/2 began to decrease as ERK1/2 was progressively degraded, but after 72 h, pERK1/2 returned to a high level (Figure 4C). To evaluate the ERK1/2 degradative activity of **B1-10J** in cancer cells, Calu-6, HCT116, and B16-F10 cancer cell lines were treated with different concentrations of **B1-10J** for 48 h, and the expression of ERK1/2 protein was quantified using Western blotting with grayscale quantification. The results demonstrated that **B1-10J** showed good degradative activity on HCT116 and Calu-6, with DC_{50} values of 102 nM and 85 nM, respectively, and weak degradative activity on B16-F10 murine melanoma cells (Figure 4D and Figure S2).

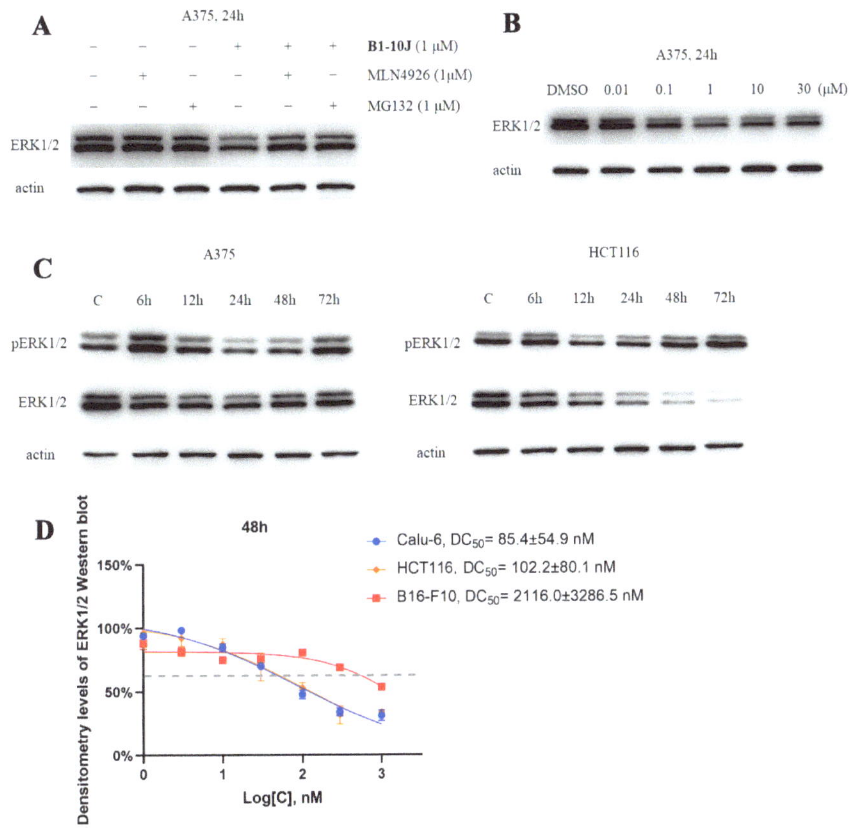

Figure 4. Characterization of B1-10J. (**A**) A375 cells were treated with/without MG132 (1 μM), MLN4924 (1 μM), and **B1-10J** (1 μM) for 24 h, followed by Western blot analysis. The control group was treated with DMSO. (**B**) A375 cells were treated with **B1-10J** for 24 h at the indicated concentrations. (**C**) Time-dependent ERK1/2 degradation upon treatment with **B1-10J** (1 μM) in A375 cells and HTC116 cells. (**D**) Cells were treated with the indicated concentrations of **B1-10J** for 48 h before Western blot and grayscale quantification. The error bars represent the mean ± s.d. of $n = 3$ biological duplicates.

2.3. *Effects of* **B1-10J** *on the Growth and Migration of Tumor Cells*

As mentioned previously, ERK1/2 is closely associated with the formation of tumors. In numerous human cancer cells, the ERK1/2 pathway is abnormally activated. Initially, the growth-inhibiting effects of **B1-10J** on three cancer cell lines were assessed, as reported in previous studies of ERK1/2 inhibitors [14,15,25,31]. A375, HCT116, and Calu-6 cells were treated with indicated concentrations of **B1-10J** for 72 h. These cancer cells displayed variable degrees of growth inhibition, with IC_{50} values of 2.1 μM for HCT116, 15.9 μM for Calu-6, and >30 μM for A375. The insensitivity of A375 cells was attributed to the lower ERK1/2 degradation ability in A375 cells, compared with that of HCT116. In addition, the growth-inhibiting effects of **B1-10J**, **B1-10**, **B1-10P**, **B2-12**, and the "warhead" part (**B-J**, Scheme 3) of **B1-10J** were evaluated in these cells (Figure 5A). Because the A375 cell line was less susceptible to the growth inhibition of **B1-10J**, it was selected to observe the effect on cell migration in a Transwell chamber. As shown in Figure 5B, A375 cell migration decreased markedly after 24 h. Using 33% acetic acid to wash crystal violet to measure the absorbance, it was found that the migration rate decreased by more than 50%. This result suggested that ERK1/2 degradation inhibited the migration of A375 cells.

A

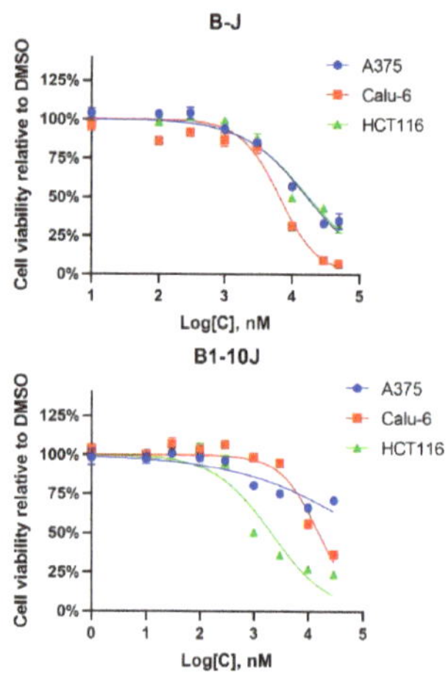

IC_{50} (μM)[a]	A375	Calu-6	HCT116
B1-10	>30	>30	>30
B-J	16.8±3.0	6.5±1.3	17.8±3.9
B1-10J	>30	15.9±2.9	2.1±0.8
B1-10P	>30	>30	2.2±0.6
B2-12	7.5±6.1	>30	11.7±10.5

[a]The error bars represent the mean ± s.d. of n = 3 biological duplicates.

B

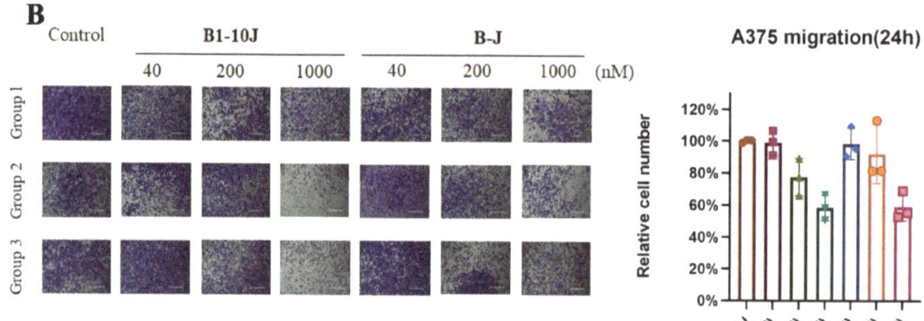

Figure 5. Characterization of antitumor activity of B1-10J. (**A**) Cytotoxicity assay. Various cell lines were treated with different concentrations of **B1-10J** for 72 h, and cell viability was detected with the CCK8 method. (**B**) A375 cells were treated with DMSO and **B1-10J** at a concentration of 1 μM for 24 h, respectively. After being stained with crystal violet, the cells were observed under a microscope and photos were taken. Crystal violet was washed with 33% acetic acid and the absorbance was measured at a wavelength of 590 nm. The error bars represent the mean ± s.d. of n = 3 biological duplicates.

2.4. *In Vivo Testing of* **B1-10J**

To investigate the in vivo antitumor activity of **B1-10J**, nude mice were subcutaneously xenografted with the HCT116 cell line. After the tumor volumes rose to ~90 mm^3, the mice were divided into a blank control group, low-, medium-, and high-dose groups of **B1-10J**, and a positive control group treated with 5-fluorouracil (5-FU), a commonly used chemotherapeutic drug in colorectal cancer treatment [32]. The mice were administrated 5-FU via intraperitoneal injection three times per week, while all other groups were dosed daily. As shown in Figure 6A, only nude mice of the 5-FU group exhibited a trend of weight loss, whereas the **B1-10J** groups exhibited similar body weight variations as the blank control group, indicating that **B1-10J** was well-tolerated. **B1-10J** was able to inhibit tumor growth in a dose-dependent manner and was more effective at a 25 mg/kg dose than 5-FU. Up until day 19, 25 mg/kg of **B1-10J** showed approximately 50% tumor growth inhibition. The size of the tumors at day 20 is depicted in Figure 6C and Figure S3. These results suggested that **B1-10J** effectively inhibited the proliferation of tumors in vivo.

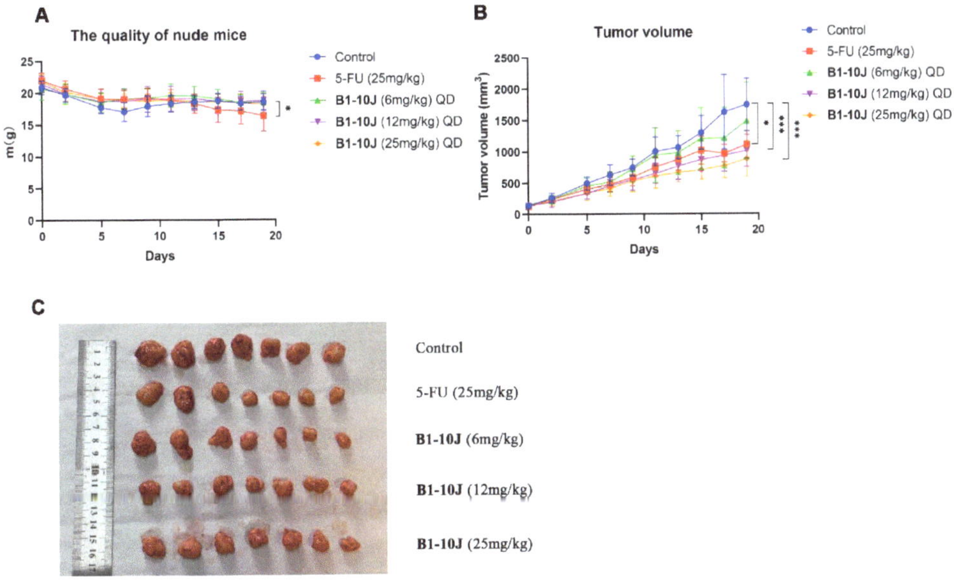

Figure 6. The antitumor activity of B1-10J in an HCT116 subcutaneous xenograft model. (**A**) The weight fluctuations of nude mice. (**B**) The plots show the average tumor volumes of HCT116 over time upon intraperitoneal injections of **B1-10J** at 6, 12, or 25 mg/kg Q.D. Average volumes are geomeans, and error bars show ± SD. * $p < 0.05$ and *** $p < 0.001$. (**C**) Photographs of nude mice and HCT116 tumors.

3. Materials and Methods

3.1. Chemistry

General Methods. All reagents and solvents used were obtained from commercial suppliers (Bidepharmatech (Shanghai, China), Energy Chemical (The Woodlands, TX, USA), Alfa (Ronkonkoma, NY, USA), etc.) without further purification, except for special cases. Reactions were monitored via TLC. Thin-layer chromatography was carried out using TLC silica gel 60 F254 plates. Flash column chromatography was performed with 200–300 mesh silica gel. The NMR spectrum was recorded on a Bruker-400 NMR spectrometer, with TMS as an internal standard and chemical shifts reported in ppm (δ). Coupling constants (J) were reported in Hz. Spin multiplicities were described as s (singlet), br (broad singlet), d (doublet), t (triplet), q (quartet), and m (multiplet). Melting point was measured by using a X-5 micro melting point meter. High-resolution mass spectra (HRMS) were obtained using a Shimadzu LCMS-IT-TOF mass spectrometer. An Agilent high-performance liquid chromatography (HPLC) system with 5 μm C18 column was used for purity analysis. The gradient was 5% CH_3CN (MeOH) to 95% CH_3CN (MeOH) over 25 min at a flow rate of 1 mL/min and the wavelength was 254 nm. All end products were analyzed via HPLC and all were over 95% pure. Additional figures for ^1H NMR, ^{13}C NMR, HRMS and HPLC of compounds used for biological testing were included in the Supporting Information.

3.2. Molecular Docking

The software Sybyl x2.0 was used for computer-aided drug design. After downloading the protein crystal structure (PDB: 6gdq), Sybyl was used for hydrogenation, dehydration, deligation, and repair to generate the docking file of the protein. Chemdraw and Chemdraw 3D were used to convert 2D small molecules into 3D forms, and energy minimization processing was performed to output a docking file for small molecules. The default docking conditions of Sybyl x2.0 were used for docking, the docking results were exported, and PyMol was used for image processing.

3.3. Cell Culture

A375, HCT116, Calu-6, and B16-F10 cells were purchased from American Type Culture Collection (ATCC) and cultured in DMEM high glucose media (SH30022.01, Cytiva, Shanghai, China) supplemented with 10% fetal bovine serum (10099-144, Gibco, Carlsbad, CA, USA) and 1% penicillin/streptomycin (Solarbio, Beijing, China) in an incubator at 37 °C with 5% CO_2.

3.4. Western Blot and Protein Degradation Assay

Cells were cultured under the indicated conditions as described in "Cell culture". Cells were seeded in 6-well plates (3516, Corning, New York, NY, USA). Compounds were dissolved in DMSO for storage (10 mM). Following compound treatment, cells were washed with PBS, scraped from the plate, and lysed using cell lysis buffer (Applygen, Beijing, China). After the protein in the cell lysate was quantified via the BCA quantification method, the samples were placed in a metal bath at 100 °C and heated for 5 min. Cell lysates were separated using a 10% SDS–PAGE gel and transferred to PVDF membranes (Merck Millipore, Burlington, MA, USA). The PVDF membrane was blocked with 5% nonfat dry milk in TBST for one hour. The PVDF membrane was transferred to the antibody diluent and incubated overnight at 4 °C. PVDF membranes were washed three times with TBST for 5 min each and then incubated with the appropriate HRP-conjugated secondary antibody for 1 h at room temperature. After washing three more times with TBST, the PVDF membrane was stained with a hypersensitive ECL chemiluminescence reagent (Biodragon, Beijing, China) and then imaged with Tanon 5200. Grayscale quantitative analysis was performed using ImageJ software. Statistical analysis was performed using GraphPad Prism 8. DC_{50} and D_{max} values were fitted using a four-parameter [inhibitor] versus response and reported directly from the Prism output. Mean ± SD and unpaired t tests were performed in GraphPad Prism.

3.5. Cell Proliferation Assay

A 96-well plate (3599, Corning, New York, NY, USA) was inoculated with 100 μL of complete medium containing 10,000 cells, and the experimental, control, and blank groups were established. Three parallel wells were used per concentration. After 10 h, the media from the experimental and control groups were replaced with a compound containing the indicated concentration of the complete medium. Cells were incubated for 72 h before 10 μL of CCK8 reagent (CA1210, Solarbio, Beijing, China) was added to each well and incubated for 1 h. Using a microplate reader (Tecan, Mannendorf, Switzerland), the absorbance at 450 nm was measured. The IC_{50} values were determined using GraphPad Prism 8 and nonlinear regression curve fitting.

3.6. Transwell Migration Assay

A total of 100 μL of FBS-free DMEM high-glucose medium containing 20,000 cells was added to the upper layer of the Transwell chamber of a 24-well plate (3422, Corning, New York, NY, USA), and a complete medium containing 10% FBS was added to the lower layer. The experimental group and the control group were established. The control group used DMSO (DMSO concentration in the culture medium was no more than 0.5% in each well), and the specified concentration of compounds was added to the medium of the experimental group. The cells were then placed in an incubator for 24 h. The cells were washed with PBS, the cells in the upper layer of the Transwell chamber were wiped off with a cotton swab, and the cells were fixed with cold methanol for 15 min. The cells were stained with crystal violet reagent and observed and photographed under a microscope. Then, crystal violet was washed with 33% acetic acid and the absorbance was measured at a wavelength of 590 nm. Graphing with GraphPad Prism 8 after counting cells was performed.

3.7. Xenograft Mouse Studies

Male BALB/c nude mice of around 8 weeks old were purchased from Charles River (Beijing, China). Nude mice were raised in SPF-grade animal rooms and had access to water and food ad libitum. HCT116 colorectal cancer cells were grown in DMEM supplemented with 10% fetal calf serum and 1% penicillin/streptomycin under standard cell culture conditions. HCT116 xenografts were prepared by subcutaneously injecting 5×10^6 cells suspended in PBS into the right flank of each male BALB/c nude mouse. Tumor growth was monitored via caliper measurement, and volume was calculated using the following equation: $0.5 \times$ max (length/width) $\times$ min (length/width) $\times$ min (length/width). When the tumors reached an average of approximately 100 mm^3, the mice were randomized into groups of five. The mice were injected intraperitoneally once a day at doses of 6, 12, 25 mg/kg compound **B1-10J** and 25 mg/kg 5-FU. Body weights and tumor volumes were measured every 2–3 days. All studies were appropriately statistically powered.

3.8. Statistical Analysis

Statistical analyses were performed using GraphPad Prism software 8. The significance analysis was conducted using a two-tailed unpaired t-test. $p < 0.05$ was considered statistically significant (* $p < 0.05$, ** $p < 0.01$, and *** $p < 0.001$).

4. Conclusions

Based on the previously reported ERK1/2 inhibitors, molecular docking was used to design multiple PROTACs containing various E3 ligands. Several effective ERK1/2 degraders, such as **B1-10**, **B1-10J**, **B1-10P**, **B2-10**, **B2-12**, and **B2-14**, were identified from 24 synthesized molecules, among which **B1-10J** was the best one; its ERK1/2 DC_{50} values were 102 nM and 85 nM for HCT116 and Calu-6, respectively. We confirmed that the degradation of ERK1/2 induced by **B1-10J** in cancer cells was via the ubiquitination pathway. **B1-10J** significantly prevented the migration of A375 cells at 1 μM, and inhibited the proliferation of HCT116 cells both in vitro and in vivo. In sum, **B1-10J** is a promising lead compound for the degradation of ERK1/2 and warrants further investigation.

Supplementary Materials: The following supporting information can be downloaded at https://www.mdpi.com/article/10.3390/ijms242216290/s1.

Author Contributions: P.P. and X.M. conceived the concept and designed the research. P.P. planned and carried out the experiments. Y.H. and T.G. participated in some experiments. P.P. wrote the manuscript. All authors reviewed and edited the manuscript. All authors have read and agreed to the published version of the manuscript.

Funding: This work was supported by the National Natural Science Foundation of China (NSFC No. 81872738) and the National Key R&D Program of China (2022YFF1203005, 2022YFC2303700).

Institutional Review Board Statement: Not applicable.

Informed Consent Statement: Not applicable.

Data Availability Statement: Data are contained within the article.

Acknowledgments: We acknowledge the China International Science and Technology Cooperation Base of Food Nutrition/Safety and Medical Chemistry, College of Biotechnology, Tianjin University of Science and Technology for providing us with high-resolution mass spectrometry analysis. We acknowledge the laboratory of M. Ma, School of Pharmaceutical Sciences, Peking University, for providing us with instruments for high-performance liquid chromatography.

Conflicts of Interest: The authors declare that they have no known competing financial interests or personal relationships that could have appeared to influence the work reported in this paper.

References

1. Johnson, G.L.; Lapadat, R. Mitogen-activated protein kinase pathways mediated by ERK, JNK, and p38 protein kinases. *Science* **2002**, *298*, 1911–1912. [CrossRef]
2. Lawrence, M.C.; Jivan, A.; Shao, C.; Duan, L.; Goad, D.; Zaganjor, E.; Osborne, J.; McGlynn, K.; Stippec, S.; Earnest, S.; et al. The roles of MAPKs in disease. *Cell Res.* **2008**, *18*, 436–442. [CrossRef] [PubMed]
3. Raman, M.; Chen, W.; Cobb, M.H. Differential regulation and properties of MAPKs. *Oncogene* **2007**, *26*, 3100–3112. [CrossRef]
4. Roskoski, R., Jr. ERK1/2 MAP kinases: Structure, function, and regulation. *Pharmacol. Res.* **2012**, *66*, 105–143. [CrossRef]
5. Pylayeva-Gupta, Y.; Grabocka, E.; Bar-Sagi, D. RAS oncogenes: Weaving a tumorigenic web. *Nat. Rev. Cancer* **2011**, *11*, 761–774. [CrossRef]
6. Davies, H.; Bignell, G.R.; Cox, C.; Stephens, P.; Edkins, S.; Clegg, S.; Teague, J.; Woffendin, H.; Garnett, M.J.; Bottomley, W.; et al. Mutations of the BRAF gene in human cancer. *Nature* **2002**, *417*, 949–954. [CrossRef]
7. Little, A.S.; Smith, P.D.; Cook, S.J. Mechanisms of acquired resistance to ERK1/2 pathway inhibitors. *Oncogene* **2013**, *32*, 1207–1215. [CrossRef]
8. Fu, L.; Chen, S.; He, G.; Chen, Y.; Liu, B. Targeting Extracellular Signal-Regulated Protein Kinase 1/2 (ERK1/2) in Cancer: An Update on Pharmacological Small-Molecule Inhibitors. *J. Med. Chem.* **2022**, *65*, 13561–13573. [CrossRef]
9. Pan, X.; Pei, J.; Wang, A.; Shuai, W.; Feng, L.; Bu, F.; Zhu, Y.; Zhang, L.; Wang, G.; Ouyang, L. Development of small molecule extracellular signal-regulated kinases (ERKs) inhibitors for cancer therapy. *Acta Pharm. Sin. B* **2022**, *12*, 2171–2192. [CrossRef]
10. Roskoski, R., Jr. Targeting ERK1/2 protein-serine/threonine kinases in human cancers. *Pharmacol. Res.* **2019**, *142*, 151–168. [CrossRef]
11. Balmanno, K.; Cook, S.J. Tumour cell survival signalling by the ERK1/2 pathway. *Cell Death Differ.* **2009**, *16*, 368–377. [CrossRef]
12. Sullivan, R.J.; Infante, J.R.; Janku, F.; Wong, D.J.L.; Sosman, J.A.; Keedy, V.; Patel, M.R.; Shapiro, G.I.; Mier, J.W.; Tolcher, A.W.; et al. First-in-Class ERK1/2 Inhibitor Ulixertinib (BVD-523) in Patients with MAPK Mutant Advanced Solid Tumors: Results of a Phase I Dose-Escalation and Expansion Study. *Cancer Discov.* **2018**, *8*, 184–195. [CrossRef]
13. Germann, U.A.; Furey, B.F.; Markland, W.; Hoover, R.R.; Aronov, A.M.; Roix, J.J.; Hale, M.; Boucher, D.M.; Sorrell, D.A.; Martinez-Botella, G.; et al. Targeting the MAPK Signaling Pathway in Cancer: Promising Preclinical Activity with the Novel Selective ERK1/2 Inhibitor BVD-523 (Ulixertinib). *Mol. Cancer Ther.* **2017**, *16*, 2351–2363. [CrossRef]
14. Blake, J.F.; Burkard, M.; Chan, J.; Chen, H.; Chou, K.J.; Diaz, D.; Dudley, D.A.; Gaudino, J.J.; Gould, S.E.; Grina, J.; et al. Discovery of (S)-1-(1-(4-Chloro-3-fluorophenyl)-2-hydroxyethyl)-4-(2-((1-methyl-1H-pyrazol-5-yl)amino)pyrimidin-4-yl)pyridin-2(1H)-one (GDC-0994), an Extracellular Signal-Regulated Kinase 1/2 (ERK1/2) Inhibitor in Early Clinical Development. *J. Med. Chem.* **2016**, *59*, 5650–5660. [CrossRef]
15. Ward, R.A.; Bethel, P.; Cook, C.; Davies, E.; Debreczeni, J.E.; Fairley, G.; Feron, L.; Flemington, V.; Graham, M.A.; Greenwood, R.; et al. Structure-Guided Discovery of Potent and Selective Inhibitors of ERK1/2 from a Modestly Active and Promiscuous Chemical Start Point. *J. Med. Chem.* **2017**, *60*, 3438–3450. [CrossRef]
16. Li, X.; Pu, W.; Zheng, Q.; Ai, M.; Chen, S.; Peng, Y. Proteolysis-targeting chimeras (PROTACs) in cancer therapy. *Mol. Cancer* **2022**, *21*, 99. [CrossRef]
17. Zhao, L.; Zhao, J.; Zhong, K.; Tong, A.; Jia, D. Targeted protein degradation: Mechanisms, strategies and application. *Signal Transduct. Target. Ther.* **2022**, *7*, 113. [CrossRef]
18. Bekes, M.; Langley, D.R.; Crews, C.M. PROTAC targeted protein degraders: The past is prologue. *Nat. Rev. Drug Discov.* **2022**, *21*, 181–200. [CrossRef]
19. Lebraud, H.; Wright, D.J.; Johnson, C.N.; Heightman, T.D. Protein Degradation by In-Cell Self-Assembly of Proteolysis Targeting Chimeras. *ACS Cent. Sci.* **2016**, *2*, 927–934. [CrossRef]
20. He, S.; Dong, G.; Cheng, J.; Wu, Y.; Sheng, C. Strategies for designing proteolysis targeting chimaeras (PROTACs). *Med. Res. Rev.* **2022**, *42*, 1280–1342. [CrossRef]
21. Liao, Y.; Niu, X.; Chen, B.; Edwards, H.; Xu, L.; Xie, C.; Lin, H.; Polin, L.; Taub, J.W.; Ge, Y.; et al. Synthesis and Antileukemic Activities of Piperlongumine and HDAC Inhibitor Hybrids against Acute Myeloid Leukemia Cells. *J. Med. Chem.* **2016**, *59*, 7974–7990. [CrossRef]
22. Wang, M.; Lin, R.; Li, J.; Suo, Y.; Gao, J.; Liu, L.; Zhou, L.; Ni, Y.; Yang, Z.; Zheng, J.; et al. Discovery of LL-K8-22: A Selective, Durable, and Small-Molecule Degrader of the CDK8-Cyclin C Complex. *J. Med. Chem.* **2023**, *66*, 4932–4951. [CrossRef] [PubMed]
23. He, M.; Cao, C.; Ni, Z.; Liu, Y.; Song, P.; Hao, S.; He, Y.; Sun, X.; Rao, Y. PROTACs: Great opportunities for academia and industry (an update from 2020 to 2021). *Signal Transduct Target Ther.* **2022**, *7*, 181. [CrossRef] [PubMed]

24. Aronov, A.M.; Tang, Q.; Martinez-Botella, G.; Bemis, G.W.; Cao, J.; Chen, G.; Ewing, N.P.; Ford, P.J.; Germann, U.A.; Green, J.; et al. Structure-guided design of potent and selective pyrimidylpyrrole inhibitors of extracellular signal-regulated kinase (ERK) using conformational control. *J. Med. Chem.* **2009**, *52*, 6362–6368. [CrossRef] [PubMed]
25. Ward, R.A.; Anderton, M.J.; Bethel, P.; Breed, J.; Cook, C.; Davies, E.J.; Dobson, A.; Dong, Z.; Fairley, G.; Farrington, P.; et al. Discovery of a Potent and Selective Oral Inhibitor of ERK1/2 (AZD0364) That Is Efficacious in Both Monotherapy and Combination Therapy in Models of Nonsmall Cell Lung Cancer (NSCLC). *J. Med. Chem.* **2019**, *62*, 11004–11018. [CrossRef]
26. Lebraud, H.; Surova, O.; Courtin, A.; O'Reilly, M.; Valenzano, C.R.; Nordlund, P.; Heightman, T.D. Quantitation of ERK1/2 inhibitor cellular target occupancies with a reversible slow off-rate probe. *Chem. Sci.* **2018**, *9*, 8608–8618. [CrossRef]
27. Tsubuki, S.; Saito, Y.; Tomioka, M.; Ito, H.; Kawashima, S. Differential inhibition of calpain and proteasome activities by peptidyl aldehydes of di-leucine and tri-leucine. *J. Biochem.* **1996**, *119*, 572–576. [CrossRef]
28. Soucy, T.A.; Smith, P.G.; Milhollen, M.A.; Berger, A.J.; Gavin, J.M.; Adhikari, S.; Brownell, J.E.; Burke, K.E.; Cardin, D.P.; Critchley, S.; et al. An inhibitor of NEDD8-activating enzyme as a new approach to treat cancer. *Nature* **2009**, *458*, 732–736. [CrossRef]
29. Pettersson, M.; Crews, C.M. PROteolysis TArgeting Chimeras (PROTACs)—Past, present and future. *Drug Discov. Today Technol.* **2019**, *31*, 15–27. [CrossRef]
30. Lavoie, H.; Gagnon, J.; Therrien, M. ERK signalling: A master regulator of cell behaviour, life and fate. *Nat. Rev. Mol. Cell Biol.* **2020**, *21*, 607–632. [CrossRef]
31. Heightman, T.D.; Berdini, V.; Braithwaite, H.; Buck, I.M.; Cassidy, M.; Castro, J.; Courtin, A.; Day, J.E.H.; East, C.; Fazal, L.; et al. Fragment-Based Discovery of a Potent, Orally Bioavailable Inhibitor That Modulates the Phosphorylation and Catalytic Activity of ERK1/2. *J. Med. Chem.* **2018**, *61*, 4978–4992. [CrossRef] [PubMed]
32. Alzahrani, S.M.; Al Doghaither, H.A.; Al-Ghafari, A.B.; Pushparaj, P.N. 5-Fluorouracil and capecitabine therapies for the treatment of colorectal cancer (Review). *Oncol. Rep.* **2023**, *50*, 175. [CrossRef] [PubMed]

International Journal of
Molecular Sciences

MDPI

Article

Activation of Chaperone-Mediated Autophagy Inhibits the Aryl Hydrocarbon Receptor Function by Degrading This Receptor in Human Lung Epithelial Carcinoma A549 Cells

Rui Xiong, Dan Shao, Sandra Do and William K. Chan *

Department of Pharmaceutics & Medicinal Chemistry, Thomas J. Long School of Pharmacy, University of the Pacific, Stockton, CA 95211, USA; r_xiong1@u.pacific.edu (R.X.); d_shao@u.pacific.edu (D.S.); s_do10@u.pacific.edu (S.D.)
* Correspondence: wchan@pacific.edu; Fax: +1-(209)-946-2410

Abstract: The aryl hydrocarbon receptor (AHR) is a ligand-activated transcription factor and a substrate protein of a Cullin 4B E3 ligase complex responsible for diverse cellular processes. In the lung, this receptor is responsible for the bioactivation of benzo[a]pyrene during tumorigenesis Realizing that the AHR function is affected by its expression level, we are interested in the degradation mechanism of AHR in the lung. Here, we have investigated the mechanism responsible for AHR degradation using human lung epithelial A549 cells. We have observed that the AHR protein levels increase in the presence of chloroquine (CQ), an autophagy inhibitor, in a dose-dependent manner Treatment with 6-aminonicotinamide (6-AN), a chaperone-mediated autophagy (CMA) activator, decreases AHR protein levels in a concentration-dependent and time-dependent manner. This decrease suppresses the ligand-dependent activation of the AHR target gene transcription, and can be reversed by CQ but not MG132. Knockdown of lysosome-associated membrane protein 2 (LAMP2), but not autophagy-related 5 (ATG5), suppresses the chloroquine-mediated increase in the AHR protein. AHR is resistant to CMA when its CMA motif is mutated. Suppression of the epithelial-to-mesenchymal transition in A549 cells is observed when the *AHR* gene is knocked out or the AHR protein level is reduced by 6-AN. Collectively, we have provided evidence supporting that AHR is continuously undergoing CMA and activation of CMA suppresses the AHR function in A549 cells.

Keywords: aryl hydrocarbon receptor; chaperone-mediated autophagy; A549; 6-AN; LAMP2; CQ

check for updates

Citation: Xiong, R.; Shao, D.; Do, S.; Chan, W.K. Activation of Chaperone-Mediated Autophagy Inhibits the Aryl Hydrocarbon Receptor Function by Degrading This Receptor in Human Lung Epithelial Carcinoma A549 Cells. *Int. J. Mol. Sci.* **2023**, *24*, 15116. https://doi.org/10.3390/ijms242015116

Academic Editor: Kang-Nan Wang

Received: 13 September 2023
Revised: 7 October 2023
Accepted: 8 October 2023
Published: 12 October 2023

1. Introduction

The aryl hydrocarbon receptor (AHR) is a biological sensor that alters gene expression in response to many environmental pollutants (such as dioxins and polychlorinated biphenyl compounds) and flavonoids [1,2]. Many tryptophan metabolites (such as 6-formylindolo (3,2-b) carbazole (FICZ)) have been shown to be the endogenous ligands of this receptor. These metabolites can be generated in the gut microbiome and subsequently activate AHR in human cells [3]. AHR suppresses the immune response in part by promoting naïve T cell differentiation into T regulatory cells [4]. It also drives the growth of many tumors [5], promotes insulin resistance [6], and is a drug target for psoriasis treatment [7]. Interestingly, AHR has been implicated as a drug target for the treatment of SARS-CoV2 infection since AHR may alter lung function in favor of supporting SARS-CoV2 infection [8,9]. Regarding the role of AHR in lung tumorigenesis, there are conflicting reports on the effect of AHR on the epithelial–mesenchymal transition (EMT). Although some researchers observed a positive correlation between the AHR action and the aggressive phenotypes of invasion and metastasis in lung and other cell types [10–13], others reported that AHR suppressed metastasis by inhibiting lung EMT [14,15]. Nonetheless, the expression levels of

AHR must correlate positively with its function in the lung. We are interested in studying the degradation mechanisms of AHR in lung epithelial cells in affecting the AHR function.

AHR exists as a complex in the cytoplasm with heat shock protein 90 (HSP90), co-chaperone protein p23, hepatitis B virus-x associated protein 2 (XAP2), and possibly proto-oncogene tyrosine-protein kinase (Src) [16–18]. Exposure of the nuclear localization sequence of AHR after ligand binding leads to nuclear translocation of the complex. The binding of the aryl hydrocarbon receptor nuclear translocator (ARNT) to AHR in the nucleus dissociates the complex [18]. The AHR–ARNT heterodimer binds to its dioxin response element (DRE) enhancer, activating the transcription of its target genes, such as cytochrome P450 1a1 (*CYP1A1*). Alternatively, AHR serves as a substrate protein that recruits its target proteins (for example, ERα) to CUL4B E3 ligase for proteasomal degradation [19].

Interestingly, AHR is degraded via autophagy in many human cell lines, namely lung A549, liver Hep3B, and breast T-47D and triple-negative MDA-MB-468 cells [20]. Proteins can be selectively degraded via selective macroautophagy and chaperone-mediated autophagy (CMA) [21,22]. Selective macroautophagy employs p62 to escort client proteins into autophagosomes by interaction with microtubule-associated protein 1 light chain (LC3) at the autophagosome membrane, followed by fusion with lysosomes. This fusion leads to lysosomal degradation of the client proteins. Many autophagy-related gene proteins, such as ATG5, are essential for LC3 lipidation at the membrane during the process of selective macroautophagy [23,24]. Alternatively, client proteins are escorted by heat shock cognate 70 kDa protein (HSC70) to lysosomes by interaction with the lysosomal membrane glycoprotein LAMP2A, leading to internalization of the client proteins for degradation. This process is called chaperone-mediated autophagy (CMA). LAMP2A plays an important role in the CMA process by mediating the lysosomal degradation of proteins in response to various stresses and keeping the normal turnover of proteins with a long biological half-life [25]. We have observed that degradation of AHR via autophagy is cell line specific: AHR undergoes selective macroautophagy in human cervical HeLa cells [20] and CMA in triple-negative breast cancer MDA-MB-468 cells [26]. Here, we provide evidence that CMA is responsible for AHR degradation in human lung epithelial A549 cells. Targeting the AHR degradation mechanism can be a viable approach in the lung since modulation of the AHR protein level via CMA alters the AHR function in A549 cells.

2. Results

2.1. Subsection

2.1.1. CQ Increases the AHR Protein Levels of A549 Cells in a Functionally Relevant Manner

CQ is a general autophagy inhibitor that inhibits the lysosomal proteases. We addressed whether AHR undergoes autophagy in A549 cells by examining AHR contents in the presence of CQ. Treatment of A549 cells with 60 and 100 μM CQ for 6 h increased the AHR protein levels to 1.5- and 2-fold, respectively (Figure 1A). To address whether a 2-fold increase in the AHR content would elicit any functional significance, we examined the effect of this fold change on the ligand-dependent upregulation of the AHR target gene *CYP1A1*. As expected, treatment with 100 μM CQ for 6 h increased the AHR protein content by 2-fold in the presence or absence of benzo[α]pyrene (BaP) (Figure 1B). Treatment with 5 μM BaP alone for 4 h reduced the AHR protein content due to proteasomal degradation of AHR after ligand binding [27,28]. We observed that BaP caused a 10-fold increase in the *CYP1A1* transcript after 4 h of treatment (Figure 1C). This upregulation was enhanced from 10- to 38-fold in the presence of 100 μM CQ, showing that a 2-fold increase in the AHR content (when compared between BaP and BaP/CQ treatments) caused a significant increase in the AHR function. Treatment with CQ alone did not change the *CYP1A1* transcript level, confirming that the upregulation of the *CYP1A1* gene transcription was mediated by BaP, not CQ. Collectively, we conclude that a 2-fold increase in the content of A549 AHR by CQ can significantly enhance the BaP-dependent AHR function.

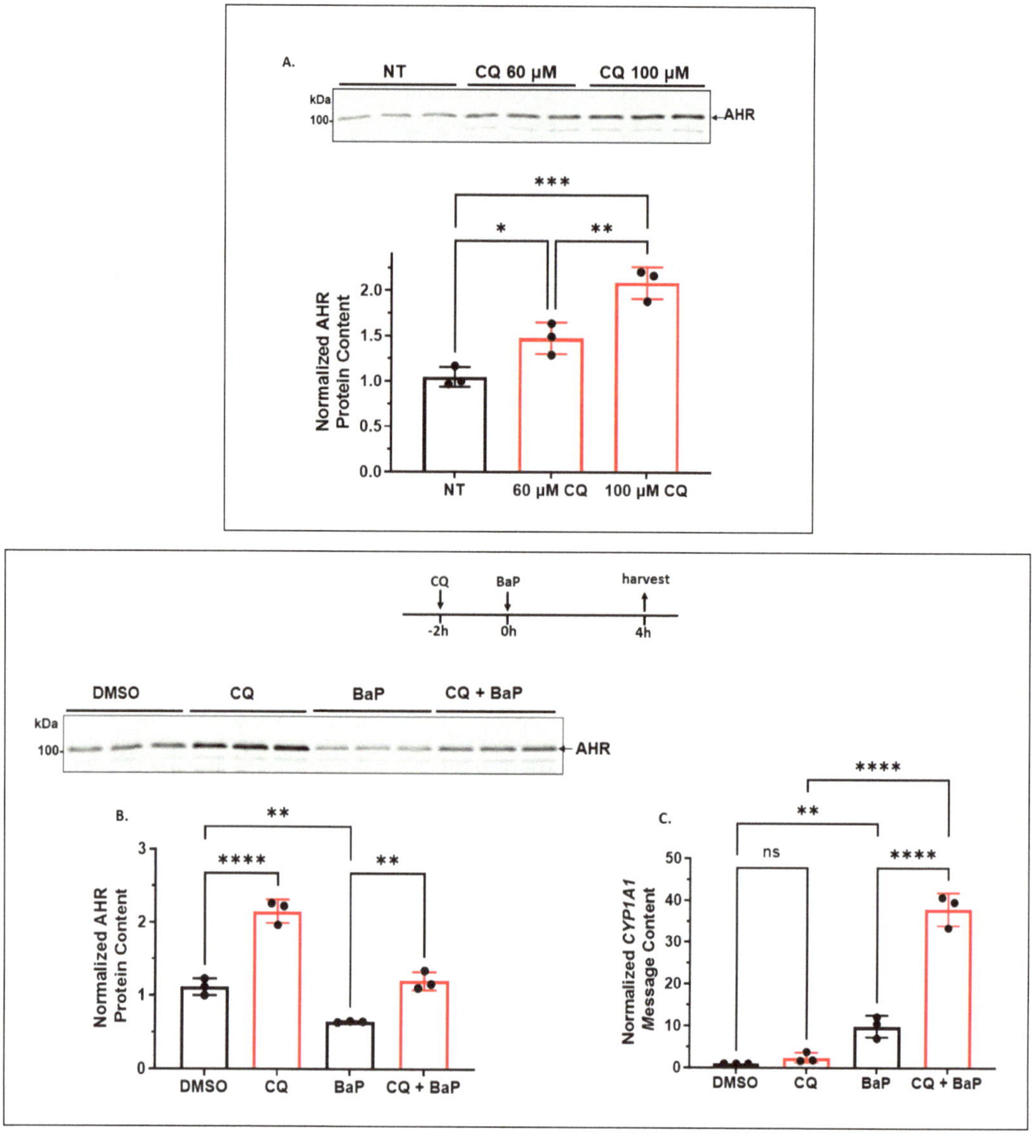

Figure 1. Inhibition of lysosomal degradation increased the AHR protein level and its ligand-dependent activation of the *CYP1A1* gene transcription in A549 cells. (**A**) Western blot results of cells treated with 0, 60 or 100 μM CQ for 6 h. The images above are biological triplicates of one experiment. (**B**) Western blot results of cells treated with 100 μM CQ for 6 h. At 2 h post-CQ treatment, cells were co-treated with DMSO or 5 μM BaP for the remaining 4 h. The images above are biological triplicates of one experiment. (**C**) RT-qPCR results of *CYP1A1* message level. Each experiment was in biological triplicate and was repeated once with similar results. The plots showed as the means with error bars (means ± SD, $n = 3$). The statistical significance of the differences between group means was evaluated by one-way ANOVA using Tukey's test for multiple comparisons. Statistical significance is indicated as follows: * $p < 0.05$, ** $p < 0.01$, *** $p < 0.001$, **** $p < 0.0001$, and $p > 0.05$ (ns, not significant).

2.1.2. 6-AN Reduces the AHR Protein Content of A549 Cells in a Dose- and Time-Dependent Manner with Functional Relevance

Next, we used activators of either selective macroautophagy or CMA to explore the autophagy mechanism for AHR degradation in A549 cells. Treatment with 6-AN, a CMA activator, caused a dose-dependent reduction in the AHR protein content (Figure 2A). This reduction was time-dependent as well, supporting that activation of CMA can effectively reduce the AHR protein content in A549 cells (Figure 2B). When compared to the reduction of the AHR protein content upon ligand (BaP) treatment, both 6-AN and BaP reduced the AHR protein content to a similar extent after 24 h of treatment. This prolonged suppression of the AHR levels is not surprising since TCDD and 3MC can cause a similar suppression of AHR for 24 h in Hepa1c1c7 cells [29]. However, unlike the BaP-induced AHR proteasomal degradation, the reduction of the AHR protein content by 6-AN was reversed in the presence of an autophagy inhibitor (CQ) but not a proteasomal inhibitor (MG132) (Figure 2C,D). Treatment with MG132 even suppressed the AHR protein content further. This is consistent with our finding that treatment of A549 cells with MG132 alone for 6 h also reduced the AHR protein levels, and this reduction was reversed in the presence of CQ (Figure 2E), which can be explained by the literature report that MG132 can induce autophagy [30,31]. In fact, lactacystin, another proteasomal degradation inhibitor, has been reported to activate autophagy partly by upregulating LC3 expression [32]. Interestingly, MG132 reversed the inhibition of AHR autophagy by CQ (Figure 2E), consistent with the notion that AHR undergoes autophagy and MG132 can activate autophagy to degrade AHR. However, both metformin (Met) and rapamycin (Rap), which are selective macroautophagy activators, did not reduce but increased the AHR protein content (Figure 2F,G). This increase can be explained by the crosstalk between selective macroautophagy and CMA in that activation of one autophagy mechanism may negatively regulate the other [33]. In any case, selective macroautophagy does not seem to be involved in the degradation of AHR in A549 cells. To address whether 6-AN can suppress AHR function, we examined the effect of 6-AN on the ligand-induced, AHR-dependent activation of the *CYP1A1* gene transcription. We observed that both BaP (5 μM) and FICZ (1 μM) upregulated the *CYP1A1* message by 10- and 20-fold, respectively, after 6 h of treatment (Figure 2H). 6-AN (100 μM) effectively suppressed the induction to less than 5-fold in both cases, showing that the reduction in the AHR protein content by 6-AN to about 40% of its content in A549 cells significantly hampered its function.

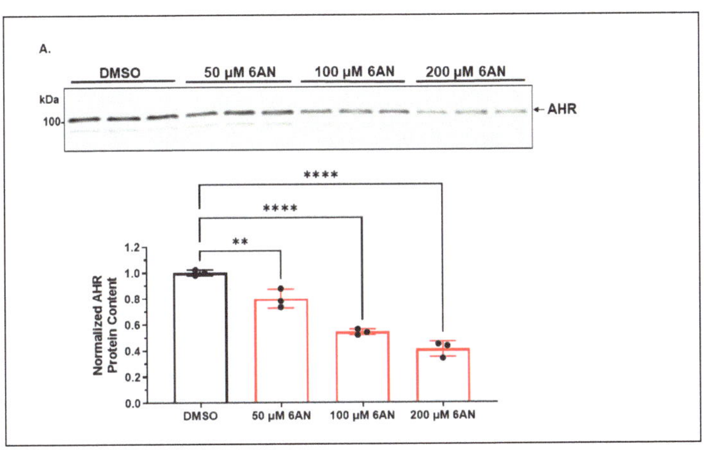

Figure 2. *Cont.*

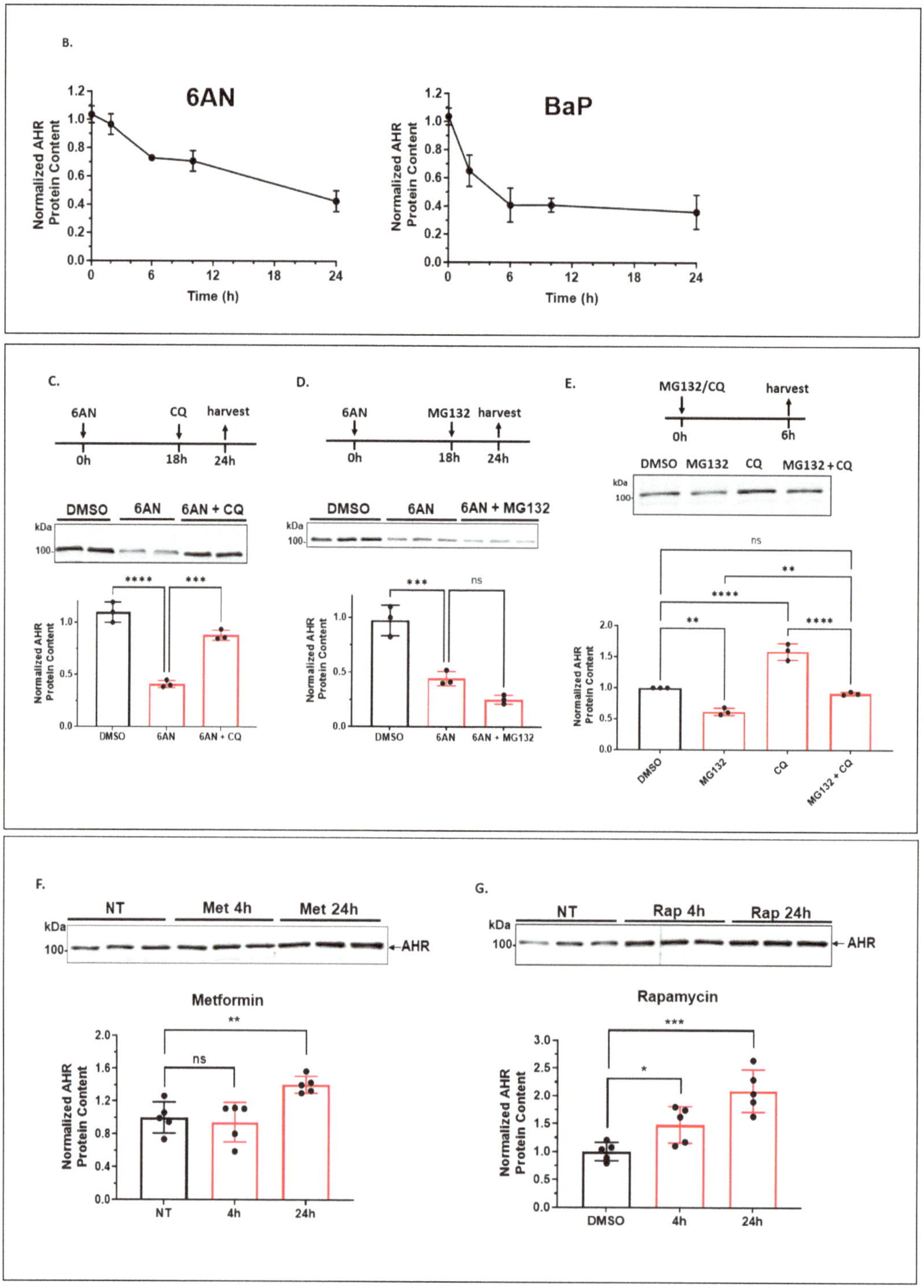

Figure 2. *Cont.*

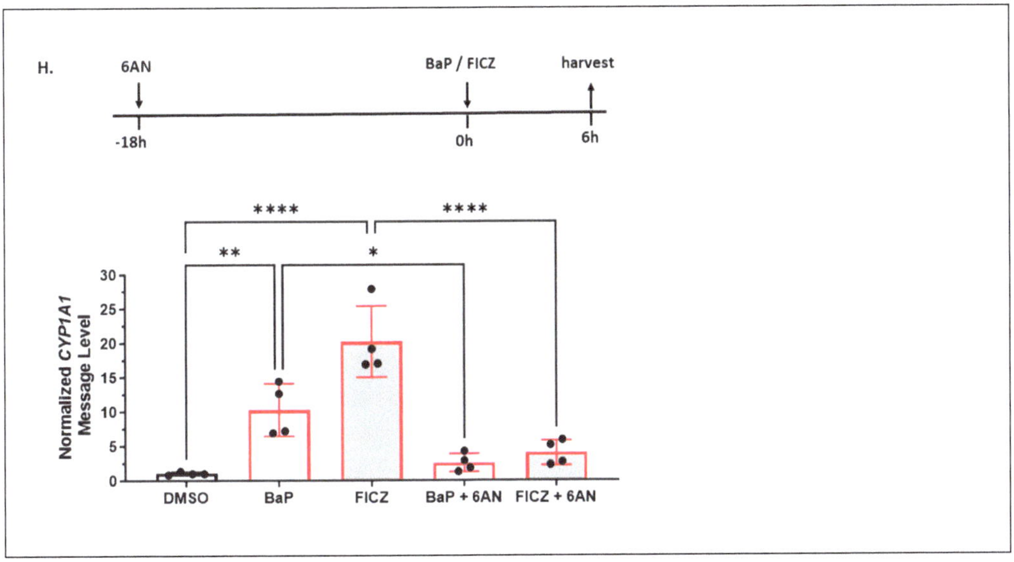

Figure 2. 6-AN decreased the AHR protein level via CMA and suppressed the ligand-induced activation of its target gene transcription in A549 cells. (**A**) Western blot results of cells treated with DMSO or 6-AN (50, 100 or 200 μM) for 24 h. The images above are biological triplicates of one experiment. (**B**) Western blot results of cells treated with 100 μM 6-AN (**B, left panel**) or 5 μM BaP (**B, right panel**) for 0, 2, 6, 10, or 24 h. Each time point represents means ± SD, $n = 3$. (**C**) Western blot results of cells pretreated with DMSO or 100 μM 6-AN for 18 h, followed by 100 μM CQ treatment for an additional 6 h. The images above the plot are biological duplicates of one experiment. This experiment was repeated one more time to generate the plot. (**D**) Western blot results of cells pretreated with DMSO or 100 μM 6-AN for 18 h followed by 10 μM MG132 treatment for an additional 6 h. The images above the plot are biological triplicates of one experiment. This experiment was repeated once with similar results. (**E**) Western blot results of cells treated with DMSO, 10 μM MG132, 100 μM CQ, or 10 μM MG132 plus 100 μM CQ for 6 h. The images are representatives of the plotted data. All plots are the means with error bars (means ± SD, $n = 3$). (**F**) Western blot results of cells treated with 4 mM metformin (Met) for 0, 4 or 24 h. The plot represents the means with error bars (means ± SD, $n = 5$). The images above are biological triplicates of one experiment. The statistical significance of the differences between group means was evaluated by one-way ANOVA using Tukey's test for multiple comparisons. (**G**) Western blot results of cells treated with 0.5 μM rapamycin (Rap) for 0, 4 or 24 h. The plot represents the means with error bars (means ± SD, $n = 5$). The images above are biological triplicates of one experiment. (**H**) RT-qPCR results of cells pretreated with DMSO or 100 μM 6-AN for 18 h, followed by BaP (5 μM) or FICZ (1 μM) treatment for another 6 h. The experiment was performed with biological triplicates and was repeated once with similar results. The plot represents the means with error bars (means ± SD, $n = 3$). The statistical significance of the differences between group means was evaluated by one-way ANOVA using Tukey's test for multiple comparisons. Statistical significance is indicated as follows: * $p < 0.05$, ** $p < 0.01$, *** $p < 0.001$, **** $p < 0.0001$, and $p > 0.05$ (ns, not significant).

2.1.3. Knockdown of LAMP2 in A549 Cells Abolishes the CQ-Mediated Increase of the AHR Protein Content

LAMP2A, a lysosomal membrane-bound protein, is responsible for the internalization of CMA substrates into lysosomes for degradation. To further address whether CMA could degrade AHR in A549 cells, we used shRNA to knock down the *LAMP2* messages, which include the *LAMP2A* message, in A549 cells to see whether the AHR protein content is affected when CMA is less active. Using LAMP2-specific shRNA/siRNA to knock down the *LAMP2A* message is a common approach to downregulate LAMP2A expression. We observed that the AHR protein content increased to 1.7-fold when LAMP2 was down-regulated (Figure 3A), suggesting that LAMP2A plays a role in the degradation of AHR. Results from the LAMP2 Western showed a stretch of the LAMP2 region that represented both LAMP2A and LAMP2B. Additionally, CQ was unable to increase the AHR protein content when LAMP2 was downregulated, supporting that AHR is degraded via CMA in A549 cells. In contrast, knocking down ATG5, which is necessary for LC3 lipidation into LC3-phosphatidylethanolamine conjugate (LC3-II) during the formation of autophagosomes, decreased the AHR protein levels to 76% when compared to the wild-type A549 cells, suggesting that ATG5 is not involved in the degradation of AHR (Figure 3B). The reduction in the AHR content could be the compensatory mechanism that might activate CMA, since this kind of crosstalk between selective macroautophagy and CMA has been reported in the literature [33]. Treatment with CQ did not change the extent of the AHR protein content increase between the wild-type and ATG5 knockdown cells, suggesting that selective macroautophagy is not involved in AHR degradation in A549 cells.

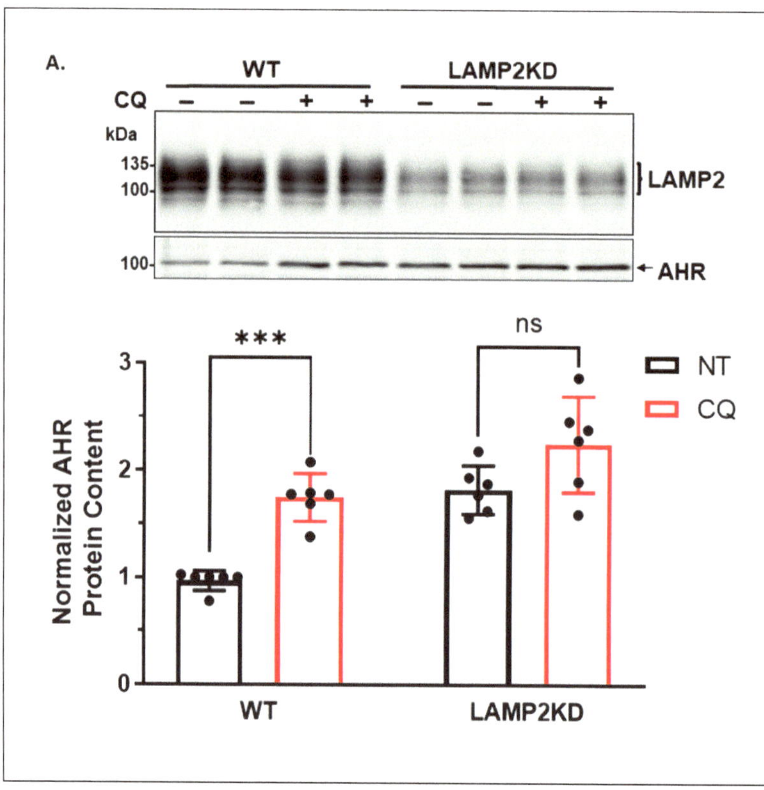

Figure 3. *Cont.*

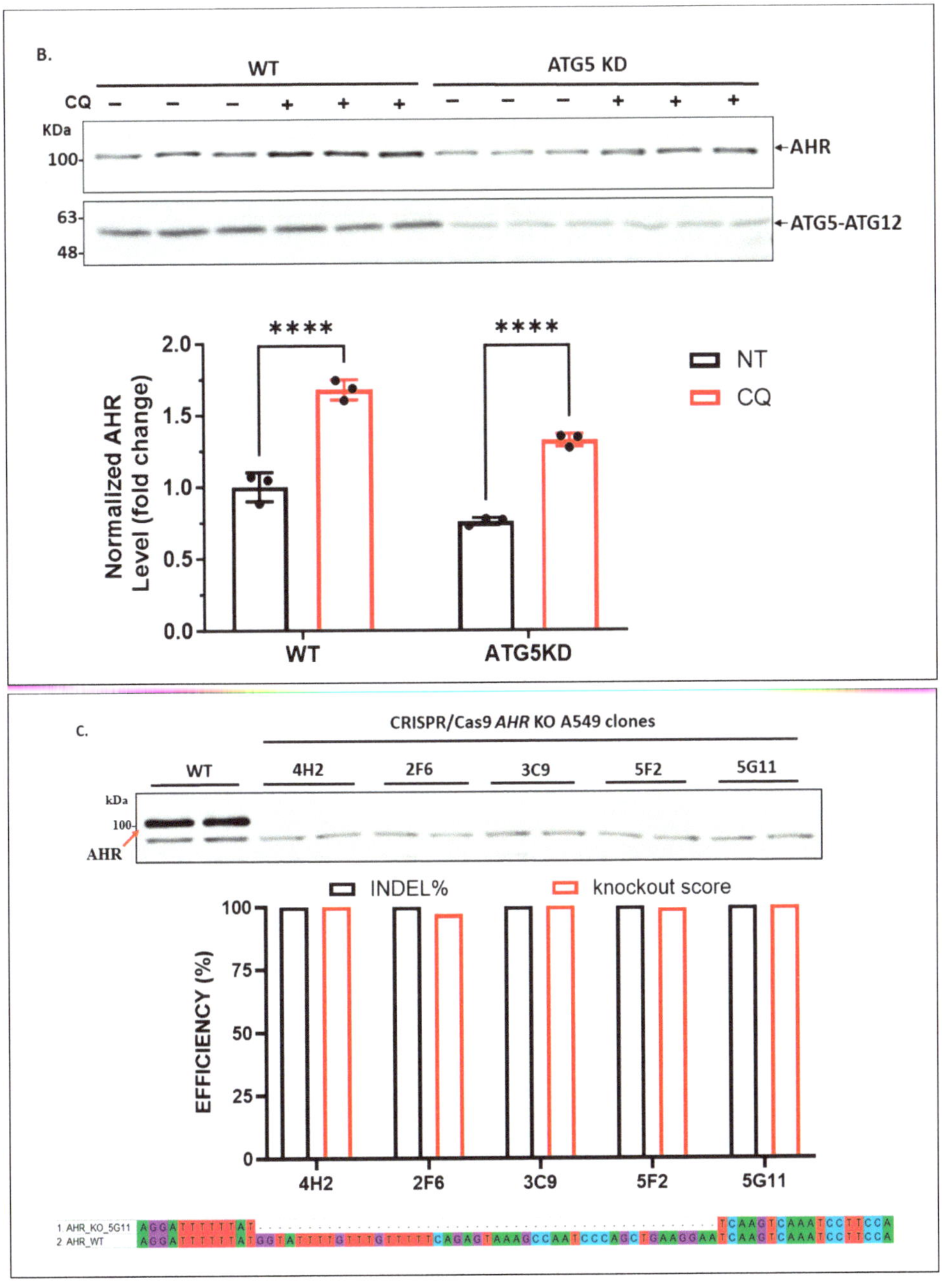

Figure 3. *Cont.*

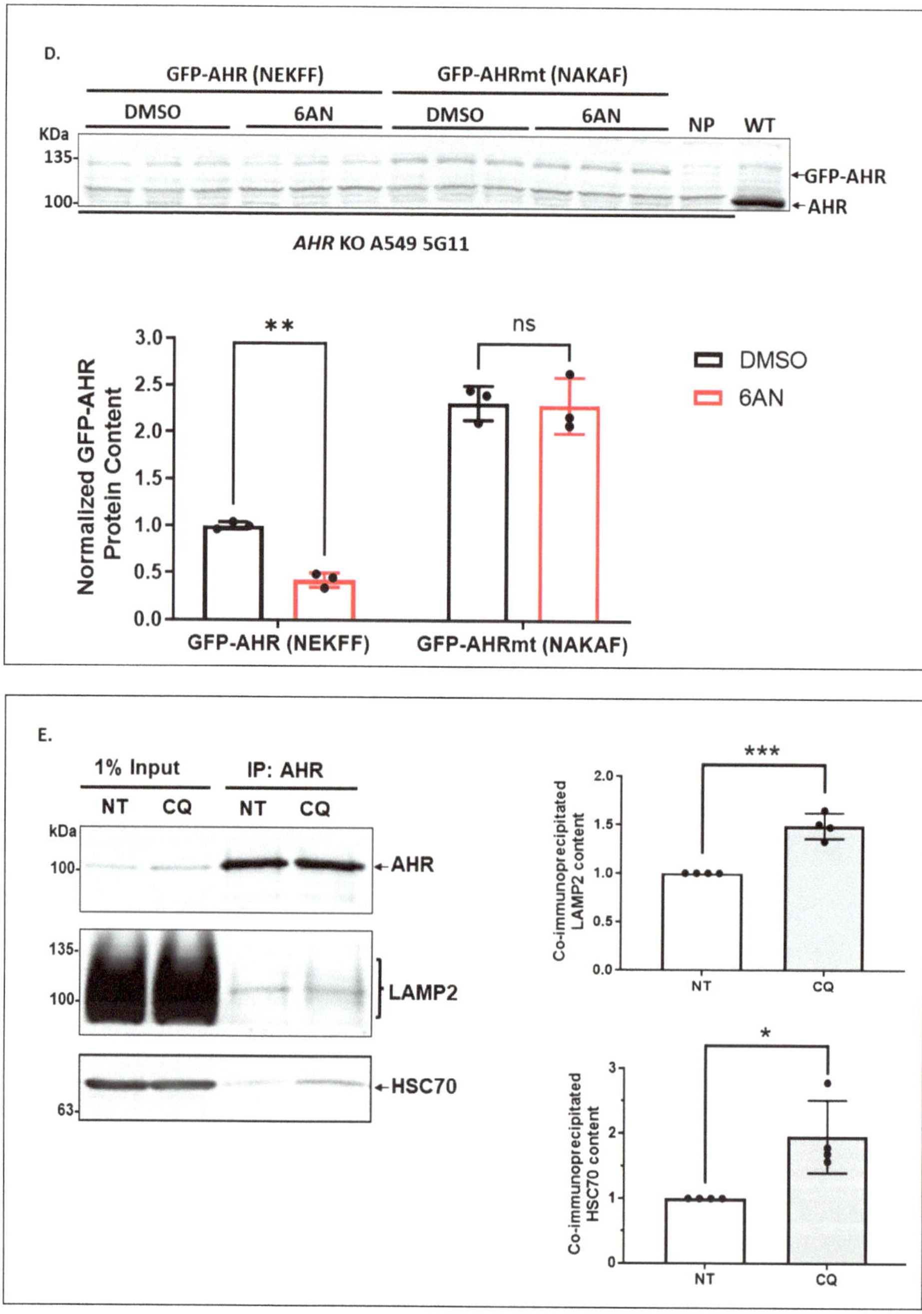

Figure 3. Lysosomal degradation of AHR required LAMP2A and the presence of its CMA motif in A549 cells. (**A**) Western blot results of A549 wild-type (WT) and LAMP2 stable knockdown (LAMP2KD) cells treated with or without 100 μM CQ for 6 h. Stable knockdown cells contain 30% of the wild-type LAMP2 content. Mature LAMP2, including LAMP2A, is highly glycosylated with a total molecular weight of about 100~130 kDa. The plot represents the means with error bars (means ± SD,

n = 6). The images above are biological duplicates of one experiment. (**B**) Western blot results of A549 wild-type (WT) and ATG5 stable knockdown (ATG5KD) cells treated with or without 100 μM CQ for 6 h. Stable knockdown cells contain 28% of the wild-type ATG5 content. Intracellular ATG5 is conjugated with ATG12. In the Western blot, a band at ~55 kDa represents the ATG5-ATG12 complex. The images are biological triplicates from one experiment, and the experiment was repeated once with similar results. (**C**) CRISPR/Cas9-mediated knockout of *AHR* gene in A549 cells. Western blot analysis of AHR protein levels in wild-type (WT) and five CRISPR/Cas9 *AHR* knockout (KO) clones. Sequencing alignment showed that clones 4H2, 2F6, 3C9, and 5F2 are heterozygous compound knockouts. The plot indicates the INDEL efficiencies and knockout scores from the online Inference of CRISPR Edits (ICE) analysis. Clone 5G11 is a homozygous knockout with 47 nucleotides deleted in the exon 2 region of AHR genomic DNA, leading to a frameshift deletion. (**D**) Western blot results of cells transiently expressing GFP-AHR wild type (NEKFF) or GFP-AHR mutant (NAKAF) treated with DMSO or 6-AN for 24 h at 48 h post-transfection. The plot represents the means with error bars (means ± SD, n = 3). The experiment was performed with biological triplicates and was repeated once with similar results. The images represent the biological triplicates of one experiment. NP represents A549 cells that have undergone the transfection process without the GFP plasmid, whereas WT is the wild-type A549 cells. (**E**) Co-immunoprecipitation results of cells treated with or without 100 μM CQ for 6 h. Anti-AHR antibody SA210 was used for immunoprecipitation of AHR (IP). The 1% input represents 1% of the total protein lysate used for immunoprecipitation. The plot represents the means with error bars (means ± SD, n = 4). The images are representative of the plotted data. The statistical significance was evaluated by one-way (**E**) or two-way (**B–D**) ANOVA using Tukey's test for multiple comparisons. Statistical significance is indicated as follows: * $p < 0.05$, ** $p < 0.01$, *** $p < 0.001$, **** $p < 0.0001$, and $p > 0.05$ (ns, not significant).

2.1.4. Degradation of AHR via CMA Is Dependent on the NEKFF Motif of AHR in A549 Cells

Earlier, we reported that NEKFF at amino acids 558–562 of the human AHR could be the CMA motif that allows interaction with HSC70, followed by recruitment to LAMP2A for degradation [26]. To further investigate the CMA-mediated AHR degradation in A549 cells, we examined the necessity of NEKFF for AHR degradation. We had previously generated a few GFP fusions of the wild-type human AHR and its mutants by altering the NEKFF sequence via site-directed mutagenesis [26]. We confirmed the sequence by sequencing the whole AHR cDNAs of the GFP-AHR (NEKFF) and the GFP-AHR mutant (NAKAF). Although the only differences in the amino acid sequence between the GFP-AHR (558-NEKFF-562) and GFP-AHR mutant (558-NAKAF-562) were E559A and F561A, we noticed two random mutations resulting in I581T and S590N when compared to the published human AHR cDNA sequence (NM_001621). To minimize any interference of the A549 AHR on the degradation of the GFP-AHR and GFP-AHR mutant due to shared machinery, we transfected the GFP plasmid into the *AHR* knockout A549 cells. First, we generated five clones of *AHR* knockout A549 cells using the clustered regularly interspaced short palindromic repeats/CRISPR-associated protein 9 (CRISPR/Cas9) strategy, namely 4H2, 2F6, 3C9, 5F2, and 5G11 (Figure 3C). Using the Synthego ICE v3.0 knockout online analysis software, all clones were 100% edited with insertions or deletions (% indel score), and clone 5G11 was determined to be a homologous bi-allelic knockout with 47 nucleotide deletions at exon 2, resulting in a frameshift mutation. Both wild-type and mutant *ahr* cDNAs were cloned downstream to the GFP cDNA, followed by transient transfection into the *AHR* knockout A549 5G11 cells. The GFP fusion of the wild-type AHR containing NEKFF showed similar suppression as the A549 AHR when treated with 100 μM 6-AN for 24 h (Figure 3D vs. Figure 2A). However, the GFP fusion of the NAKAF mutant was resistant to degradation upon 6-AN treatment, supporting that NEKFF is the CMA motif and AHR undergoes the CMA-mediated degradation. This observation is consistent with our immunoprecipitation results that treatment with CQ enhanced the interactions of AHR with HSC70 and LAMP2 in vitro (Figure 3E). These interactions were also observed in MDA-MB-468 cells [26].

2.1.5. Autophagy of AHR Is Ongoing in the Background While AHR Is Undergoing Rapid Degradation via the Ubiquitin–Proteasome System after Treatment with an AHR Ligand in A549 Cells

It is well accepted that upon ligand treatment, AHR undergoes proteasomal degradation within hours of treatment [28]. We were interested in how autophagy of AHR is affected by ligand treatment in A549 cells. As expected, AHR was degraded to 40% of its wild-type content within 2 h of 5 μM BaP treatment via 26S proteasome since cotreatment with 10 μM MG132 completely reversed the degradation (Figure 4A). Cotreatment of BaP-treated cells with 100 μM CQ, however, increased the AHR protein levels to 1.8-fold when compared to cells treated with BaP alone. This fold change was close to the increase we observed when we treated A549 cells with 100 μM CQ for 6 h (Figure 1A), suggesting that AHR underwent the usual rate of degradation via autophagy while proteasomal degradation of AHR was triggered by a ligand. Interestingly, cotreatment of A549 cells with 5 μM BaP and 10 μM MG132 for an additional 4 h (a total of 6 h) showed that the AHR protein levels pronouncedly dropped to less than 30% content (Figure 4A,B). Considering that MG132 can activate autophagy, we examined the possibility that AHR underwent autophagy between the second and sixth hours of MG132 treatment. We exposed the BaP-treated cells to both MG132 and CQ for 6 h and observed only a 2-fold increase of the AHR protein levels (from 29% to 58%) (unpublished data). The strong reduction in the AHR protein levels from the second to the sixth hours of MG132 treatment cannot be fully explained by merely the lysosomal degradation of AHR. The AHR protein levels were similar from 6 h up to 24 h after BaP treatment (Figure 2B, right). Interestingly, similar inhibition of AHR autophagy by CQ was observed within the 6 h to 24 h period since a 1.5 to 1.8-fold increase in the AHR levels was observed when we compared BaP and BaP/CQ treatment groups (Figure 4A–D). Figure 4E contains the representative Western blot images of Figure 4A–D, which showed that from 6 h to 24 h, the AHR levels increased in the presence of CQ, whereas MG132 did not have any effect. Collectively, we concluded that autophagy is likely involved in maintaining AHR levels after ligand treatment.

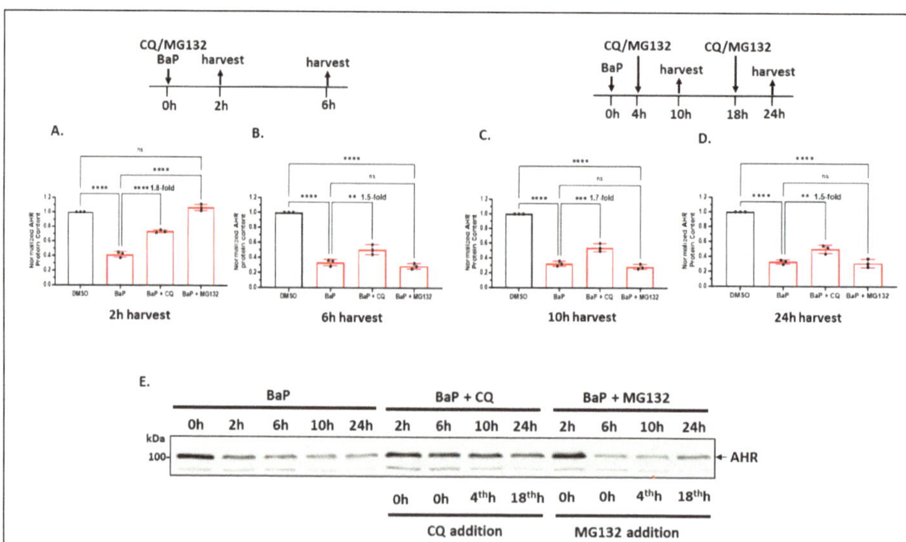

Figure 4. Lysosomal degradation of AHR was ongoing in the background while AHR underwent rapid proteasomal degradation after ligand treatment in A549 cells. Cells were treated with 5 μM

BaP (or DMSO), 100 μM CQ (or water), and 10 μM MG132 (or DMSO) in different combinations, followed by Western blot analysis of the AHR content. Treatment conditions are diagrammed above the plots in (**A–D**). The plots represent the means with error bars (means ± SD, $n = 3$). The experiment was performed with biological triplicates and was repeated once with similar results. The statistical significance of the differences between group means was evaluated by one-way ANOVA using Tukey's test for multiple comparisons. (**E**) A representative sample of the Western blot images of (**A–D**) in one Western blot analysis. Statistical significance is indicated as follows: ** $p < 0.01$, *** $p < 0.001$, **** $p < 0.0001$, and $p > 0.05$ (ns, not significant).

2.1.6. Autophagy Is Not Involved in the Quick-Onset Degradation of AHR Triggered by a Low Dose of Geldanamycin (GA) But Is Involved in Controlling the AHR Levels after Both Low and High Doses of GA Treatment in A549 Cells

It is known that treatment with a low dose of GA (0.1 μM) causes proteasomal degradation of AHR [34]. We examined whether autophagy might also be involved in this GA-mediated degradation. Our results showed that treatment with a low dose of GA (0.1 μM) for 2 h caused a pronounced reduction in the AHR content to 17% in A549 cells, which could be partially reversed to 41% content by 10 μM MG132 (Figure 5A). The fact that MG132 could reverse 100% of the BaP-mediated degradation of AHR (Figure 4A) and, to a much lesser extent, the GA-mediated degradation of AHR (Figure 5A), suggested that the GA effect on AHR degradation is more complex than merely proteasomal degradation. On the other hand, inhibition of autophagy by CQ (100 μM) did not alter the low AHR levels caused by 0.1 μM GA. Realizing that a low dose of GA (0.5 μM) could inhibit autophagy by suppressing the Atg7, Beclin-1, and ULK1 protein levels, as reflected by a reduction in the autophagic flux [35], AHR degradation via autophagy might have been inhibited in the presence of 0.1 μM GA. Although two additional hours of GA exposure (which was 4 h treatment) did not further suppress AHR but rather caused a slight increase in content (28% vs. 17%), MG132 restored more AHR content when compared to 2 h (66% vs. 41%). These results suggested that there might be newly synthesized AHR between 2 and 4 h that was degraded by proteasomal degradation (Figure 5B). In addition, CQ also increased the AHR content from 28% to 43%, suggesting autophagy was also at play between the second and fourth hours of GA treatment.

It has been reported that treatment with a high dose of GA (2–10 μM) for 24 h causes degradation of IKK via the ATG5-dependent autophagy in 293, HeLa, and B and T cells [36]. The activity of CMA is doubled when IMR-90 cells are treated with 2 μM of GA [37]. Therefore, we examined whether autophagy and proteasomal degradation might play a role in the sustained low levels of AHR many hours after a high dose of GA treatment. We treated the cells with 1 μM GA (high dose) for 24 h, followed by exposure to CQ, MG132, or the DMSO vehicle in the last 6 h before harvesting in A549 cells. We used 1 μM because a higher dose (2 μM) of GA caused apparent cell death after 24 h. We observed that CQ increased the AHR levels significantly from 29 to 58%, whereas MG132 did not alter the AHR levels in a statistically significant manner (Figure 5C,D), showing that autophagy, but not proteasomal degradation, is essential for maintaining the AHR protein levels. The slight reduction in AHR levels by MG132 in this case is probably due to the activation of autophagy by MG132, as demonstrated elsewhere [30,31]. Results from the time-dependent experiment showed that a high dose of GA suppressed the AHR protein levels more effectively than a low dose of GA (Figure 5E), supporting that a high dose of GA causes more AHR degradation, possibly through autophagy. However, we cannot rule out the possibility that a higher dose of GA may also cause an increase in the proteasomal degradation of AHR. Collectively, AHR undergoes autophagy, which can be triggered by a high dose of GA and blocked by CQ (Figure 5C,E).

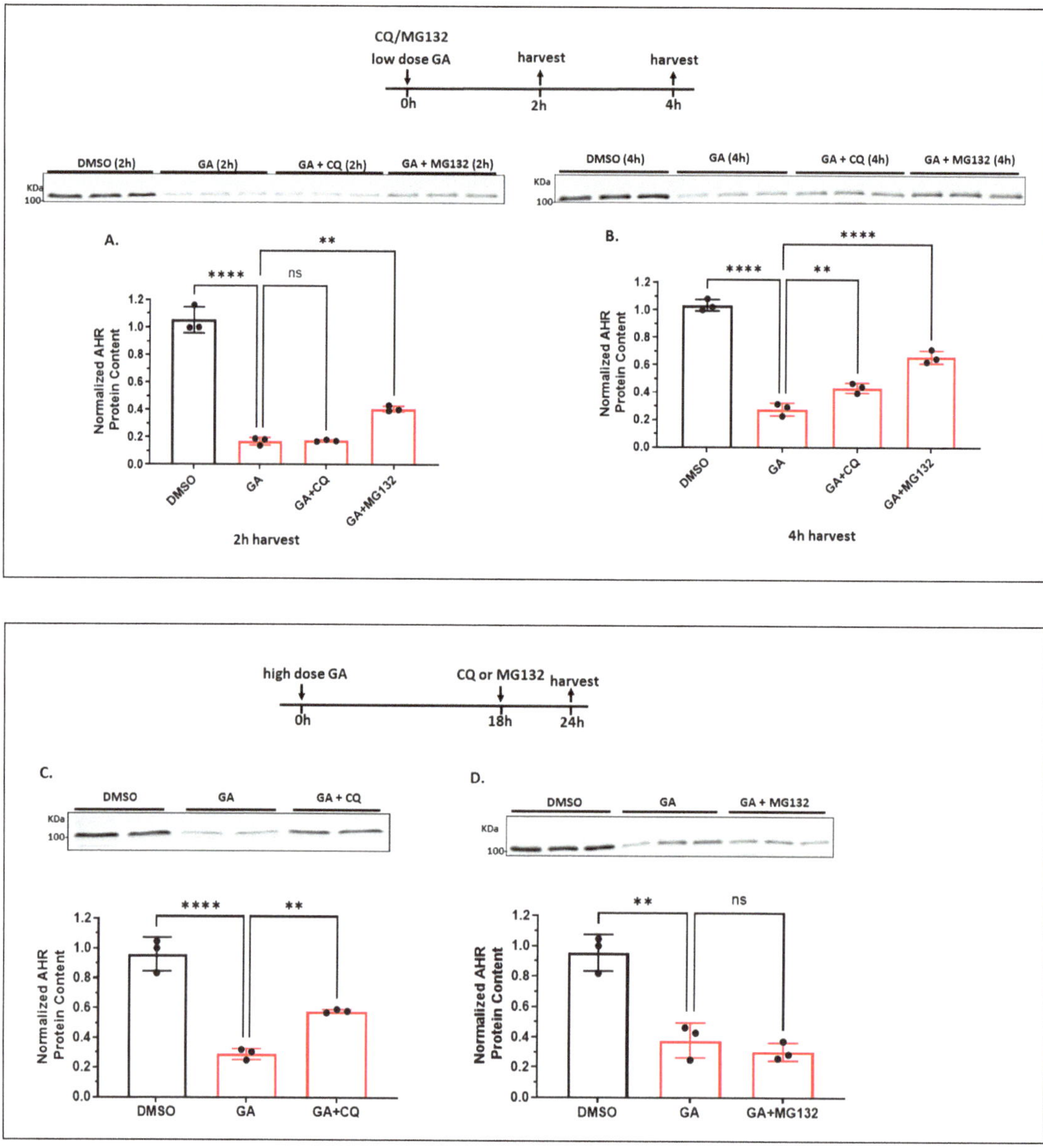

Figure 5. *Cont.*

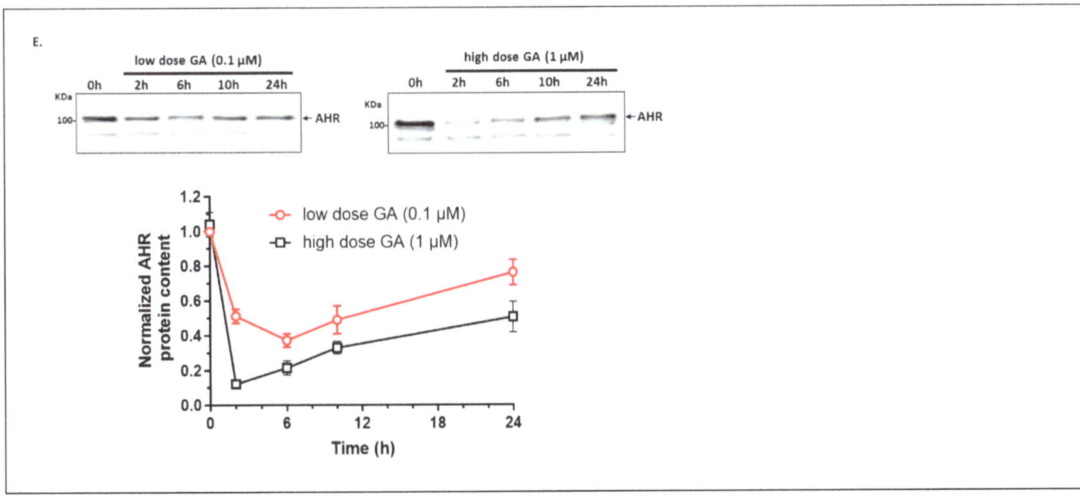

Figure 5. Lysosomal degradation of AHR occurred after A549 cells were treated with low or high doses of GA. Western blot results of cells treated with DMSO, 0.1 μM GA (low dose), 0.1 μM GA plus 100 μM CQ, or 0.1 μM GA plus 10 μM MG132 for 2 h (**A**) or 4 h (**B**). The images represent the biological triplicates of one experiment. (**C**) Western blot results of cells pretreated with 1 μM GA (high dose) for 18 h, followed by 100 μM CQ treatment for another 6 h. The images represent the biological duplicates of one experiment. (**D**) Western blot results of cells pretreated with 1 μM GA (high dose) for 18 h, followed by 10 μM MG132 treatment for another 6 h. The images represent the biological triplicates of one experiment. For (**A–D**), all plots represent the means with error bars (means ± SD, $n = 3$); all experiments were performed with biological triplicates and were repeated once with similar results; the statistical significance of the differences between group means are evaluated by one-way ANOVA using Tukey test for multiple comparisons. (**E**) A plot showing the Western blot results of cells treated with either a low dose of GA (0.1 μM) or a high dose of GA (1 μM) for 0, 2, 6, 10, or 24 h. Each time point represents means ± SD, $n = 3$. Statistical significance is indicated as follows: ** $p < 0.01$, **** $p < 0.0001$, and $p > 0.05$ (ns, not significant).

2.1.7. 6-AN Is Not an AHR Ligand

Although our data supported that 6-AN activates CMA to degrade AHR, we examined whether 6-AN is also an AHR ligand, knowing that the binding of an AHR ligand normally causes AHR protein degradation. For this experiment, we used rat H4G1.1c3 cells stably expressing the DRE-driven GFP protein [26]. Treating these cells with a prototypical AHR ligand such as BaP and FICZ for 12 h caused the GFP expression that could be captured by fluorescence microscopy (Figure 6). This GFP fluorescence was suppressed in the presence of an AHR antagonist CH223191, supporting that the fluorescence corresponded to the AHR ligand-activated GFP expression. We observed that 100 μM 6-AN, which caused AHR protein degradation at that concentration, did not show any more fluorescence than the DMSO vehicle control, supporting that 6-AN is not an AHR ligand. We also determined that both MG132 (10 μM) and GA (0.1 μM) did not cause GFP expression and are not AHR ligands; however, all of them (6-AN, MG132, and GA) can reduce AHR protein levels (Figures 2A,E and 5A).

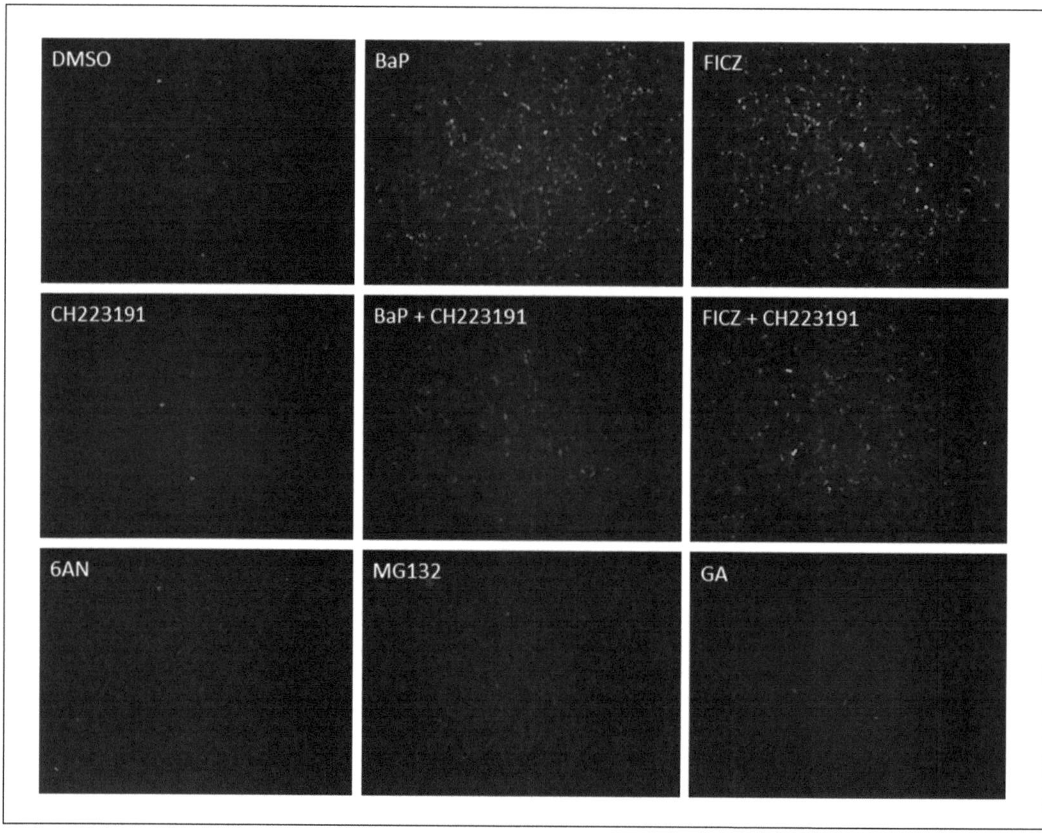

Figure 6. 6-AN, (S)-MG132, and GA are not AHR ligands. Treatment with 6-AN, MG132, GA, BaP, FICZ, and CH223191 in rat H4G1.1c3 stable cells carrying a DRE-driven GFP cDNA. Cells were treated for 12 h with 0.2% DMSO, 100 μM 6-AN, 10 μM MG132, 100 nM GA, 5 μM BaP, 1 μM FICZ, 10 μM CH223191, 5 μM BaP plus 10 μM CH223191, or 1 μM FICZ plus 10 μM CH223191. The experiment was repeated two more times with similar results. Magnification: 4×.

2.1.8. AHR Promotes Migration of A549 Cells in a Wound Healing Assay

There have been conflicting reports on the role of AHR in cell migration and metastasis, particularly on lung epithelial cells, in which some researchers showed suppression of migration and metastasis by AHR using the knockdown approach in A549 cells [14]. Here, we used our *AHR* knockout (5G11) A549 cells to determine the role of AHR on cell migration in a wound healing assay. We observed that A549 cells migrated slower when the AHR protein was absent (Figure 7A). Promoting CMA by 100 μM 6-AN treatment of A549 cells for up to 48 h showed similar retardation of migration as observed in *AHR* knockout 5G11 cells. This inhibition of cell migration by 6-AN was expected since 6-AN was shown to suppress the migration of acute myelogenous leukemia cells by inhibiting enzymes involved in the pentose phosphate pathway [38]. However, this suppression of migration by 6-AN was abolished in *AHR* knockout 5G11 cells, supporting that this 6-AN effect on A549 cells is AHR-dependent.

2.1.9. AHR Promotes EMT in A549 Cells

Next, we examined the role of AHR on the EMT of A549 cells. We measured the transcript and protein levels of an epithelial marker, E-cadherin, and the mesenchymal markers vimentin and N-cadherin in wild-type and *ahr* KO 5G11 A549 cells. We observed that the transcript levels of *E-cadherin* and *Vimentin* were 2.6- and 0.5-fold, respectively, in *AHR* knockout 5G11 cells when compared to the wild-type A549 cells (Figure 7B), although the reduction in the *vimentin* transcript was not significant. The Western blot results showed that E-cadherin and vimentin proteins had the same trend as the transcripts, with a clear increase in E-cadherin and a decrease in vimentin in a statistically significant manner (Figure 7C). These results clearly supported the fact that AHR favors the mesenchymal phenotype in A549 cells. However, there was no change in the *N-Cadherin* transcript and protein levels in the presence or absence of AHR (Figure 7B,C). Treatment of A549 cells with 100 μM 6-AN for 24 h did not seem to alter the E-cadherin and vimentin protein levels in A549 cells, and lacking AHR in 5G11 did not reveal any trend of the 6-AN effect that was AHR-dependent (Figure 7C). Realizing that there should be about 50% of AHR content after this 6-AN treatment, it is conceivable that 50% of AHR content might be sufficient to maintain the levels of these markers, and 6-AN might very well elicit some AHR-independent effect that complicated the picture. Next, we were interested to see any effect on these markers when the AHR content was increased < 2-fold by knocking down LAMP2 in A549 cells (Figure 3A). We observed that vimentin was upregulated in LAMP2 knockdown A549 cells when compared to the wild-type A549 cells (Figure 7D), consistent with the predicted pattern when AHR was upregulated. However, E-cadherin was also upregulated, showing a more complex picture that might be complicated by the inhibition of the CMA-mediated degradation of an EMT inhibitor when LAMP2 was downregulated. Downregulation of LAMP2 might also trigger partial EMT, causing cells to co-express both epithelial and mesenchymal markers. During partial EMT, the retained epithelial markers (such as E-cadherin) were shown to cluster cancer cells before migration [39]. Interestingly, upregulation of E-cadherin expression might not be sufficient to block invasion, such as in the case of pancreatic cancer [40]. When we repeated the LAMP2 knockdown experiment using *AHR* knockout 5G11 cells, we observed that E-cadherin and vimentin were essentially unchanged when LAMP2 was knocked down, whereas a slight increase in N-cadherin levels was observed (Figure 7E). When comparing these results with the *AHR* knockout results in Figure 7C, knocking down LAMP2 appeared to suppress E-cadherin and increase vimentin levels, revealing a mesenchymal phenotype that was LAMP2-dependent but not AHR-dependent.

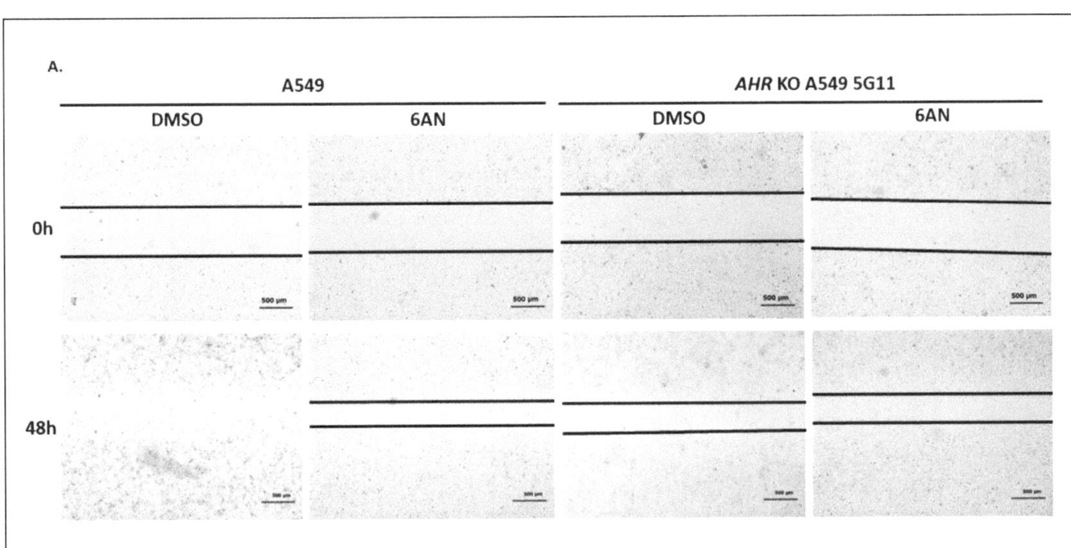

Figure 7. *Cont.*

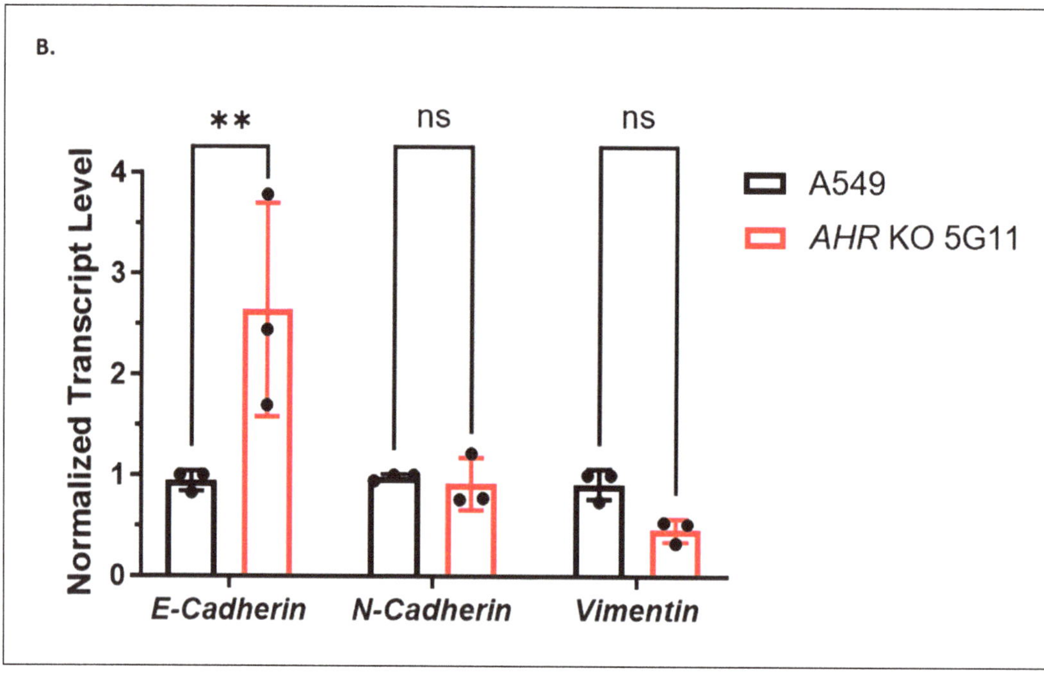

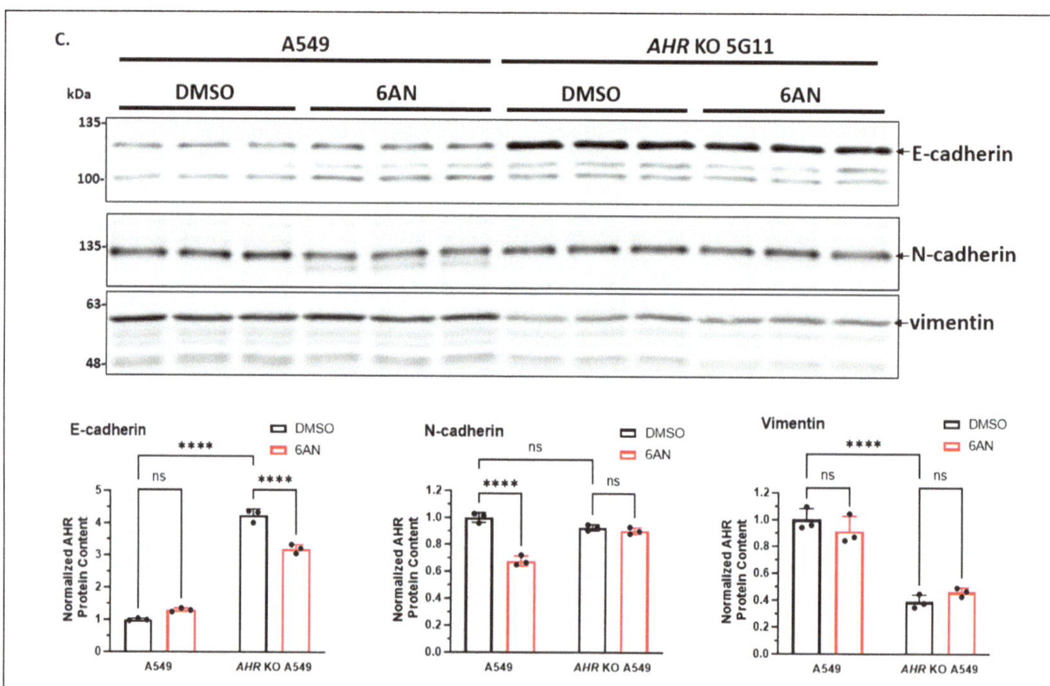

Figure 7. *Cont.*

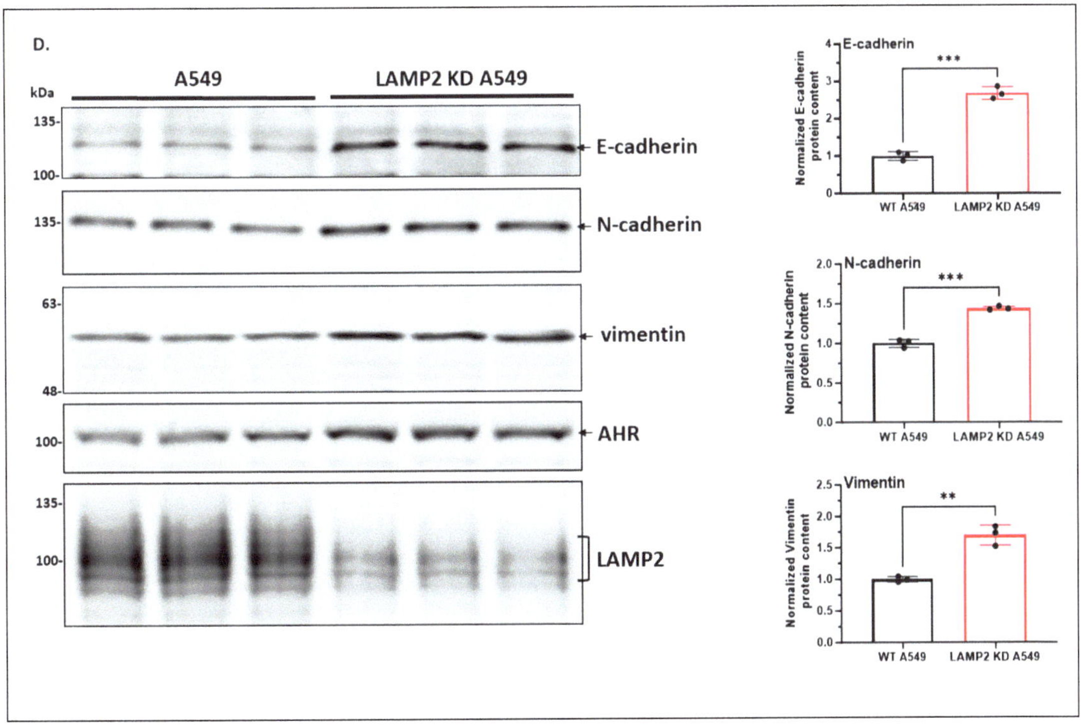

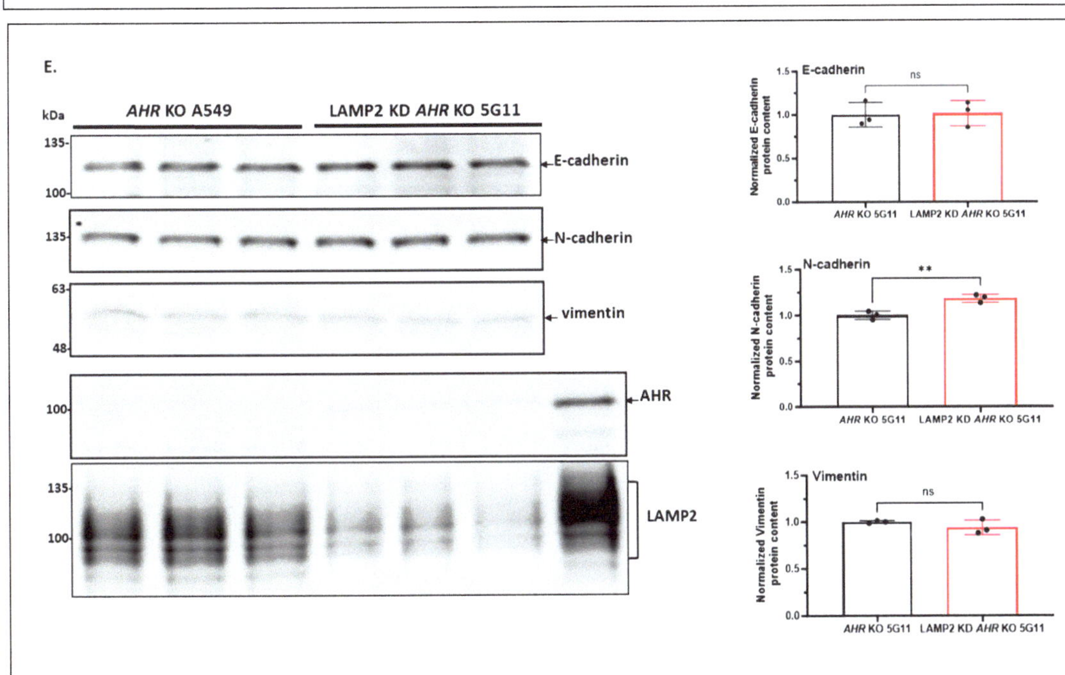

Figure 7. *Cont.*

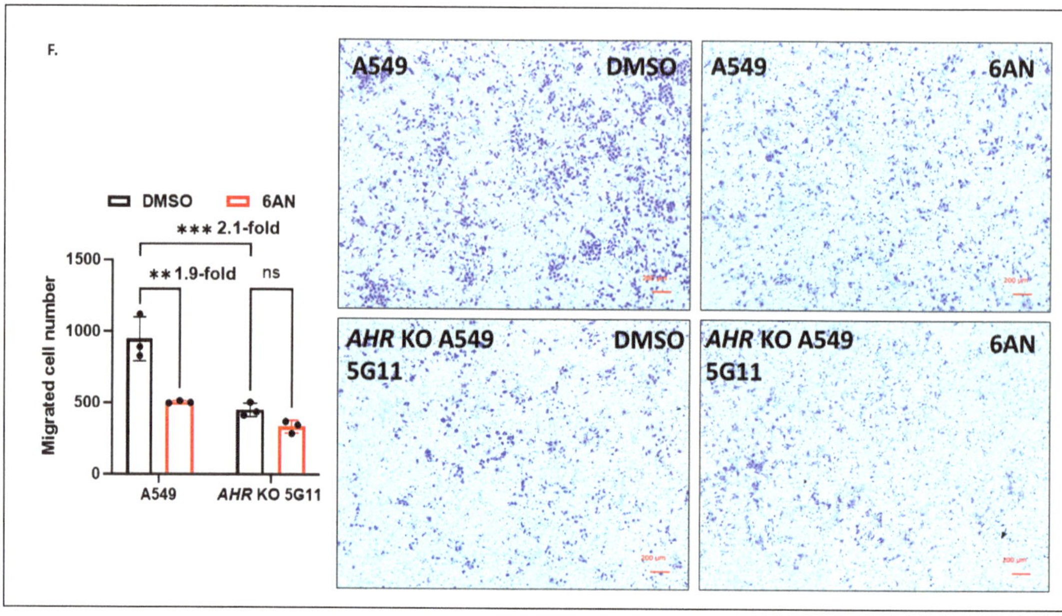

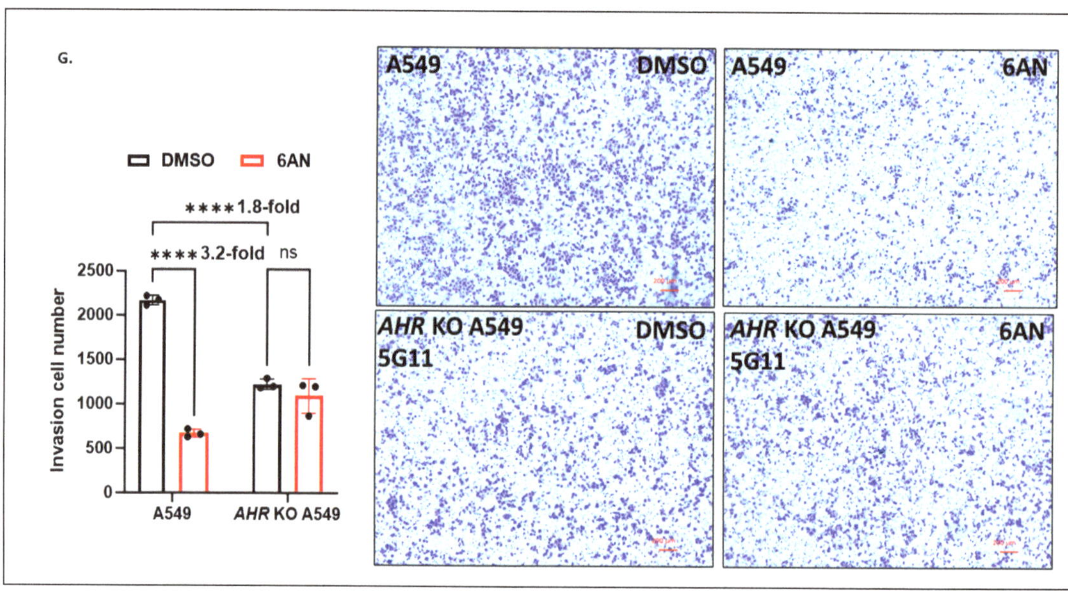

Figure 7. Suppression of the AHR protein level slowed down EMT in A549 cells. (**A**) Wound healing assay results showing the migration capacities of the wild-type and *AHR* knockout (KO, 5G11) A549 cells. Both cell lines were treated with DMSO or 100 μM 6-AN for 48 h. The microscopy images of wound closure were captured at 0 and 48 h after wounding. The black lines represent the area lacking cells. Scale bars, 500 μm. The experiment was repeated two more times with similar results. Magnification: 4×. (**B**) RT-qPCR results of the message levels of *E-cadherin*, *N-cadherin*, and *vimentin* in wild-type and *AHR* knockout (KO, 5G11) A549 cells. The plot represents the means with error bars (means ± SD, n = 3). The statistical significance of the differences between group means was evaluated

by two-way ANOVA using Sidak's test for multiple comparisons. Western blot results of E-cadherin, N-cadherin, and vimentin in (**C**) *AHR* knockout (KO, 5G11), (**D**) LAMP2 stable knockdown (LAMP2KD), and (**E**) LAMP2 stable knockdown (LAMP2 KD) of *AHR* knockout (KO) 5G11 A549 cells are shown as biological triplicates in one blot. Corresponding images in A549 wild-type cells are shown as the controls. 6-AN condition in C was performed with 100 μM 6-AN for 24 h. (**F**) Transwell migration assay results of the wild-type and *AHR* knockout (KO, 5G11) A549 cells treated with DMSO or 6-AN for 24 h. (**G**) Transwell invasion assay results of wild-type and *AHR* knockout (KO, 5G11) A549 cells treated with DMSO or 100 μM 6-AN for 24 h. Representative images of migrated and invaded cells were captured. Scale bars, 200 μm. The numbers of migrated and invaded cells were quantitated by ImageJ v1.53k software. All plots (**B–G**) represent the means with error bars (means ± SD, $n = 3$) and were repeated once with similar results. The statistical significance of the differences between group means was evaluated by two-way (**B,C,F,G**) ANOVA using Tukey's test for multiple comparisons. The statistical significance of the differences between group means was evaluated by unpaired *t*-test (**D,E**). Statistical significance is indicated as follows: ** $p < 0.01$, *** $p < 0.001$, **** $p < 0.0001$, and $p > 0.05$ (ns, not significant).

To directly address the migration and invasion potential mediated by AHR, we conducted Transwell assays to determine the migration and invasion potential of the wild-type and *AHR* knockout A549 cells. We observed that AHR significantly promoted cell migration and invasion by 2.1- and 1.8-fold, respectively, when we compared the *AHR* knockout 5G11 with the wild-type A549 cells (Figure 7F,G). 6-AN similarly suppressed the migration and invasion of A549 cells by 1.9- and 3.2-fold, respectively, in an AHR-dependent manner since this 6-AN effect was abrogated in the *AHR* knockout cells. Collectively, AHR clearly drives the invasion and migration of A549 cells. Activating the CMA degradation of AHR by 6-AN effectively slows down the invasion and migration of A549 cells in a mechanism that cannot be fully explained by E-cadherin and vimentin levels.

3. Discussion

Without the addition of exogenous ligand, AHR is subjected to autophagy in several human cell lines. Two autophagic mechanisms, namely selective macroautophagy and CMA, can selectively degrade client proteins such as human AHR. Here, we provide evidence that AHR undergoes CMA regularly in A549 cells without the addition of any exogenous ligand (Figure 8). This CMA-mediated AHR degradation can be activated by 6-AN in a dose- and time-dependent manner. Down-regulation of LAMP2A or mutation of the CMA motif of AHR abrogates this effect, strongly supporting the role of CMA in degrading AHR in A549 cells. Importantly, altering the degradation of AHR clearly affects its activities in the ligand-activated gene transcription and modulation of the EMT of A549 cells. In the case of HeLa cells, downregulation of LC3 impairs the degradation of AHR, supporting that selective macroautophagy is responsible for AHR protein degradation [20]. Interestingly, unlike A549 cells, treatment with 100 μM 6-AN for 24 h does not change the AHR protein levels in HeLa cells [20], revealing that selective macroautophagy, but not CMA, degrades AHR in a different human cell type. However, we also observed that AHR undergoes LAMP2A-mediated degradation in the triple-negative MDA-MB-468 human breast cancer cells [26], revealing that AHR can undergo different autophagy mechanisms in a cell-specific manner. There is precedent in the literature that maintaining cellular protein levels via autophagy can be important. For example, CMA is responsible for the degradation of SMAD3 [41], Erk3 [42], Dicer [43], and oxidized PRL2 [44]. Analogous to human AHR, Tau and α-synuclein have been reported to undergo both selective macroautophagy and CMA [45–47]. Although AHR can be degraded via either CMA or selective macroautophagy in a cell-line-specific manner, it is yet unclear how different human cells select which of the two autophagy mechanisms, or both, to degrade AHR.

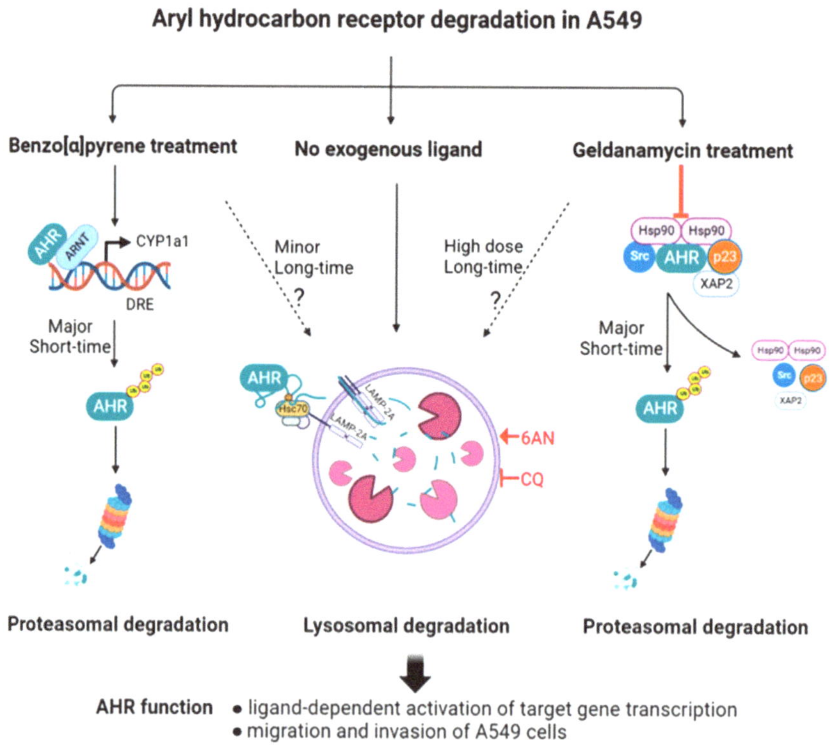

Figure 8. A diagram of CMA-mediated degradation of AHR in A549 cells. This figure was created with BioRender.com, www.biorender.com (accessed on 8 September 2023).

Targeting autophagy-dependent AHR degradation can potentially be an effective therapeutic approach. For example, activation of AHR in the gut can be beneficial for the treatment of inflammatory bowel disease [48]. Thus, AHR agonists might be effective for this treatment. Interestingly, P140, a 21mer phosphopeptide derived from U1-70K spliceosomal protein, suppresses CMA and is proposed to be a mechanism for the treatment of inflammatory bowel disease [49]. Suppression of CMA might increase the AHR protein levels in the gut and thereby elicit a synergistic effect of AHR activation when used in combination with an AHR agonist—an interesting regimen for the treatment of inflammatory bowel disease.

The carboxy terminus of Hsc70-interacting protein (CHIP), a co-chaperone of HSC70, contains a U-box ring-finger motif found in ubiquitin ligases. This CHIP has been implicated in promoting proteasomal degradation of CFTR [50], tau [51], and hypoxia-inducible factor 1 alpha (HIF-1α) [52]. It is well known that binding HSC70 to the CMA motif of proteins escorts proteins to LAMP2A at the lysosomal membrane, followed by the internalization of proteins for degradation via CMA. Interestingly, HIF-1α also undergoes CMA, and CHIP is required for the interaction of HSC70 and HIF-1α interaction [53]. Like HIF-1α, AHR has been shown to interact with CHIP in vitro [54]. It is conceivable that CHIP may be involved in the degradation of AHR via CMA.

Although it has been widely accepted that AHR ligands and GA cause degradation of AHR via the ubiquitin–proteasome pathway, it appears that autophagy might also be involved in determining the AHR levels after ligand or GA treatment in A549 cells (Figure 8). Clearly, proteasomal degradation is primarily responsible for the degradation of AHR within 2 h of ligand or low-dose GA treatment in A549 cells. However, in time, autophagy becomes active in degrading AHR in A549 cells, suggesting that degradation of

AHR via autophagy may be temporarily interrupted soon after a low dose of GA treatment, which activates proteasomal degradation of AHR. A high dose of GA (1 μM), however, promotes AHR degradation via autophagy in A549 cells.

Glutathione peroxidase 4 (GPX4) is known to be a CMA substrate. Inhibition of CMA increases GPX4 protein levels, leading to the inhibition of ferroptosis in mouse hippocampal HT-22 cells [55]. It has been reported that increased AHR protein levels inhibit ferroptosis in human lung adenocarcinoma PC-9 cells by increasing the expression of SLC7A1, a molecule that suppresses ferroptosis by reducing the reactive oxygen species content [56]. Interestingly, AHR is also a CMA substrate, and like GPX4, inhibition of the CMA-mediated degradation of AHR can be a viable approach for the regulation of ferroptosis.

Resveratrol has been reported to inhibit ligand-activated AHR transcriptional activity in a manner that does not act as a typical AHR antagonist since no binding of resveratrol to AHR was observed [57]. Nevertheless, resveratrol inhibits AHR function. Interestingly, resveratrol also inhibits the transforming growth factor beta 1-mediated EMT by downregulating vimentin and upregulating E-cadherin in A549 cells [58]. This is the same outcome as observed in the *AHR* knockout A549 cells, suggesting that inhibition of the AHR function by resveratrol may be, in part, responsible for EMT inhibition in A549 cells.

Although there are conflicting data in the literature showing the effect of AHR on the EMT in A549 cells, we clearly observed that AHR promotes the EMT in these cells via the CRISPR/Cas9 knockout approach. When the *ahr* gene is disrupted in A549 cells with no AHR protein production, E-cadherin is clearly upregulated, whereas vimentin is also clearly downregulated, supporting the epithelial phenotype when AHR is absent. Wound healing, invasion, and migration experiments all unambiguously support the fact that AHR drives the EMT in A549 cells. Our finding is consistent with other researchers reporting that knocking down AHR in A549 cells suppresses invasion and migration potential [56]. Additionally, AHR drives non-small cell lung cancer tumorigenesis when the AHR protein is stabilized after deubiquitination by ubiquitin carboxy-terminal hydrolase isozyme L3 [59], and this AHR action of cancer tumorigenesis may involve Jak2/STAT3 signaling [60].

When we attempted to modulate AHR protein levels by either 6-AN or LAMP2 knockdown, we were unable to see any consistent trend of E-cadherin, vimentin, and N-cadherin expression that was AHR-dependent. Naturally, we must keep in perspective the non-AHR dependent effects of 6-AN and LAMP2 knockdown on EMT marker expression. In the case of LAMP2 knockdown, mechanisms such as inhibition of the transcriptional activation of an EMT inducer that favors EMT and upregulation of some EMT inhibitor that favors the epithelial phenotype could be involved. Examining more than three markers of tumor invasion may be necessary in this case. For example, the levels of matrix metalloproteinase (MMPs), actin cytoskeleton proteins, and cell–extracellular matrix interaction molecules can be measured to possibly provide a better picture of how AHR drives EMT [61,62].

Exploring how modulation of the AHR protein levels may alter cancer stem cell-like properties and associated gene expression can be insightful since overexpression of AHR has been implicated in an aggressive tumor phenotype in non-small cell lung cancer [60]. Nevertheless, treatment with 6-AN clearly suppresses the invasion and migration of A549 cells in an AHR-dependent manner. Although looking at the EMT marker levels alone fails to explain how a reduction in AHR levels (by 6-AN) favors the epithelial phenotype of A549 cells, modulation of AHR protein levels can be a viable approach in controlling AHR's function.

4. Materials and Methods

4.1. Reagents and Antibodies

CQ, BaP, GA, 6-AN, puromycin, CH223191, crystal violet, PMSF, and leupeptin were purchased from Sigma-Aldrich (St. Louis, MO, USA). (S)-MG132 and FICZ were purchased from Cayman Chemical (Ann Arbor, MI, USA). pLKO.1 Lentiviral LAMP2 shRNA plasmids and pLKO.1 Lentiviral ATG5 shRNA plasmids were purchased from Dharmacon (Lafayette, CO, USA). pGFP^2-N2-AHR(NEKFF) and pGFP^2-N2-mutant AHR(NAKAF) plasmids were previously generated by our lab [26]. pCMV-VSV-G was a gift from Bob Weinberg (Addgene plasmid # 8454; RRID: Addgene_8454). pCMV-dR8.2 dvpr was a gift from Bob Weinberg (Addgene plasmid #8455; RRID: Addgene_8455). EndoFectin transfection reagent was purchased from GeneGopoeia (Rockville, MD, USA). TRI Reagent and a Direct-zol RNA miniprep kit were purchased from Zymo Research (Irvine, CA, USA). MMLV high-performance reverse transcriptase was purchased from Epicentre (Madison, WI, USA). iTaq SYBR green supermix was purchased from Bio-Rad (Hercules, CA, USA). Dynabeads Protein G was purchased from Invitrogen (Carlsbad, CA, USA). EnGen Spy Cas9 NLS was purchased from NEB (Ipswich, MA, USA), multi-guide sgRNA was purchased from Synthego (Redwood, CA, USA), Lipofectamine RNAiMAX was purchased from Thermo Fisher Scientific (Waltham, MA, USA), and QuickExtract DNA extraction solution was purchased from Lucigen (Middleton, WI, USA). PCR Master Mix was purchased from Promega (Madison, WI, USA), and Falcon 24-well cell culture inserts with transparent PET membranes (8.0 μm pore size) and Matrigel matrix were purchased from Corning Inc. (Corning, NY, USA). FBS (HyClone), DMEM (HyClone), and Opti-MEM reduced serum medium were purchased from Thermo Fisher Scientific (Waltham, MA, USA). GlutaMAX-I and penicillin–streptomycin were purchased from Invitrogen (Carlsbad, CA, USA). Rabbit anti-AHR polyclonal antibody (SA210) was purchased from Enzo Life Sciences (Farmingdale, NY, USA). Mouse anti-LAMP2 monoclonal antibody (H4B4), mouse anti-Hsc70 monoclonal antibody (B-6), mouse anti-E-cadherin monoclonal antibody (G-10), mouse anti-Vimentin monoclonal antibody(V9), and mouse anti-N-cadherin monoclonal antibody (13A9) were purchased from Santa Cruz Biotechnology (Dallas, TX, USA). Rabbit anti-ATG5 polyclonal antibody (2630) was purchased from Cell Signaling Technology (Danvers, MA, USA). Donkey anti-rabbit and donkey anti-mouse secondary antibody conjugated with IRDye 680 or 800CW, Revert 700 Total Protein Stain, and nitrocellulose membrane were purchased from LI-COR Bioscience (Lincoln, NE, USA).

4.2. Cell Culture

A549 cells were a gift from Dr. John Livesey (University of the Pacific) and were authenticated by ATCC before being used for experiments in this paper. AD293 cells were purchased from Agilent Technologies (Santa Clara, CA, USA). Rat H4G1.1c3 stable cells carrying a DRE-driven GFP cDNA were a gift from Dr. Michael Denison (University of California, Davis). All the cell lines were cultured in DMEM supplemented with 10% fetal bovine serum (FBS), 1% penicillin–streptomycin, and 1% GlutaMAX-I at 37 °C, 5% CO_2.

4.3. Preparation of Whole Cell Extract and Western Blot Analysis

A549 cells were scraped in cold PBS and centrifuged at 400× g for 5 min to collect cell pellets. The collected cells were washed once with cold PBS and then lysed in lysis buffer (25 mM HEPES pH7.4, 0.4 M KCl, 1 mM EDTA, 1 mM DTT, 10% glycerol, 1% NP40, 1 mM PMSF, and 2 μg/mL leupeptin). After 3 cycles of freezing and thawing, cell lysates were kept on ice for 30 min and then centrifuged at 16,000× g for 15 min at 4 °C. The supernatants were used as whole-cell extracts. Total protein concentrations were measured by BCA assay. Proteins in 20 μg of whole cell extract from each sample were separated by 12% SDS-PAGE and then transferred to nitrocellulose membranes via the wet transfer method. Total protein staining was determined using LI-COR Revert 700 Total Protein Stain for normalization. Non-specific binding was blocked in a blocking buffer (PBS, 0.1% Tween-20, and 5% BSA) for 1 h. The primary antibodies and their dilution are as follows: 1:2000 for anti-AHR (SA210); 1:1000 for anti-ATG5 (2630); 1:200 for anti-LAMP2 (H4B4); 1:200 for anti-Hsc70 (B-6); 1:200 for anti-E-cadherin (G-10); 1:200 for anti-Vimentin (V9); and 1:200 for anti-N-cadherin (13A9). After washing with PBST (PBS and 0.1% Tween-20) 5 times, the nitrocellulose membrane was incubated in 1:10,000 dilution of donkey secondary antibody conjugated with IRDye 680 or 800 CW. Results were obtained and quantified using a LI-COR Odyssey CLx imaging system (Lincoln, NE, USA).

4.4. RNA Extraction and Reverse Transcription-Quantitative Polymerase Chain Reaction (RT-qPCR)

Total RNA was extracted from A549 cells using TRI Reagent (Zymo Research) and an RNA miniprep kit (Direct-zol) according to the manufacturer's recommendations. cDNA was reverse transcribed from 1 μg of RNA using MMLV high-performance reverse transcriptase (Epicentre) into a final volume of 20 μL cDNA solution, and 1 μL of it was used as the qPCR template. qPCR was performed with iTaq SYBR green supermix (Bio-rad, USA) on a CFX Connect real-time PCR operating system (Bio-rad, USA) according to the following protocol: 95 °C for 2 min, 40 cycles of 95 °C for 15 s, and 60 °C for 1 min. Relative gene expression was analyzed by the $2^{-\Delta\Delta Cq}$ method [63], and *18s* rRNA was used as an internal control for normalization. The primer sequences were as follows: *18s* forward: 5′-CGCCCCCTCGATGCTCTTAG-3′ and reverse: 5′-CGGCGGTCATGGAATAAC-3′; *CYP1A1* forward: 5′-GGCCACATCCGGGACATCACAGA-3′ and reverse: 5′-TGGGGATG GTGAAGGGGACGAA-3′; *E-Cadherin* forward: 5′-GCCTCCTGAAAAGAGAGTGGAAG-3′ and reverse: 5′-TGGCAGTGTCTCTCCAAATCCG-3′; *Vimentin* forward: 5′-TGTCCAAA TCGATGTGGATGTTTC-3′ and reverse: 5′-TTGTACCATTCTTCTGCCTCCTG-3′; *N-Cadherin* forward: 5′-ACAGTGGCCACCTACAAAGG-3′ and reverse: 5′-CCGAGATGGGGTTGATAATG-3′.

4.5. Generation of ATG5, LAMP2 Stable Knockdown A549 Cells Using Lentivirus

Lentivirus containing ATG5 or LAMP2 shRNA was prepared as follows: AD293 cells (7×10^5) in 5 mL of growth media (10% fetal bovine serum and 2 mM GlutaMAX-I in DMEM) were seeded in a 25 cm^2 flask. Cells were incubated at 37 °C and 5% CO_2 overnight. Then, AD293 cells were transfected using 10 μL EndoFectin transfection reagent with 5 μg plasmids (2.5 μg of pLKO.1 specific shRNA plasmid, 1.875 μg of the pCMV-dR8.2 dvpr packaging plasmid, and 0.625 μg of the pCMV-VSV-G envelope plasmid). Fresh complete medium was replaced 15 h after transfection. After 24 h, the medium containing lentiviral particles was transferred to a 15 mL tube and stored at 4 °C. Another 5 ml of fresh complete medium was added to the cells, and the medium containing lentiviral particles was harvested after 24 h of incubation. The combined medium was centrifuged at 400× *g* for 5 min to pellet any AD293 cells, and the supernatant was used for infection. Stable ATG5 or LAMP2 knockdown cell lines were generated as follows: A549 cells were seeded in a 25 cm^2 flask to achieve 50–70% confluent the next day. Fresh complete medium containing 8 μg/mL polybrene was replaced. A total of 500 μL of medium containing lentiviral particles was added into the flask. After 24 h, the medium was replaced with fresh complete medium containing 1.5 μg/mL of puromycin for stable cell line selection. This was replaced with fresh medium containing puromycin every 2–3 days. ATG5 knockdown stable A549 cells were generated using pLKO.1 Lentiviral (TRC) ATG5 shRNA #5 (TRCN0000151963) plasmid; LAMP2 knockdown stable A549 cells were generated using pLKO.1 Lentiviral (TRC) LAMP2 shRNA #4 (TRCN0262) plasmid.

4.6. CRISPR/Cas9-Mediated AHR Knockout in A549 Cells

Three different single guide RNAs (sgRNA) targeting exon 2 of the human *AHR* gene were used to knock out the *AHR* gene in A549 cells. The sequences were as follows: sgRNA1: 5′- GCTGAAGGAATCAAGTCAAA-3′; sgRNA2: 5′- ACAAGATGTTAT-TAATAAGT-3′; and sgRNA3:5′- GAGAGCCAAGAGCTTCTTTG-3′. Cas9 nuclease NLS and sgRNAs were introduced as ribonucleoprotein (RNP) complex into A549 cells through transfection using Lipofectamine RNAiMAX according to the manufacturer's recommendations. In brief, 3 μM Cas9 nuclease NLS was combined with 3 μM sgRNAs (1 μM each of 3 sgRNAs) to form RNPs in 12.5 μL volume with the Opti-MEM. We gently mixed the reaction and incubated it at room temperature for 10 min. A total of 1.2 μL of transfection reagent RNAiMAX was diluted in 12.5 μL of the Opti-MEM and was added directly into the RNP tube. The RNPs/liposome complexes were mixed gently and incubated at room temperature for 20 min. Meanwhile, a 3.2×10^5 cells/mL A549 cell suspension was prepared, and 125 μL of it was added into each well of a 96-well plate, followed by mixing it with

25 μL of RNPs/liposome complexes. The transfected cells were incubated at 37 °C and 5% CO_2 for 72 h. Then, the isolation of single cells from the knockout cell pool was accomplished through limiting dilution according to the protocol from Synthego. A total of 0.5–1 cell/100 μL of the diluted cell suspension was seeded into each well of a 96-well plate. To genotype clones, genomic DNA was isolated using QuickExtract DNA extraction solution (Lucigen). PCR was performed to amplify the edited region using PCR Master Mix (Promega) with the following primers: OL921 forward 5′-TCGGAAGAATTTAACCCATTCCCT-3′ and OL922 reverse 5′-TGCAGCCACTGAAATGATGC-3′. A DNA fragment ~500 bp was observed by agarose gel electrophoresis and was purified for Sanger sequencing (Functional Biosciences, WI, USA). The sequencing data were uploaded to the online Inference of CRISPR Edits (ICE) v3.0 analysis tool (http://ice.synthego.com, accessed on 8 September 2023) for knockout analysis.

4.7. Transient Transfection

A549 cells were seeded in a 6-well plate at 90–95% confluency at the time of transfection. Plasmid DNA, EndoFectin transfection reagent, and Opti-MEM were equilibrated to room temperature before use. Cells were transfected with 4 μg of plasmid and 8 μL of transfection reagent. Both plasmids and transfection reagents were diluted in the Opti-MEM. Then, the diluted transfection reagent and the diluted DNA were combined and kept at room temperature for 20 min to allow DNA–transfection reagent complexes to form. The combined complexes were added to each well and mixed gently. The cells were harvested for analysis after 48 h of incubation at 37 °C and 5% CO_2.

4.8. Co-Immunoprecipitation

A549 cells were cultured in a 75 cm^2 flask and reached 90–95% confluency at the time of the experiment. Cells were treated with or without CQ for 6 h and then were harvested to be lysed in lysis buffer (25 mM HEPES pH7.4, 0.15 M KCl, 1 mM EDTA, 1 mM DTT, 10% glycerol, 10 mM N-Ethylmaleimide, 1 mM PMSF, and 2 μg/mL leupeptin). About 2 mg of the whole cell extract was incubated with rabbit anti-AHR antibody (SA210) for 30 min at room temperature. Then, the samples were added to the pre-equilibrated Dynabeads Protein G (Invitrogen) and q.s. to 1 mL with IP buffer (25 mM HEPES pH7.4, 1 mM EDTA, 1 mM DTT, 10% glycerol, 150 mM NaCl, 0.05% Tween-20, and 1 mg/mL BSA). The samples were rotated at 60 rpm in a cold room for 16–18 h. The magnetic beads-Ab-Ag complex was washed 3 times with cold IP buffer on the magnet. The magnetic beads-Ab-Ag complex was resuspended in 30 μL of electrophoresis sample buffer and boiled at 95 °C for 3 min to free the bound protein. A total of 1% of the whole cell extract was used as an input control. All the samples were analyzed by Western blot with antibodies against AHR, LAMP2, and Hsc70.

4.9. Ligand Dependent, DRE-Driven Expression of GFP in H4G1.1c3 Cells

H4G1.1c3 cells (3×10^5) were seeded into each well of a 24-well plate. After incubation at 37 °C and 5% CO_2 for 24 h, the cells reached 70–85% confluence. A total of 1 mL of fresh complete medium was exchanged for each well. The cells were then treated with DMSO (0.2%), 6-AN (100 μM), MG132 (10 μM), GA (100 nM), BaP (5 μM), FICZ (1 μM), CH223191 (10 μM), BaP plus CH223191, and FICZ plus CH223191 and incubated at 37 °C and 5% CO_2 for 12 h. Fresh complete medium was exchanged after treatment. Fluorescence images were acquired by a Keyence BZ-X700 fluorescence microscope in 4× objective.

4.10. Wound Healing Assay

A549 cells were seeded in 6-well plates and formed monolayers at the time of wounding. A sterile 1 mL pipette tip was used to scratch across the monolayers to form a linear wound. Then, the disassociated cells and debris were removed by washing with PBS. Cells were treated with DMSO or 6-AN for 48 h. Representative images were taken at 0 and 48 h after the treatments at the same position under an inverted microscope (BZ-X700,

KEYENCE, Itasca, IL, USA) with a camera. The scale bar on the representative images is 500 μm.

4.11. Transwell Migration and Invasion Assay

Falcon 24-well cell culture inserts with transparent PET membrane (8.0 μm pore size) and Corning Matrigel matrix (1:5 dilution) were used to determine the cell migration and invasion capability. For migration assay, 5×10^4 cells were seeded into the upper chamber, and 700 μL of complete DMEM medium was added in the lower chamber and placed in 24-well plates. For the invasion assay, 1×10^5 cells were seeded into the upper chamber, which was coated with Matrigel before use. After 24 h incubation, the cells on the inserts were fixed with methanol for 10 min and stained with 0.1% crystal violet for 5 min. Then, the cells on the top side of the membrane were removed with cotton swabs carefully. Only the cells that migrated or invaded through the membrane to the bottom of inserts were imaged using an inverted microscope (BZ-X700, KEYENCE, Itasca, IL, USA) with a camera. Three fields, which covered about 80% of the well, were randomly captured and were analyzed by ImageJ v1.53k software.

4.12. Statistical Analysis

GraphPad Prism 9 software (La Jolla, CA, USA) was used for statistical analysis. The statistical significance of the differences between group means was evaluated by one-way or two-way ANOVA using Tukey's or Sidak's test for multiple comparisons. Statistical significance is indicated as follows: * $p < 0.05$, ** $p < 0.01$, *** $p < 0.001$, **** $p < 0.0001$, and $p > 0.05$ (ns, not significant). The two-tailed unpaired t-test was used to determine the statistical significance in Figure 7D,E.

5. Conclusions

AHR undergoes CMA continuously in A549 cells. Activation of CMA by 6-AN degrades AHR, causing downregulation of AHR functions, such as the ligand-dependent activation of gene transcription and promotion of the EMT in A549 cells.

Author Contributions: Conceptualization, R.X. and W.K.C.; methodology, R.X. and W.K.C.; software, R.X., D.S., S.D. and W.K.C.; validation, R.X., D.S., S.D. and W.K.C.; formal analysis, R.X., D.S., S.D. and W.K.C.; investigation, R.X., D.S. and W.K.C.; resources, R.X. and W.K.C.; data curation, R.X., D.S., S.D. and W.K.C.; writing—original draft preparation, R.X. and W.K.C.; writing—review and editing, R.X., D.S., S.D. and W.K.C.; supervision, W.K.C.; project administration, W.K.C.; funding acquisition, W.K.C. All authors have read and agreed to the published version of the manuscript.

Funding: This work was supported by the National Institutes of Health grant R15ES023104 (WKC).

Institutional Review Board Statement: Not applicable.

Informed Consent Statement: Not applicable.

Data Availability Statement: Data are available upon request.

Conflicts of Interest: The authors declare no conflict of interest.

References

1. Zhang, L.; Hatzakis, E.; Nichols, R.G.; Hao, R.; Correll, J.; Smith, P.B.; Chiaro, C.R.; Perdew, G.H.; Patterson, A.D. Metabolomics reveals that aryl hydrocarbon receptor activation by environmental chemicals induces systemic metabolic dysfunction in mice. *Environ. Sci. Technol.* **2015**, *49*, 8067–8077. [CrossRef] [PubMed]
2. Goya-Jorge, E.; Jorge Rodriguez, M.E.; Veitia, M.S.; Giner, R.M. Plant occurring flavonoids as modulators of the aryl hydrocarbon receptor. *Molecules* **2021**, *26*, 2315. [CrossRef] [PubMed]
3. Hubbard, T.D.; Murray, I.A.; Perdew, G.H. Indole and tryptophan metabolism: Endogenous and dietary routes to Ah receptor activation. *Drug Metab. Dispos.* **2015**, *43*, 1522–1535. [CrossRef] [PubMed]
4. Quintana, F.J.; Basso, A.S.; Iglesias, A.H.; Korn, T.; Farez, M.F.; Bettelli, E.; Caccamo, M.; Oukka, M.; Weiner, H.L. Control of T_{reg} and Th17 cell differentiation by the aryl hydrocarbon receptor. *Nature* **2008**, *453*, 65–71. [CrossRef] [PubMed]

5. Perdew, G.H.; Esser, C.; Snyder, M.; Sherr, D.H.; van den Bogaard, E.H.; McGovern, K.; Fernandez-Salguero, P.M.; Coumoul, X.; Patterson, A.D. The Ah receptor from toxicity to therapeutics: Report from the 5th AHR Meeting at Penn State University, USA, June 2022. *Int. J. Mol. Sci.* **2023**, *24*, 5550. [CrossRef]
6. Haque, N.; Ojo, E.S.; Krager, S.L.; Tischkau, S.A. Deficiency of adipose aryl hydrocarbon receptor protects against diet-induced metabolic dysfunction through sexually dimorphic mechanisms. *Cells* **2023**, *12*, 1748. [CrossRef]
7. Smith, S.H.; Jayawickreme, C.; Rickard, D.J.; Nicodeme, E.; Bui, T.; Simmons, C.; Coquery, C.M.; Neil, J.; Pryor, W.M.; Mayhew, D.; et al. Tapinarof is a natural AhR agonist that resolves skin inflammation in mice and humans. *J. Investig. Dermatol.* **2017**, *137*, 2110–2119. [CrossRef] [PubMed]
8. Giovannoni, F.; Li, Z.; Remes-Lenicov, F.; Davola, M.E.; Elizalde, M.; Paletta, A.; Ashkar, A.A.; Mossman, K.L.; Dugour, A.V.; Figueroa, J.M.; et al. AHR signaling is induced by infection with coronaviruses. *Nat. Commun.* **2021**, *12*, 5148. [CrossRef]
9. Shi, J.; Du, T.; Wang, J.; Tang, C.; Lei, M.; Yu, W.; Yang, Y.; Ma, Y.; Huang, P.; Chen, H.; et al. Aryl hydrocarbon receptor is a proviral host factor and a candidate pan-SARS-CoV-2 therapeutic target. *Sci. Adv.* **2023**, *9*, eadf0211. [CrossRef] [PubMed]
10. Duan, Z.; Li, Y.; Li, L. Promoting epithelial-to-mesenchymal transition by D-kynurenine via activating aryl hydrocarbon receptor. *Mol. Cell. Biochem.* **2018**, *448*, 165–173. [CrossRef] [PubMed]
11. Moretti, S.; Nucci, N.; Menicali, E.; Morelli, S.; Bini, V.; Colella, R.; Mandarano, M.; Sidoni, A.; Puxeddu, E. The aryl hydrocarbon receptor is expressed in thyroid carcinoma and appears to mediate epithelial-mesenchymal transition. *Cancers* **2020**, *12*, 145. [CrossRef]
12. Gao, H.; Ye, G.; Lin, Y.; Chi, Y.; Dong, S. Benzo[a]pyrene at human blood equivalent level induces human lung epithelial cell invasion and migration via aryl hydrocarbon receptor signaling. *J. Appl. Toxicol.* **2020**, *40*, 1087–1098. [CrossRef] [PubMed]
13. Selvam, P.; Cheng, C.M.; Dahms, H.U.; Ponnusamy, V.K.; Sun, Y.Y. AhR mediated activation of pro-inflammatory response of RAW 264.7 cells modulate the epithelial-mesenchymal transition. *Toxics* **2022**, *10*, 642. [CrossRef]
14. Tsai, C.H.; Li, C.H.; Cheng, Y.W.; Lee, C.C.; Liao, P.L.; Lin, C.H.; Huang, S.H.; Kang, J.J. The inhibition of lung cancer cell migration by AhR-regulated autophagy. *Sci. Rep.* **2017**, *7*, 41927. [CrossRef] [PubMed]
15. Nothdurft, S.; Thumser-Henner, C.; Breitenbucher, F.; Okimoto, R.A.; Dorsch, M.; Opitz, C.A.; Sadik, A.; Esser, C.; Holzel, M.; Asthana, S.; et al. Functional screening identifies aryl hydrocarbon receptor as suppressor of lung cancer metastasis. *Oncogenesis* **2020**, *9*, 102. [CrossRef]
16. Enan, E.; Matsumura, F. Identification of c-Src as the integral component of the cytosolic Ah receptor complex, transducing the signal of 2,3,7,8-tetrachlorodibenzo-*p*-dioxin (TCDD) through the protein phosphorylation pathway. *Biochem. Pharmacol.* **1996**, *52*, 1599–1612. [CrossRef] [PubMed]
17. Meyer, B.K.; Perdew, G.H. Characterization of the AhR-hsp90-XAP2 core complex and the role of the immunophilin-related protein XAP2 in AhR stabilization. *Biochemistry* **1999**, *38*, 8907–8917. [CrossRef] [PubMed]
18. Kazlauskas, A.; Poellinger, L.; Pongratz, I. Evidence that the co-chaperone p23 regulates ligand responsiveness of the dioxin receptor. *J. Biol. Chem.* **1999**, *274*, 13519–13524. [CrossRef]
19. Ohtake, F.; Baba, A.; Takada, I.; Okada, M.; Iwasaki, K.; Miki, H.; Takahashi, S.; Kouzmenko, A.; Nohara, K.; Chiba, T.; et al. Dioxin receptor is a ligand-dependent E3 ubiquitin ligase. *Nature* **2007**, *446*, 562–566. [CrossRef] [PubMed]
20. Yang, Y.; Chan, W.K. Selective autophagy maintains the aryl hydrocarbon receptor levels in HeLa cells: A mechanism that is dependent on the p23 co-chaperone. *Int. J. Mol. Sci.* **2020**, *21*, 3449. [CrossRef] [PubMed]
21. Cuervo, A.M.; Knecht, E.; Terlecky, S.R.; Dice, J.F. Activation of a selective pathway of lysosomal proteolysis in rat liver by prolonged starvation. *Am. J. Physiol.* **1995**, *269*, C1200–C1208. [CrossRef] [PubMed]
22. Rogov, V.; Dotsch, V.; Johansen, T.; Kirkin, V. Interactions between autophagy receptors and ubiquitin-like proteins form the molecular basis for selective autophagy. *Mol. Cell* **2014**, *53*, 167–178. [CrossRef] [PubMed]
23. Suzuki, K.; Kirisako, T.; Kamada, Y.; Mizushima, N.; Noda, T.; Ohsumi, Y. The pre-autophagosomal structure organized by concerted functions of APG genes is essential for autophagosome formation. *EMBO J.* **2001**, *20*, 5971–5981. [CrossRef] [PubMed]
24. Mizushima, N. The ATG conjugation systems in autophagy. *Curr. Opin. Cell Biol.* **2020**, *63*, 1–10. [CrossRef]
25. Li, P.; Ji, M.; Lu, F.; Zhang, J.; Li, H.; Cui, T.; Li Wang, X.; Tang, D.; Ji, C. Degradation of AF1Q by chaperone-mediated autophagy. *Exp. Cell Res.* **2014**, *327*, 48–56. [CrossRef]
26. Chen, J.; Yang, Y.; Russu, W.A.; Chan, W.K. The aryl hydrocarbon receptor undergoes chaperone-mediated autophagy in triple-negative breast cancer cells. *Int. J. Mol. Sci.* **2021**, *22*, 1654. [CrossRef] [PubMed]
27. Davarinos, N.A.; Pollenz, R.S. Aryl hydrocarbon receptor imported into the nucleus following ligand binding is rapidly degraded via the cytosplasmic proteasome following nuclear export. *J. Biol. Chem.* **1999**, *274*, 28708–28715. [CrossRef] [PubMed]
28. Roberts, B.J.; Whitelaw, M.L. Degradation of the basic helix-loop-helix/Per-ARNT-Sim homology domain dioxin receptor via the ubiquitin/proteasome pathway. *J. Biol. Chem.* **1999**, *274*, 36351–36356. [CrossRef] [PubMed]
29. Giannone, J.V.; Li, W.; Probst, M.; Okey, A.B. Prolonged depletion of AH receptor without alteration of receptor mRNA levels after treatment of cells in culture with 2,3,7,8-tetrachlorodibenzo-*p*-dioxin. *Biochem. Pharmacol.* **1998**, *55*, 489–497. [CrossRef] [PubMed]
30. Bao, W.; Gu, Y.; Ta, L.; Wang, K.; Xu, Z. Induction of autophagy by the MG132 proteasome inhibitor is associated with endoplasmic reticulum stress in MCF7 cells. *Mol. Med. Rep.* **2016**, *13*, 796–804. [CrossRef]
31. Harhouri, K.; Navarro, C.; Depetris, D.; Mattei, M.G.; Nissan, X.; Cau, P.; De Sandre-Giovannoli, A.; Levy, N. MG132-induced progerin clearance is mediated by autophagy activation and splicing regulation. *EMBO Mol. Med.* **2017**, *9*, 1294–1313. [CrossRef]

32. Shen, Y.F.; Tang, Y.; Zhang, X.J.; Huang, K.X.; Le, W.D. Adaptive changes in autophagy after UPS impairment in Parkinson's disease. *Acta Pharmacol. Sin.* **2013**, *34*, 667–673. [CrossRef] [PubMed]
33. Massey, A.C.; Kaushik, S.; Sovak, G.; Kiffin, R.; Cuervo, A.M. Consequences of the selective blockage of chaperone-mediated autophagy. *Proc. Natl. Acad. Sci. USA* **2006**, *103*, 5805–5810. [CrossRef] [PubMed]
34. Song, Z.; Pollenz, R.S. Ligand-dependent and independent modulation of aryl hydrocarbon receptor localization, degradation, and gene regulation. *Mol. Pharmacol.* **2002**, *62*, 806–816. [CrossRef] [PubMed]
35. Kimura, T.; Uesugi, M.; Takase, K.; Miyamoto, N.; Sawada, K. Hsp90 inhibitor geldanamycin attenuates the cytotoxicity of sunitinib in cardiomyocytes via inhibition of the autophagy pathway. *Toxicol. Appl. Pharmacol.* **2017**, *329*, 282–292. [CrossRef] [PubMed]
36. Qing, G.; Yan, P.; Xiao, G. Hsp90 inhibition results in autophagy-mediated proteasome-independent degradation of IkappaB kinase (IKK). *Cell Res.* **2006**, *16*, 895–901. [CrossRef] [PubMed]
37. Finn, P.F.; Mesires, N.T.; Vine, M.; Dice, J.F. Effects of small molecules on chaperone-mediated autophagy. *Autophagy* **2005**, *1*, 141–145. [CrossRef] [PubMed]
38. Chen, Y.; Xu, Q.; Ji, D.; Wei, Y.; Chen, H.; Li, T.; Wan, B.; Yuan, L.; Huang, R.; Chen, G. Inhibition of pentose phosphate pathway suppresses acute myelogenous leukemia. *Tumour Biol.* **2016**, *37*, 6027–6034. [CrossRef] [PubMed]
39. Aiello, N.M.; Maddipati, R.; Norgard, R.J.; Balli, D.; Li, J.; Yuan, S.; Yamazoe, T.; Black, T.; Sahmoud, A.; Furth, E.E.; et al. EMT subtype influences epithelial plasticity and mode of cell migration. *Dev. Cell* **2018**, *45*, 681–695. [CrossRef] [PubMed]
40. Liu, X.; Huang, H.; Remmers, N.; Hollingsworth, M.A. Loss of E-cadherin and epithelial to mesenchymal transition is not required for cell motility in tissues or for metastasis. *Tissue Barriers* **2014**, *2*, e969112. [CrossRef]
41. Liu, H.; Yong, Y.; Li, X.; Ye, P.; Tao, K.; Peng, G.; Mo, M.; Guo, W.; Chen, X.; Luo, Y.; et al. Chaperone-mediated autophagy regulates cell growth by targeting SMAD3 in glioma. *Neurosci. Bull.* **2022**, *38*, 637–651. [CrossRef] [PubMed]
42. Jin, Y.; Pan, Y.; Zheng, S.; Liu, Y.; Xu, J.; Peng, Y.; Zhang, Z.; Wang, Y.; Xiong, Y.; Xu, L.; et al. Inactivation of EGLN3 hydroxylase facilitates Erk3 degradation via autophagy and impedes lung cancer growth. *Oncogene* **2022**, *41*, 1752–1766. [CrossRef]
43. Su, C.M.; Hsu, T.W.; Chen, H.A.; Wang, W.Y.; Huang, C.Y.; Hung, C.C.; Yeh, M.H.; Su, Y.H.; Huang, M.T.; Liao, P.H. Chaperone-mediated autophagy degrade Dicer to promote breast cancer metastasis. *J. Cell. Physiol.* **2023**, *238*, 829–841. [CrossRef] [PubMed]
44. Li, Q.; Yue, T.; Du, X.; Tang, Z.; Cui, J.; Wang, W.; Xia, W.; Ren, B.; Kan, S.; Li, C.; et al. HSC70 mediated autophagic degradation of oxidized PRL2 is responsible for osteoclastogenesis and inflammatory bone destruction. *Cell Death Differ.* **2023**, *30*, 647–659. [CrossRef] [PubMed]
45. Caballero, B.; Wang, Y.; Diaz, A.; Tasset, I.; Juste, Y.R.; Stiller, B.; Mandelkow, E.M.; Mandelkow, E.; Cuervo, A.M. Interplay of pathogenic forms of human tau with different autophagic pathways. *Aging Cell* **2018**, *17*, e12692. [CrossRef] [PubMed]
46. Caballero, B.; Bourdenx, M.; Luengo, E.; Diaz, A.; Sohn, P.D.; Chen, X.; Wang, C.; Juste, Y.R.; Wegmann, S.; Patel, B.; et al. Acetylated tau inhibits chaperone-mediated autophagy and promotes tau pathology propagation in mice. *Nat. Commun.* **2021**, *12*, 2238. [CrossRef]
47. Sahoo, S.; Padhy, A.A.; Kumari, V.; Mishra, P. Role of ubiquitin-proteasome and autophagy-lysosome pathways in alpha-synuclein aggregate clearance. *Mol. Neurobiol.* **2022**, *59*, 5379–5407. [CrossRef]
48. Dvorak, Z. Pelargonidin and its glycosides as dietary chemopreventives attenuating inflammatory bowel disease symptoms through the aryl hydrocarbon receptor. *Eur. J. Nutr.* **2020**, *59*, 3863–3864. [CrossRef]
49. Retnakumar, S.V.; Geesala, R.; Bretin, A.; Tourneur-Marsille, J.; Ogier-Denis, E.; Maretzky, T.; Nguyen, H.T.T.; Muller, S. Targeting the endo-lysosomal autophagy pathway to treat inflammatory bowel diseases. *J. Autoimmun.* **2022**, *128*, 102814. [CrossRef]
50. Meacham, G.C.; Patterson, C.; Zhang, W.; Younger, J.M.; Cyr, D.M. The Hsc70 co-chaperone CHIP targets immature CFTR for proteasomal degradation. *Nat. Cell Biol.* **2001**, *3*, 100–105. [CrossRef]
51. Kim, J.; de Haro, M.; Al-Ramahi, I.; Garaicoechea, L.L.; Jeong, H.H.; Sonn, J.Y.; Tadros, B.; Liu, Z.; Botas, J.; Zoghbi, H.Y. Evolutionarily conserved regulators of tau identify targets for new therapies. *Neuron* **2023**, *111*, 824–838. [CrossRef]
52. Bento, C.F.; Fernandes, R.; Ramalho, J.; Marques, C.; Shang, F.; Taylor, A.; Pereira, P. The chaperone-dependent ubiquitin ligase CHIP targets HIF-1alpha for degradation in the presence of methylglyoxal. *PLoS ONE* **2010**, *5*, e15062. [CrossRef] [PubMed]
53. Ferreira, J.V.; Fofo, H.; Bejarano, E.; Bento, C.F.; Ramalho, J.S.; Girao, H.; Pereira, P. STUB1/CHIP is required for HIF1α degradation by chaperone-mediated autophagy. *Autophagy* **2013**, *9*, 1349–1366. [CrossRef] [PubMed]
54. Morales, J.L.; Perdew, G.H. Carboxyl terminus of hsc70-interacting protein (CHIP) can remodel mature aryl hydrocarbon receptor complexes and mediate ubiquitination of both the AhR and the 90 kDa heat-shock protein (hsp90) in vitro. *Biochemistry* **2007**, *46*, 610–621. [CrossRef] [PubMed]
55. Wu, Z.; Geng, Y.; Lu, X.; Shi, Y.; Wu, G.; Zhang, M.; Shan, B.; Pan, H.; Yuan, J. Chaperone-mediated autophagy is involved in the execution of ferroptosis. *Proc. Natl. Acad. Sci. USA* **2019**, *116*, 2996–3005. [CrossRef] [PubMed]
56. Peng, Y.; Ouyang, L.; Zhou, Y.; Lai, W.; Chen, Y.; Wang, Z.; Yan, B.; Zhang, Z.; Zhou, Y.; Peng, X.; et al. AhR promotes the development of non-small cell lung cancer by inducing SLC7A11-dependent antioxidant function. *J. Cancer* **2023**, *14*, 821–834. [CrossRef]
57. Ciolino, H.P.; Yeh, G.C. Inhibition of aryl hydrocarbon-induced cytochrome P-450 1A1 enzyme activity and CYP1A1 expression by resveratrol. *Mol. Pharmacol.* **1999**, *56*, 760–767.
58. Wang, H.; Zhang, H.; Tang, L.; Chen, H.; Wu, C.; Zhao, M.; Yang, Y.; Chen, X.; Liu, G. Resveratrol inhibits TGF-beta1-induced epithelial-to-mesenchymal transition and suppresses lung cancer invasion and metastasis. *Toxicology* **2013**, *303*, 139–146. [CrossRef]

59. Ouyang, L.; Yan, B.; Liu, Y.; Mao, C.; Wang, M.; Liu, N.; Wang, Z.; Liu, S.; Shi, Y.; Chen, L.; et al. The deubiquitylase UCHL3 maintains cancer stem-like properties by stabilizing the aryl hydrocarbon receptor. *Signal Transduct. Target Ther.* **2020**, *5*, 78. [CrossRef]
60. Xiong, J.; Zhang, X.; Zhang, Y.; Wu, B.; Fang, L.; Wang, N.; Yi, H.; Chang, N.; Chen, L.; Zhang, J. Aryl hydrocarbon receptor mediates Jak2/STAT3 signaling for non-small cell lung cancer stem cell maintenance. *Exp. Cell Res.* **2020**, *396*, 112288. [CrossRef]
61. Gerashchenko, T.S.; Novikov, N.M.; Krakhmal, N.V.; Zolotaryova, S.Y.; Zavyalova, M.V.; Cherdyntseva, N.V.; Denisov, E.V.; Perelmuter, V.M. Markers of Cancer Cell Invasion: Are They Good Enough? *J. Clin. Med.* **2019**, *8*, 1092. [CrossRef] [PubMed]
62. Jin, U.H.; Karki, K.; Cheng, Y.; Michelhaugh, S.K.; Mittal, S.; Safe, S. The aryl hydrocarbon receptor is a tumor suppressor-like gene in glioblastoma. *J. Biol. Chem.* **2019**, *294*, 11342–11353. [CrossRef] [PubMed]
63. Livak, K.J.; Schmittgen, T.D. Analysis of relative gene expression data using real-time quantitative PCR and the $2^{-\Delta\Delta CT}$ Method. *Methods* **2001**, *25*, 402–408. [CrossRef] [PubMed]

International Journal of
Molecular Sciences

MDPI

Article

Activation of G-Protein-Coupled Estrogen Receptor 1 (GPER1) Reduces Progression of Vulvar Carcinoma Cells

Johanna Loris, Lena Hanesch, Gerd Bauerschmitz, Julia Gallwas and Carsten Gründker *

Department of Gynecology and Obstetrics, University Medical Center Göttingen, 37075 Göttingen, Germany; johanna.loris@stud.uni-goettingen.de (J.L.); lena.hanesch@stud.uni-goettingen.de (L.H.); gerd.bauerschmitz@med.uni-goettingen.de (G.B.); julia.gallwas@med.uni-goettingen.de (J.G.)
* Correspondence: grundker@med.uni-goettingen.de; Tel.: +49-551-3969810

Abstract: Whether G protein-coupled estrogen receptor 1 (GPER1) is tumor-promoting or tumor-suppressive depends in part on tumor entity. Little is known about the function of GPER1 in vulvar carcinoma. In this work, we aim to clarify what role GPER1 plays in vulvar cancer, tumor-promoting or tumor-suppressive. Localization of GPER1 in A431 and CAL-39 vulvar carcinoma cells was examined by immunofluorescence. Using a tissue microarray of vulvar neoplasias, the correlation between GPER1 expression and grade of malignancy was investigated. A431 and CAL-39 cells were treated either with GPER1 agonist G1 or antagonist G36. Proliferation was quantified by BrdU assay and viability examined using Resazurin assay. Morphological changes were analyzed by microscopy and measured using ImageJ. Cell migration was analyzed by gap closure assay. Clonogenic potential was tested by colony and sphere formation. Expression of estrogen receptors was examined by Western blot. GPER1 was found consistently expressed in vulvar neoplasia tissues. The immune-reactive score was found to be significantly higher in tissue samples of lymph node metastases and neoplasias with grade 3. In A431 and CAL-39 vulvar carcinoma cells, GPER1 expression was mainly found in the cytoplasm and nuclei. Treatment of A431 and CAL-39 cells with GPER1 agonist G1 resulted in a decrease in proliferation and migration. In addition, colony formation and tumor sphere formation were reduced. Furthermore, morphological signs of necrosis and reduction in cell viability after G1 treatment were observed. The GPER1 antagonist G36 did not have significant effects on vulvar carcinoma cells. Neither agonist G1 nor antagonist G36 treatment resulted in altered expression of estrogen receptors. Activation of GPER1 with GPER1 agonist G1 reduces the tumorigenic potential of the vulvar carcinoma cells. It can be deduced from this that GPER1 appears to have a tumor-suppressive effect in vulvar carcinoma.

Keywords: G-protein-coupled estrogen receptor; GPER1; vulvar carcinoma; tumor suppressor; oncogene

Citation: Loris, J.; Hanesch, L.; Bauerschmitz, G.; Gallwas, J.; Gründker, C. Activation of G-Protein-Coupled Estrogen Receptor 1 (GPER1) Reduces Progression of Vulvar Carcinoma Cells. *Int. J. Mol. Sci.* **2023**, *24*, 13705. https://doi.org/10.3390/ijms241813705

Academic Editor: Farzad Pakdel

Received: 31 July 2023
Revised: 31 August 2023
Accepted: 4 September 2023
Published: 5 September 2023

1. Introduction

Vulvar carcinomas are uncommon, accounting for 3–5% of all gynecological cancers [1–3]. Their etiology is diverse with squamous cell carcinoma (SCC) being by far the most common subtype. Around 15–25% of SCC are induced by high-risk human papillomavirus (HPV) occurring in younger women with increasing incidence, whereas most vulvar carcinomas are HPV-negative and associated often with lichen sclerosus, primarily affecting postmenopausal women [1,4]. Vulvar cancers are regarded to be nonhormone-dependent [5]. Their treatment is predominantly surgical, although concurrent chemoradiation is an effective alternative, particularly for advanced tumors [6]. Management should be individualized and carried out by a multidisciplinary team experienced in the treatment of these tumors. When treatment options are considered, the most appropriate treatment of the primary lesion and the inguinal lymph nodes should be considered independently of each other to maximize the chance of cure while minimizing morbidity. The 5-year survival rates vary between 86% if the tumor has not spread to lymph nodes or nearby tissues and 54% in case of lymph node involvement [7]. As the diagnosing of lymph node metastasis in earlier stages has not improved since the 1980s, a reconsideration of the screening and care model is needed [8].

The G-protein-coupled estrogen receptor 1 (GPER1, GPR30) is a seven transmembrane-domain G protein-coupled receptor that is mainly located in the endoplasmic reticulum [9]. In gynecological neoplasms, the receptor is described to be located in cytoplasm und nucleus [10–12]. The receptor is a part of cellular pathways of angiogenesis, migration, proliferation, invasion, and apoptosis [9,13–15]. GPER1 is also responsible for nongenomic, membrane-initiated estrogen effects [16]. According to Girgert et al. [16], the Hippo, FOXO3a, and HOTAIR pathways depend on regulation by GPER1. Furthermore calcium signaling, cAMP, epidermal-growth-factor (EGF) receptor, and IxB pathways are influenced by GPER1 [16]. For the experiments in this study, GPER1 analogs G1 and G36 were used. G1 is a nonsteroidal compound that is a highly selective and potent agonist of GPER1 [9]. It does not bind to the estrogen receptors (ER) α and β [9,17]. G36 is a nonsteroidal, selective antagonist of GPER1, which does not bind to ERα or ERβ either. Further, it blocks activation of phosphoinositide-3-kinase and prevents activation of extracellular signal-regulated kinase by G1 or estrogen [17].

GPER1 is expressed by many hormone-sensitive tumor entities and rapidly activates signaling cascades mediated by estrogen, making it a potential target for carcinoma treatment [13]. GPER1 has been detected in tissues including the testis, ovaries, breast, endometrium, and lung [15]. GPER1 is likely to modulate carcinogenesis [14]. The tissue type seems to define the effects of the receptor [15]. In cutaneous neoplasms, the effects of estrogen signaling are largely unexplored [18,19]. Bai et al. [13] reported the function of GPER1 as an oncogene in the skin, but the treatments with estrogens and its antagonists led to conflicting results [12,18]. GPER1 has a tumor-suppressive function in ovarian carcinoma [20]. In breast carcinoma, GPER1 is often described as an oncogene [21–25], but Han et al. [20] observed a tumor-suppressive effect. Tumor-suppressive [14,26–28] and oncogenic effects [11,29,30] are discussed in cervical carcinoma. In vulvar carcinoma, Lan et al. [12] suggest a function of GPER1 as an oncogene. Due to the limited evidence concerning the importance of GPER1 in the progression of vulvar carcinoma and the significance of the receptor as a therapeutic target, especially in gynecological neoplasms, it is further investigated in this study.

Using a tissue microarray of vulvar neoplasia, the correlation between GPER1 expression and grade of malignancy was explored. GPER1 was detected in A431 and CAL-39 vulvar carcinoma cells using immune cytochemical staining. Via Western blotting, the effects of treatment with GPER1 agonist G1 and antagonist G36 on estrogen receptor expression in vulvar carcinoma cell lines A431 and CAL-39 was examined. To investigate the impact of GPER1 on proliferation and viability of vulvar carcinoma cells, the BrdU and Resazurin assays were used. Effects of G1 and G36 on cell morphology of vulvar carcinoma cells A431 and CAL-39 were observed. Gap closure assay was used to analyze the migration of vulvar carcinoma cells A431 and CAL-39 after treatment with G1 and G36. To examine the ability of the vulvar carcinoma cells A431 and CAL-39 to form colonies and tumor spheres and to migrate and proliferate after being treated with G1 and G36, the colony formation and sphere formation assays were used.

2. Results

2.1. Correlation of GPER1 Expression and Grade of Malignancy within a Tissue Microarray of Vulvar Neoplasia

In the evaluated microarray, GPER1 was stained green and cell nuclei were stained blue (Figure 1A–C). Each microarray was evaluated for its GPER1 stained area and its GPER1 staining intensity. The immune-reactive score gives a range of 0–9 (0 = negative, 1–3 = mild, 4–6 = moderate, 7–9 = strong) as a product of multiplication between positive-stained tumor proportion score (0–3; 0 = no positive stained tumor tissue, 1 = <30%

positive stained tumor tissue, 2 = 30–60% positive stained tumor tissue, 3 = >60% positive stained tumor tissue) and staining intensity score (0–3; 0 = no color reaction, 1 = mild color reaction, 2 = moderate color reaction, 3 = intense color reaction), and then subjected to statistical analysis one-way ANOVA (Figure 1A). The immune-reactive score was found to be significantly higher in tissue samples of lymph node metastases (M = 6.5 ± 2.5; n = 2) compared to squamous cell carcinomas (M = 2.68 ± 0.207; n = 72; p = 0.0065), papillary hyperplasias of squamous cell carcinomas (M = 2 ± 0.447; n = 6; p = 0.0047), hyperplasias (M = 2.75 ± 0.75; n = 4; p = 0.0315), and chronic mucosal inflammations (M = 2.313 ± 0.373; n = 16; p = 0.0042). The immune-reactive score was significantly higher in neoplasias with grade 3 (M = 6.5 ± 2.5; n = 2) compared to neoplasias with grade 1 (M = 2.65 ± 0.22; n = 54; p = 0.0017) and grade 2 (M = 1.95 ± 0.285; n = 20; p = 0.0004). The tissue microarray was analyzed for the presence of GPER1 revealing that, among other findings, in hyperplasia an accumulation of GPER1 was observed in the epithelial layer (Figure 1B), while in squamous cell carcinoma samples an accumulation was observed in the stroma with infiltrating tumor cells (Figure 1C).

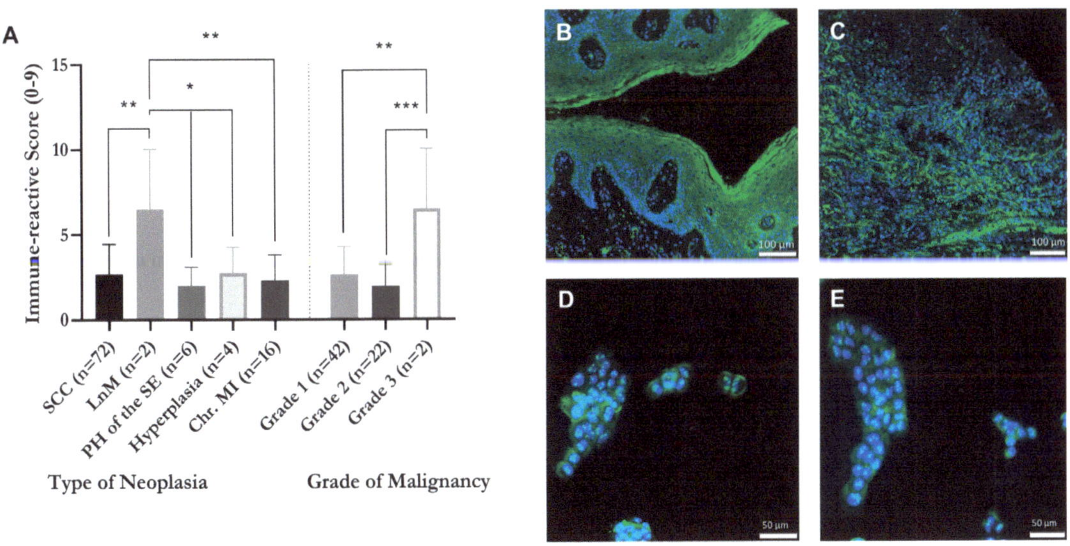

Figure 1. Correlation of GPER1 expression and grade of malignancy within tissue samples of vulvar neoplasia, based on FITC staining (green) of GPER1. Immune-reactive score (**A**) (0–9; staining intensity (0–3) multiplied with area size (0–3)). Detection of GPER1 epithelial in a vulva hyperplasia (**B**) and within the stroma with infiltrating tumor cells of a squamous cell carcinoma of the vulva (**C**). Magnification 10× (**B**,**C**). SCC = squamous cell carcinoma, LnM = lymph node metastasis, PH = papillary hyperplasia, Chr. MI = chronic mucosal inflammation. Mean with SEM, one-way ANOVA, n = 2–72. Detection of GPER1 within vulvar carcinoma cells. Cell lines A431 (**D**) and CAL-39 (**E**). Cell nuclei stained with DAPI (blue) and GPER1 stained with FITC (green). Magnification 20× (**D**,**E**). * $p < 0.05$, ** $p < 0.01$, *** $p < 0.001$.

2.2. Detection of GPER1 in Vulvar Carcinoma Cells in Cytoplasm and Nucleus

In A431 and CAL-39, vulvar carcinoma cell GPER1 was stained green while the cell nuclei were stained blue. In both vulvar carcinoma cell lines, GPER1 was visible to the same extent in the nucleus and cytoplasm (Figure 1D,E).

2.3. G1 Acts through GPER1 in Vulvar Carcinoma Cells

To show that the effect of G1 is elicited by GPER1, cells were treated with G1, both without and with increasing concentrations of GPER1 antagonist G36. The proliferation assay showed that G1 treatment (1.25 μM) significantly decreased proliferation of the vulvar carcinoma cells A431 and CAL-39 (A431: M = 37.39 ± 3.88% vs. ethanol control (= 100%); $n = 3$; $p < 0.001$ (Figure 2A); CAL-39: M = 39.91 ± 3.39% vs. ethanol control (= 100%); $n = 3$; $p < 0.01$ (Figure 2B)). This effect could be inhibited by the GPER1 antagonist G36 in a dose-dependent manner. Cotreatment of A431 cells with 1.25 μM G1 with 1.25 μM G36 resulted in a slight increase in proliferation to 40.28 ± 3.34% as compared with G1 alone ($n = 3$) (Figure 2A). After treatment with 1.25 μM G1 in combination with 2.5 μM G36, proliferation was significantly increased to 72.34 ± 6.62% vs. G1 alone ($n = 3$; $p = 0.05$) If the cells were treated with 1.25 μM G1 and 5 μM G36, proliferation was significantly increased to 100.65 ± 12.64% vs. G1 alone ($n = 3$; $p = 0.001$). After simultaneous treatment of CAL-39 cells with 1.25 μM G1 and 1.25 μM G36, the proliferation rate was 40.43 ± 4.38% ($n = 3$) (Figure 2B). After treatment with 1.25 μM G1 in combination with 2.5 μM G36, proliferation was increased to 67.44 ± 15.04% vs. G1 alone ($n = 3$). The maximum blocking effect was achieved with 5 μM G36 (M = 85.18 ± 10.71% vs. control; $n = 3$; $p < 0.05$).

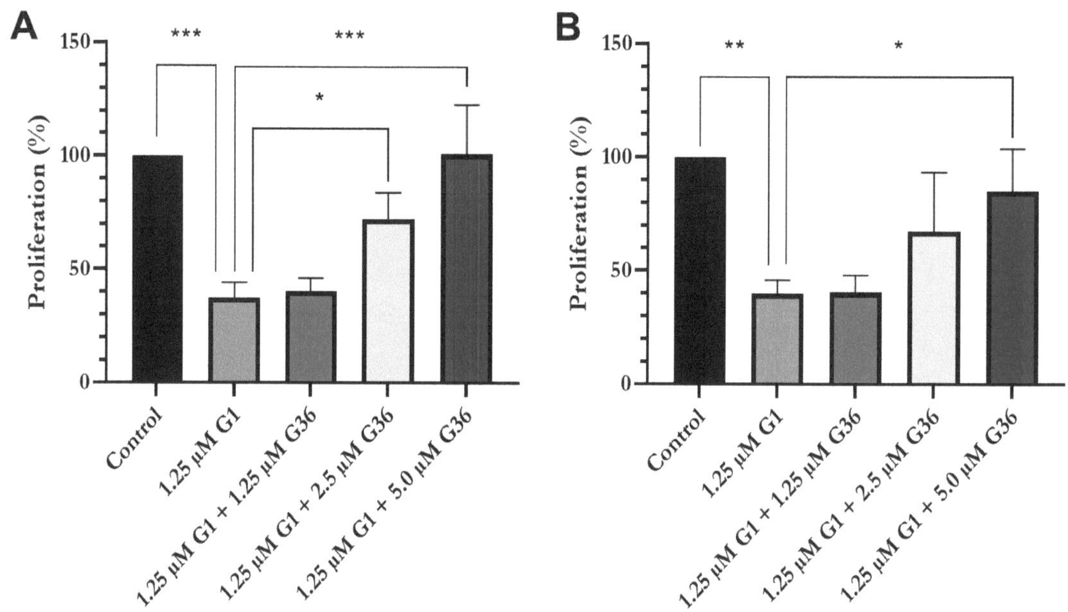

Figure 2. Proliferation after treatment with G1 alone and in combination with GPER1 antagonist G36. Effects of treatments with G1 or with G1 in combination with increasing concentrations of G36 on proliferation of A431 (**A**) and CAL-39 (**B**). Mean with SEM, one-way ANOVA, $n = 3$. * $p < 0.05$, ** $p < 0.01$, *** $p < 0.001$.

2.4. No Impact of GPER1 Agonist G1 and Antagonist G36 Treatment on Expression of Estrogen Receptors in Vulvar Carcinoma Cells

In the A431 vulvar carcinoma cell lines (Figure 3A), there were no changes in the expression of GPER1, ERα, and ERβ after treatment with G1 or G36, compared with control. In the CAL-39 cell line (Figure 3B), the expression of GPER1, ERα, and ERβ also remained unchanged after treatment with G1 or G36. Densiometric evaluation of 3–4 independent experiments each showed no significant effects (Figure S1, Table S1). Additional bands are nonspecific detections of the antibodies.

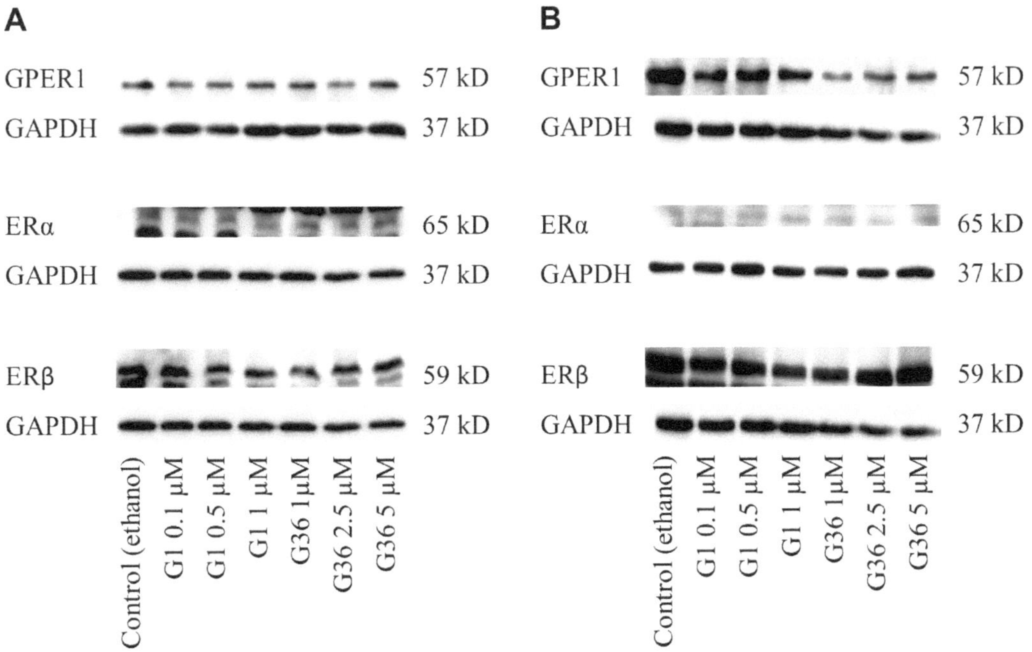

Figure 3. Effects of GPER1 agonist G1 and antagonist G36 on estrogen receptor expression in vulvar carcinoma cells. Expression of GPER1, ERα, and ERβ after treatment with G1 or G36 on vulvar carcinoma cells A431 (**A**) and CAL-39 (**B**). n = 3–4.

2.5. Reduced Proliferation and Viability after Treatment with GPER1 Agonist G1

The results of the BrdU assay showed that GPER1 agonist G1 treatment decreased proliferation of the vulvar carcinoma cells A431 and CAL-39. A slight inhibitory effect was observed in A431 cells treated with 1 μM G1 (M = 90.026 ± 6.95; $n = 4$; $p = 0.3328$) but was dose-dependently significant with 2.5 μM (M = 71.475 ± 2.387; $n = 4$; $p = 0.0003$) and 5 μM (M = 47.267 ± 3.785; $n = 4$; $p < 0.0001$) (Figure 4A). In CAL-39 cells, the inhibition of proliferation was already significant at treatment with 1 μM G1 (M = 70.021 ± 7.221; $n = 4$; $p < 0.0001$) and dose-dependently stronger at 2.5 μM (M = 24.226 ± 3.601; $n = 4$; $p < 0.0001$) and 5 μM (M = 16.885 ± 3.056; $n = 4$; $p < 0.0001$) (Figure 4B). The GPER1 antagonist G36 did not have any significant effect on the vulvar carcinoma cells at all doses compared to the ethanol control (Figure 4A,B). Inhibition of GPER1 with G36 at concentrations of 0.5–5 μM had no significant effect on the viability of the carcinoma cells in either cell line (Figure 4C,D). The GPER1 agonist G1 led to a dose-dependent decrease in cell viability in both cell lines. In A431, a slight reduction in cell viability was achieved at doses of 0.5 μM G1 (M = 93.72 ± 1.745; $n = 5$; $p = 0.3809$) but a significant reduction at doses of 1.25 μM (M = 64.4 ± 10.037; $n = 3$; $p < 0.0001$). This effect was dose-dependently amplified in G1 for concentrations 2.5 μM (M = 13.05 ± 1.429; $n = 4$; $p < 0.0001$) and 3.75 μM (M = 3.167 ± 0.338; $n = 3$; $p < 0.0001$) (Figure 4E). In CAL-39 cells, a significant reduction in cell viability was already achieved at 0.5 μM G1 (M = 72.25 ± 9.962; $n = 4$; $p = 0.0069$) but stronger at 1.25 μM (M = 28.333 ± 8.685; $n = 3$; $p < 0.0001$) and 2.5 μM (M = 15.3 ± 5.081; $n = 4$; $p < 0.0001$) with the minimum measured viability being almost reached at 3.75 μM (M = 13.167 ± 0.731; $n = 3$; $p < 0.0001$) (Figure 4F). Compared to the control in both cell lines, no significant effect of ethanol treatment on cell viability was observed (Figure 4C–F).

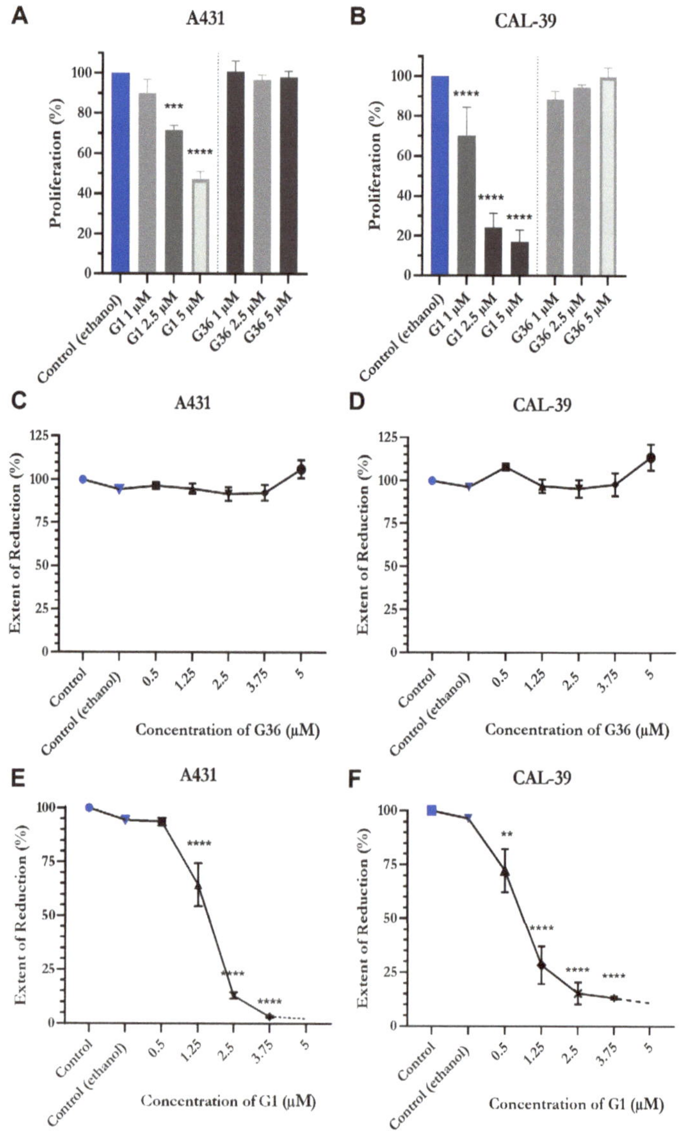

Figure 4. Proliferation and viability after treatment with GPER1 agonist G1. Proliferation after treatment of the vulva carcinoma cells A431 (**A**) and CAL-39 (**B**) with G1 or G36 (controls = blue). Effects of treatments with G1 or G36 on viability of A431 (**C**,**E**) and CAL-39 (**D**,**F**). Mean with SEM (**A**–**F**), one-way ANOVA (**A**–**F**), $n = 3$ (**A**,**B**), $n = 5$–6 (**C**), $n = 3$–5 (**D**–**F**). ** $p < 0.01$, *** $p < 0.001$, **** $p < 0.0001$.

2.6. Signs of Necrosis after Treatment with GPER1 Agonist G1

In the A431 vulvar carcinoma cells, signs of necrosis such as release of cell contents, karyolysis, cell separation from the cell association, and cell swelling were observed after treatment with G1 at time-points t2 (48 h after treatment) and t3 (72 h after treatment). No similar effects were observed after treatment with ethanol (control) and G36 (Figure 5A). The statistical analysis of the cell's length–width ratio did not show any significant differences at all time points and treatments. The ratio was about 1.5 (Figure 5B). The size of the cells remained nonsignificant compared to the ethanol control after G36 treatment, but it increased significantly at t2 (M = 330,370.667 ± 42,431.936; $n = 6$; $p = 0.0032$) and t3 (M = 1,273,521 ± 198,552.721; $n = 6$; $p < 0.0001$) after G1 treatment (Figure 5C).

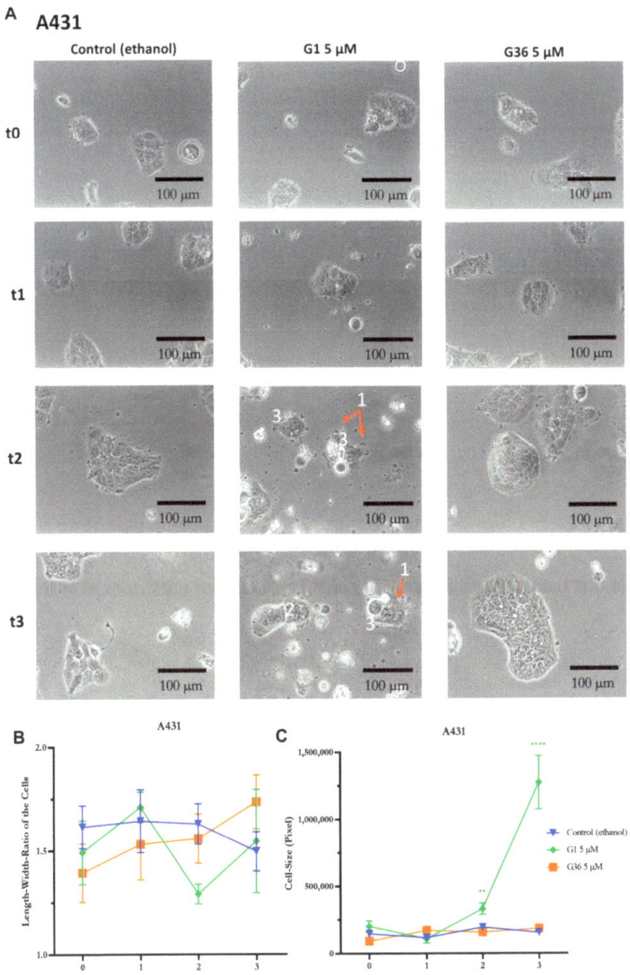

Figure 5. Signs of necrosis after treatment with GPER1 agonist G1 in vulvar carcinoma A431. Photos of the cancer cells after treatment with G1 or G36 vs. control (ethanol) at time t0 (treatment)–t3 (72 h after treatment), interval 24 h (**A**). Necrosis signs: 1 = cell contents are released; 2 = karyolysis; 3 = cell detaches from cell association, cell swelling (**C**). Changes after treatment in length-to-width ratio (**B**) and cell size (**C**). Mean with SEM, one-way ANOVA, 6–12 cells per treatment, $n = 3$ (**B**,**C**). Magnification 20×. ** $p < 0.01$, **** $p < 0.0001$.

In the CAL-39 cell line, the signs of necrosis—release of cell contents, karyolysis, chromatin condensation and clumping to nuclear membrane, cell separation from the cell association, and cell swelling—were observed after G1 treatment, starting slightly at t2 and unambiguous at t3 (Figure 6A). No similar effects were observed after ethanol (control) and G36 treatment (Figure 6C). The length–width ratio of the cells remained not significantly different after ethanol (control) and G36 treatment at all time points. After G1 treatment, a significantly lower length–width ratio was achieved at t3 (M = 1.63 ± 0.081; $n = 12$;

$p = 0.0069$) compared to the ethanol control (Figure 6B). The cell size was significantly increased compared to the ethanol control after G1 treatment at time-points t1 (M = 374,616.125 ± 95,908.505; $n = 8$; $p = 0.0482$), t2 (M = 499,313.7 ± 37,611.973, $n = 10$, $p < 0.0001$), and t3 (M = 503,506.111 ± 85,248.282; $n = 9$; $p = 0.0149$) (Figure 6C).

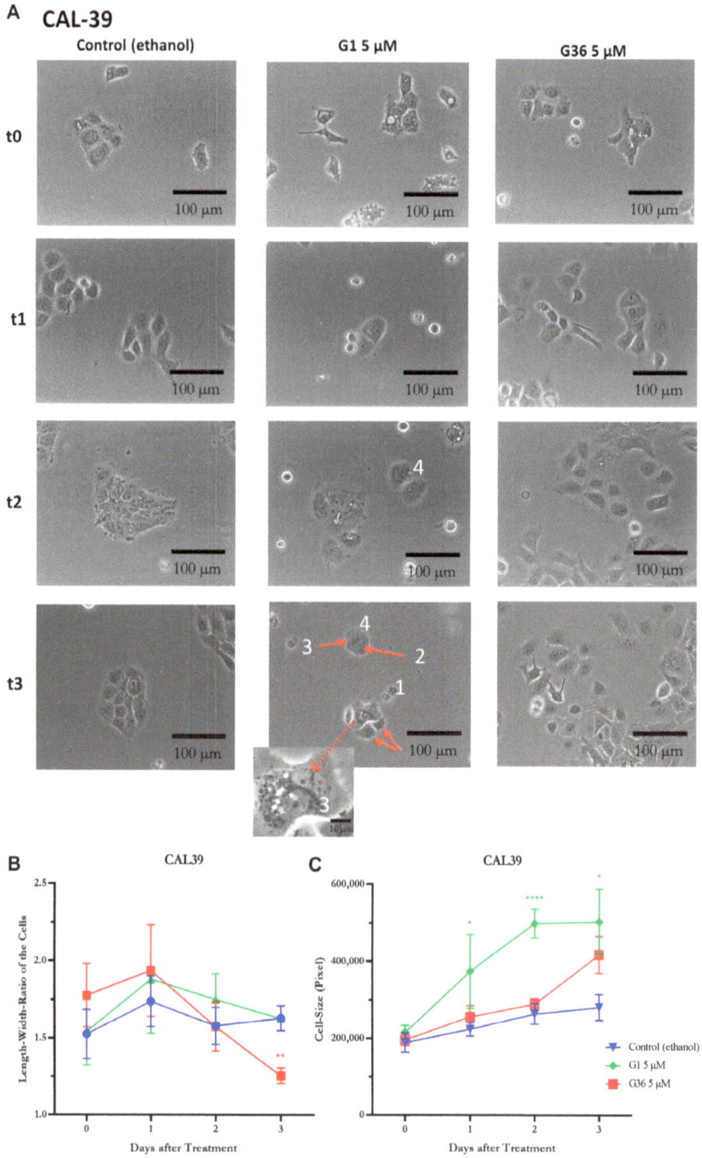

Figure 6. Signs of necrosis after treatment with GPER1 agonist G1 in vulvar carcinoma cells CAL-39. Photos of the cancer cells after treatment with G1/G36 vs. control (ethanol) at time t0 (treatment)–t3 (72 h after treatment), interval 24 h (**A**). Necrosis signs: 1 = cell contents are released; 2 = karyolysis; 3 = chromatin condenses and clumps to nuclear membrane; 4 = cell detaches from cell association, cell swelling (**C**). Changes after treatment in length-to-width ratio (**B**) and cell size (**C**). Mean with SEM, one-way ANOVA, 6–12 cells per treatment, $n = 3$ (**B**,**C**). Magnification 20×. * $p < 0.05$, ** $p < 0.01$, **** $p < 0.0001$.

2.7. Effects of GPER1 Agonist G1 and Antagonist G36 on Migration of Vulvar Carcinoma Cells

G36 treatment at all concentrations did not result in any significant changes in gap size or migration compared to the ethanol control at both time-points t1 and t2 on A431 cells. Treatment with 1 μM G1 did not affect the migration of A431 cells. G1 inhibited the migration of A431 cells at t1 at 2.5 μM (M = 71.008 ± 4.082, n = 6; p = 0.0210) and 5 μM (M = 79.438 ± 4.251, n = 5; p = 0.0005). This inhibition was further amplified at t2 after treatment with G1 at concentrations of 2.5 μM (M = 41.833 ± 1.644, n = 6; p = 0.0001) and 5 μM (M = 65.593 ± 3.078, n = 5; p < 0.0001) (Figure 7B), as demonstrated in the photographs (Figure 7A). The migration of the CAL-39 vulvar carcinoma cells was not significantly affected by G1 and G36 treatment at t1 compared to the ethanol control. At t2, treatment with G1 slowed gap closure at 2.5 μM (M = 50.488 ± 11.119, n = 5; p = 0.0255) and 5 μM (M = 68.572 ± 7.828, n = 6; p = 0.0002). There was no significant effect at lower concentrations. G36 treatment at 1 μM and 5 μM had no observable effects while at 2.5 μM (M = 49.803 ± 4.933, n = 7; p = 0.0145) it reduced migration compared to the ethanol control (Figure 8B). The photographs of the treatments with significant effects on migration are shown in Figure 8A.

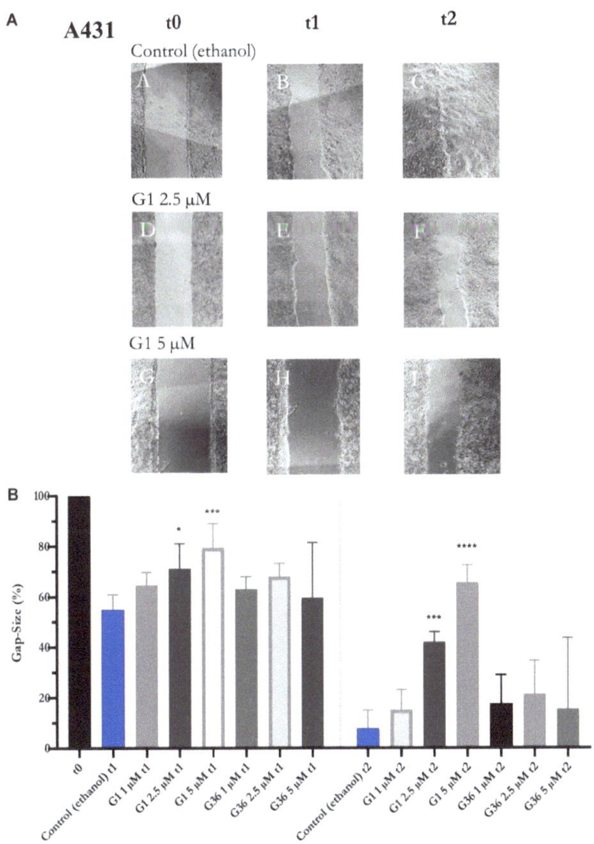

Figure 7. Inhibition of migration of vulvar carcinoma cells A431 by GPER1 agonist G1. Photos of the gaps after treatment with G1 vs. control (ethanol) at time t0–t2, interval 10 h (**A**). Presentation of the effects of treatment with G1 or G36 vs. control (ethanol) on migration (**B**). Mean with SEM, one-way ANOVA, compared to control (ethanol), n = 5–7 (B). Magnification 4×. * p < 0.05, *** p < 0.001, **** p < 0.0001.

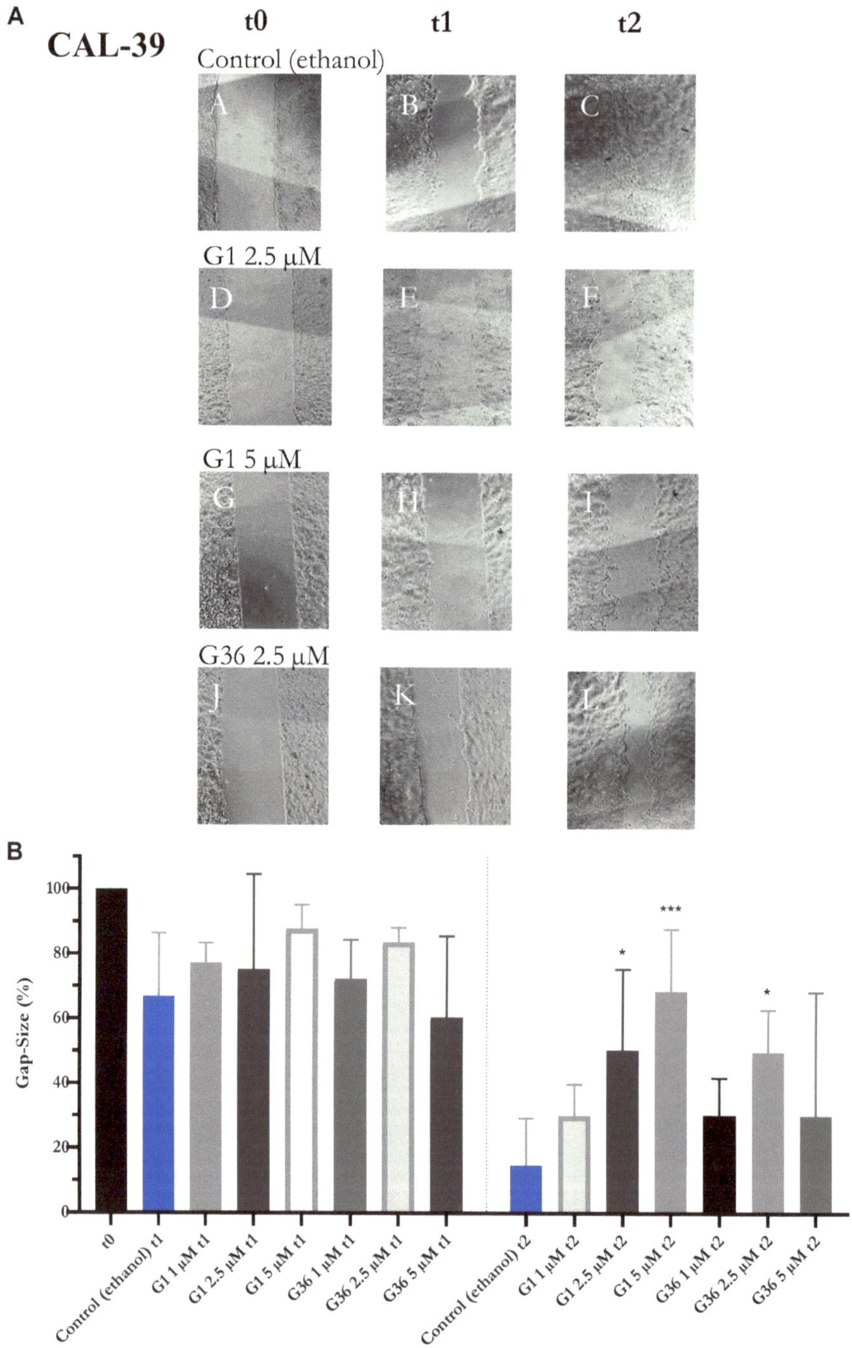

Figure 8. Inhibition of migration of vulvar carcinoma cells CAL-39 by GPER1 agonist G1 and antagonist G36. Photos of the gaps after treatment with G1 or G36 vs. control (ethanol) at time t0–t2, interval 10 h (**A**). Effects of treatments with G1 or G36 vs. control (ethanol) on migration (**B**). Mean with SEM, one-way ANOVA, $n = 4$–7 (**B**). Magnification 4×. * $p < 0.05$, *** $p < 0.001$.

2.8. Effects of GPER1 Agonist G1 and Antagonist G36 on Colony Formation and Sphere Formation

In the vulvar carcinoma cell line A431, the colony size was significantly reduced compared to the ethanol control following treatments with G1 at concentrations of 0.1 µM (M = 21.21 ± 1.503, n = 1611; $p < 0.0001$), 0.5 µM (M = 22.1 ± 1.501, n = 1617; $p < 0.0001$), 1 µM (M = 23.23 ± 2.294, n = 545; $p = 0.0336$), 2.5 µM (M = 0.09968 ± 5.229, n = 93; $p < 0.0001$), and 5 µM (M = 0 ± 5.593, n = 81; $p < 0.0001$). The colony size was significantly increased compared to the ethanol control following treatment with G36 at concentrations of 1 µM (M = 36.46 ± 1.339, n = 2312; $p < 0.0001$) and 5 µM (M = 51.58 ± 1.365, n = 2168; $p < 0.0001$), with the strongest effect at 2.5 µM (M = 56.49 ± 1.375, n = 2119; $p < 0.0001$) (Figure 9A). After treatment with G1, the colony number was significantly increased compared to the ethanol control at concentrations of 0.1 µM (M = 536 ± 25.239, n = 3; $p = 0.0394$) and 0.5 µM (M = 538 ± 27.429, n = 3; $p = 0.0359$), but was dose-dependent significantly reduced at concentrations of 1 µM (M = 90.333 ± 44.867, n = 6; $p < 0.0001$), 2.5 µM (M = 0.167 ± 0.167, n = 6; $p < 0.0001$), and 5 µM (M = 0 ± 0, n = 6; $p < 0.0001$). The colony number was not significantly different from the ethanol control following treatment with G36 at all concentrations (Figure 9B).

The size of tumor spheres was compared to the ethanol control after treatment of G1 at 1 µM on day 8 (M = 21,048.245 ± 3584.927, n = 53; $p = 0.0567$) and day 16 (M = 44,125.283 ± 8145.794, n = 46; $p = 0.0508$) tendentiously reduced but significantly reduced on day 12 (M = 32,251.216 ± 5503.460, n = 51; $p = 0.0073$). At 2.5 µM G1, the size of tumor spheres was significantly reduced on day 4 (M = 10514.057 ± 1105.903, n = 88; $p = 0.0086$), day 8 (M = 11,154.221 ± 1203.315, n = 77; $p < 0.0001$), day 12 (M = 10,156.815 ± 1330.249, n = 74; $p < 0.0001$), and day 16 (M = 10,156.815 ± 1330.249, n = 65; $p < 0.0001$). The size of tumor spheres was also significantly reduced at 5 µM G1 on day 4 (M = 7662.938 ± 678.615, n = 112; $p < 0.0001$), day 8 (M = 6507.705 ± 544.729, n = 98; $p < 0.0001$), day 12 (M = 5922.547 ± 642.367, n = 64; $p < 0.0001$), and day 16 (M = 4698.797 ± 374.268, n = 59; $p < 0.0001$) (Figure 9C).

During the observation period, the number of tumor spheres did not differ between ethanol (control), G1, and G36 treatment (Figure 9D,F). The size of tumor spheres was significantly reduced after treatment with G36 at a concentration of 1 µM on day 8 (M = 12,784.092 ± 1625.408, n = 65, $p < 0.0001$), day 12 (M = 31,820.967 ± 3700.883, n = 92; $p = 0.0014$), and day 16 (M = 44,043.552 ± 5480.897, n = 87; $p = 0.0183$), after treatment with 2.5 µM on day 8 (M = 17,626.044 ± 1580.861, n = 91; $p < 0.0001$), day 12 (M = 28,805.060 ± 3518.959, n = 84; $p = 0.0003$), and day 16 (M = 35,438.506 ± 4760.253, n = 89; $p = 0.0005$), and after treatment with 5 µM on day 12 (M = 29,327.087 ± 3657.152, n = 69; $p = 0.0006$) and day 16 (M = 36,558.309 ± 5034.015, n = 68; $p = 0.0016$) (Figure 9E).

The colony size of CAL-39 vulvar carcinoma cells was significantly reduced after treatment with 0.1 µM G1 (M = 13.07 ± 2.404, n = 273; $p = 0.0005$) compared to the ethanol control. After treatment with 0.5 µM G1 (M = 15.06 ± 6.224, n = 35; $p = 0.9973$), a nonsignificantly reduced colony size was observed. The reduction in colony size was significantly increased compared to the ethanol control after treatment with G1 at concentrations of 1 µM (M = 0.4167 ± 3.836, n = 96; $p < 0.0001$), 2.5 µM (M = 0 ± 3.855, n = 95; $p < 0.0001$), and 5 µM (M = 0 ± 3.855, n = 95; $p < 0.0001$). At concentrations of 1 µM and 2.5 µM, GPER1 antagonist G36 had no effect on colony size. At 5 µM, G36 (M = 31.34 ± 1.95, T n = 829; $p < 0.0001$) increased the colony size significantly compared to the ethanol control (Figure 10A). A slight decrease in colony number compared to the ethanol control was observed after treatment with G1 at a concentration of 0.1 µM (M = 92 ± 31.974, n = 3; $p = 0.5022$) and significantly reduced colony number after treatment with 0.5 µM (M = 11.667 ± 7.688, n = 3; $p = 0.0053$), 1 µM (M = 0.333 ± 0.333, n = 6; $p < 0.0001$), 2.5 µM (M = 0 ± 0, n = 6; $p < 0.0001$), and 5 µM (M = 0 ± 0, n = 6; $p < 0.0001$). Treatment with G36 at all the concentrations used did not significantly affect the number of colonies (Figure 10B). Regardless of treatment with G36, G1, or ethanol (control), the tumor spheres of CAL-39 cells that formed were consistently dissolved on the ultralow attachment plate. Accordingly, without a significant difference compared to the ethanol control, the number of tumor spheres steadily decreased after treatment with G1 or G36 at the concentrations noted (Figure 10D,F). During the period examined, there were no significant differences in the size of the tumor spheres after treatment with G1 and G36 when compared to ethanol control (Figure 10C,E).

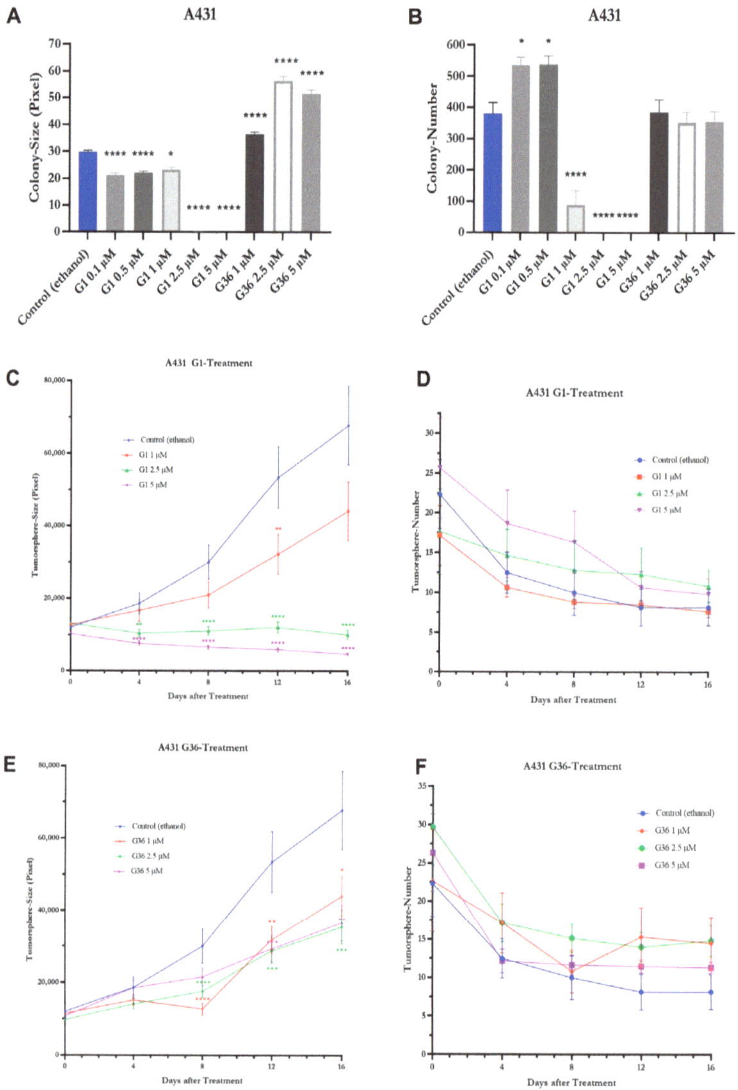

Figure 9. Effects of GPER1 agonist G1 and antagonist G36 on colony formation and tumor sphere formation in vulvar carcinoma cells A431. Presentation of the effects of treatment with G1/G36 vs. control (ethanol) on colony size (**A**) and number (**B**), and on tumor sphere size (**C**,**E**) and number (**D**,**F**). Mean with SEM (**A**–**F**), one-way ANOVA (**A**–**F**), n = 81–2312 (**A**), n = 3–9 (**B**), n = 45–160 (**C**,**E**), n = 5 (**D**,**F**). * $p < 0.05$, ** $p < 0.01$, *** $p < 0.001$, **** $p < 0.0001$.

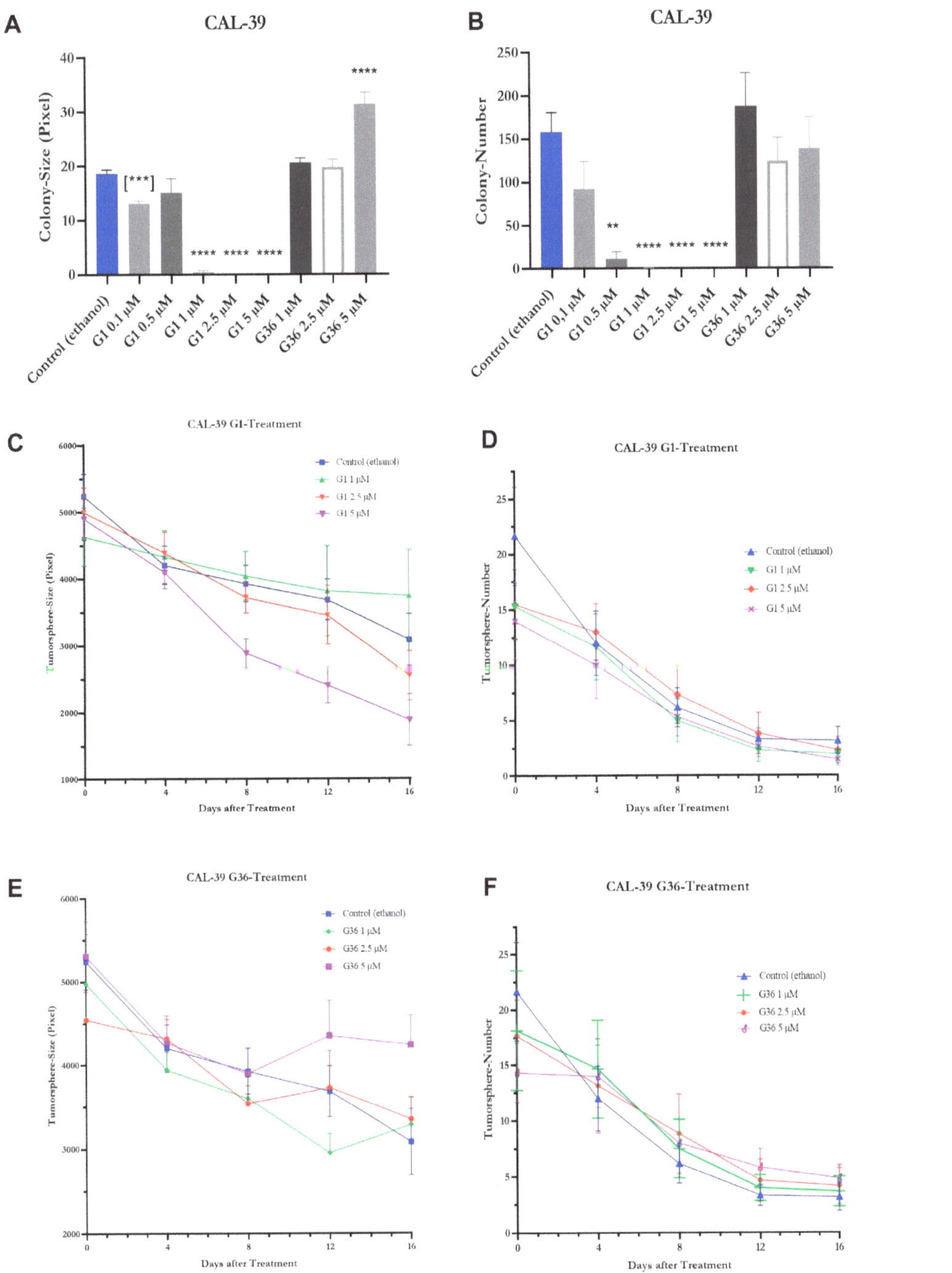

Figure 10. Effects of GPER1 agonist G1 and antagonist G36 on colony formation and tumor sphere formation in vulvar carcinoma cells CAL-39. Plots of the effects of treatment with G1/G36 vs. control (ethanol) on the colony size (**A**) and number (**B**), and on tumor sphere size (**C**,**E**) and number (**D**,**F**). Mean with SEM (**A–F**), one-way ANOVA (**A–F**), (***) = unpaired *t*-test (**A**), n = 35–1121 (**A**), n = 3–9 (**B**), n = 11–164 (**C**,**E**), n = 5 (**D**,**F**). ** $p < 0.01$, *** $p < 0.001$, **** $p < 0.0001$.

3. Discussion

GPER1 is often involved in the progression of neoplasms with varying impacts. Its role as a tumor suppressor or an oncogene appears to depend on the tissue type and tumor entity [15]. The precise effects of GPER1 in carcinogenesis remain unclear and are subject to discussion within specific tissues [10,27]. This study aims to examine the function of GPER1 in vulvar carcinoma.

Using immune cytochemistry, which allows the visualization of molecular markers [31], it has been demonstrated that GPER1 is expressed in the cytoplasm and nuclei of vulvar carcinoma cells A431 and CAL-39. The cytoplasmic localization of GPER1 in the A431 cell line has also been validated by Lan et al. [12]. GPER1 has been localized in the cytoplasm and nucleus of cervical neoplasms [10] and at the cell membrane, particularly in the area of invasion fronts [11].

Tissue microarrays enable the quantification and localization of proteins [32]. The staining of the microarray indicates that GPER1 is consistently expressed in vulvar carcinoma, as evidenced by the presence of GPER1 staining in all samples. The receptor is primarily observed in the epithelial layers or in the stroma with infiltrating tumor cells of the tissue samples. Additionally, a high immune-reactive value, which corresponds to elevated GPER1 expression, showed a positive correlation with a high tumor grade. Hence, similar findings have been observed in cervical carcinoma [26,29]. Accordingly, this high GPER1 expression is prognostically associated with lower 5-year survival rates in patients with cervical carcinoma [11]. Likewise, in breast carcinoma, a correlation has been observed between high GPER1 expression, increased cell proliferation, invasion, migration [23], and worsened prognosis [21].

To analyze the impact of GPER1 activation or inhibition on vulvar carcinoma cells, various aspects such as cell viability, proliferation, migration, colony formation, and sphere formation were examined following treatment with the GPER1 agonist G1 or the GPER1 antagonist G36. Although G1 is considered to be a selective GPER1 agonist [9], receptor-independent effects have been demonstrated by showing that G1 is able to interact directly with tubulin [33–35] at the same site as colchicine [33]. Therefore, we checked whether the biological effect of G1 occurs through GPER1 in vulvar carcinoma cells. We could demonstrate that the effect of G1 was abolished in a dose-dependent manner in the presence of GPER1 antagonist G36 (not shown). This implies that G1 mediates its effects via binding to GPER1. This does not exclude additional receptor-independent effects.

The BrdU assay was used to assess the proliferation of the cells [36]. Results revealed a dose-dependent decrease in proliferation at 1 μM G1 for CAL-39 and 2.5 μM for A431. Han et al. [20] reported similar findings for the G1 impact in ovarian carcinoma cells. GPER1 activation also leads to reduced proliferation in breast carcinoma cells [20], while in contrast, Girgert et al. [37] observed decreased growth in triple-negative breast carcinoma cells following GPER1 knockdown. Correspondingly, inhibiting GPER1 resulted in decreased proliferation in experiments involving oral squamous cell carcinoma [13].

During the evaluation of cell viability using the Resazurin assay [38], activating GPER1 with at least 0.5 μM G1 in CAL-39 and 1.25 μM in A431 resulted in a decline in cell viability. Han et al. [20] also observed reduced cell viability in ovarian carcinoma cells CaOV3 and CaOV4 after treatment with G1, as assessed by an MTT assay. In contrast, Lan et al. [12] found that treatment of A431 cells with 0.01–0.5 μM G1 led to a dose-dependent increase in viability. Furthermore, a decrease in cell viability was observed after treatment with 5 μM GPER1 antagonist G15 [12]. Since the same A431 cell line was used by Lan et al. [12] as in this study, cell-line-specific factors can be ruled out as the cause of the contradictory results. It is possible this different effect can be explained with the use of GPER1 antagonist G15 instead of G36, as G15 has the ability to bind to ERα and ERβ, which is not the case for G36 [17]. This could have activated other signaling pathways than treatment with G36.

The migration ability of vulvar carcinoma cells was investigated using the gap closure assay [39]. For A431 cells, it was dose-dependently reduced by GPER1 activation with at least 2.5 μM G1 after 10 h, while CAL-39 cells exhibited a reduction after 20 h. Likewise, the migration ability of ovarian carcinoma cells was reduced by G1 in a wound-healing scratch-assay [20].

The colony formation ability is influenced by cell proliferation and migration. The impact of GPER1 activation/inhibition on this ability of vulvar carcinoma cells was assessed using the colony formation assay [40]. In A431 cells, the activation of GPER1 with 0.1 μM and 0.5 μM G1 led to an increase in colony number, followed by a dose-dependent decrease from 1 to 5 μM G1. Lower doses may have caused colony disintegration, resulting in reduced colony size but increased count. At 1 μM G1, the tumor-suppressive effect became more prominent, oppressing the formation of new colonies. This is supported by the observation that even at 0.1 μM G1 colony size was reduced compared to the control, but a significant decrease first was achieved at 2.5 μM G1. In CAL-39 cells, treatment with G1 resulted in a dose-dependent decrease in colony number and size. These results support a tumor-suppressive function of GPER1 activation.

To observe cell proliferation and their ability to form tumor spheres, the sphere formation assay was used [41]. Treatment with G1 and G36 did not affect the number of tumor spheres. In A431 cells, however, the size of tumor spheres decreased in a dose-dependent manner after treatment with G1. Thus, it can be inferred that GPER1 activation has a tumor-suppressive effect in this assay. In CAL-39 cell line, both the number and size of tumor spheres decreased without significant differences during the observation period in the G1-treated, G36-treated, and control. It is likely that CAL-39 cells faced challenges in forming and maintaining stable tumor spheres in the nonadherent environment, which explains the lack of treatment differences.

To gain a better understanding of the effects of G1 and G36 on the cells, cell morphology was observed after treatment. In A431 cells, signs of necrosis such as cell content release, significant cell swelling, cell detachment from the cell cluster, and karyolysis were observed after G1 treatment. Additionally, chromatin condensation at the nuclear membrane was seen in CAL-39 cells. Consequently, GPER1 activation leads to necrosis, a nonprogrammed cell death. These observations support the tumor-suppressive effect of GPER1 shown in other experiments. Hernandez-Silva et al. [10] also reported an increase in necrosis and apoptosis after G1 treatment in cervical neoplasia.

By conducting Western blot analysis, changes in protein expression of estrogen receptors in vulvar carcinoma cells A431 and CAL-39 were examined after GPER1 activation/inhibition. However, no significant alterations in expression were observed, suggesting that the effects of G1 and G36 treatment shown in this study were not caused by changes in the expressions of GPER1, ERα, or ERβ.

In the gap closure assay, migration of CAL-39 vulvar carcinoma cells was inhibited after treatment with 2.5 μM G36, similar to the inhibitory effect observed after G1 treatment. An increased or decreased expression of estrogen receptors cannot be a reason for that. In the colony formation assay, it was noticed that G36 treatment increased the size of colonies in A431 cells at all doses and in CAL-39 cells at only 5 μM, while the number of colonies remained unchanged compared to the control. This suggests that growth occurred within existing colonies. However, it is contradictory that no changes in cell size or morphology were observed compared to the control in the examination of cell morphology and the BrdU assay showed no effects from G36 treatment on proliferation. Surprisingly, the size of A431 tumor spheres decreased compared to the control after G36 treatment. This decrease was independent of the dosage. Further, the size difference to the control remained relatively constant instead of increasing with higher doses as expected with inhibition. The splitting of the spheres cannot explain these observations since the number of tumor spheres did not increase either. The inhibition of GPER1 using G36 did not yield significant results in the other experiments of this study.

4. Materials and Methods

4.1. Cell Culture

The human vulvar cancer cell lines A431 and CAL-39 were obtained from American Type Cell Collection (ATCC; Manassas, VA, USA). They were cultured in Minimum Essential Medium (MEM; Thermo Fisher Scientific, Waltham, MA, USA) supplemented with 0.1% Transferrin (Sigma-Aldrich, St. Louis, MO, USA), 1% Penicillin/Streptomycin (*p/S* Thermo Fisher Scientific), 10% fetal calf serum (FCS; Pan Biotech, Aidenbach, Germany), and 26 IU Insulin (Sanofi, Paris, France), in a humidified atmosphere with 5% CO_2 at 37 °C

4.2. Drugs

GPER1 agonist G1 and GPER1 antagonist G36 were purchased from Biomol (Hamburg, Germany). G1 and G36 were dissolved in ethanol. Each control was treated with 0.03 v/v% ethanol.

4.3. Proliferation

The BrdU assay was implemented with a kit (Roche Applied Science, Penzberg, Germany). Five thousand cells per well were seeded in 96-well plates (Corning Life Sciences, Tewksbury, MA, USA). They were treated with G1 or G36 after 24 h and incubated for 50 h The cells were treated with 10 µL BrdU Labeling Reagent 1:100 in MEM. After another 22 h of incubation, the BrdU amount was measured by a multidetection microplate reader (BioTek Instruments, Winooski, VT, USA) and GEN5 3.12 software (BioTek Instruments).

4.4. Viability

For the Resazurin assay, 600 cells of A431 and 1200 cells of CAL-39 per well were seeded in 96-well plates (Corning Life Sciences) in Dulbecco's Minimum Essential Medium w/o phenol-red (DMEM; Thermo Fisher Scientific). A greater quantity of CAL-39 cells was utilized to ensure that the color change occurred simultaneously in both cell lines, enabling a simultaneous execution of the assay. The treatment with G1 or G36 was carried out 24 h later. After incubation for 72 h, 20 µL Resazurin (Thermo Fisher Scientific) was added to every well. Ten hours later, the relative reduction in Resazurin was measured at 570 nm and 630 nm using a multidetection microplate reader (BioTek Instruments) and analyzed using GEN5 1.08 software (BioTek Instruments). The relative reduction was calculated with Excel (Microsoft, Redmond, WA, USA).

4.5. Cell Morphology

In a 6-well plate (Greiner Bio-One, Frickenhausen, Germany), 30,000 cells per well of A431 and 40,000 cells per well of CAL-39 were seeded. The cells were treated 24 h later with 5 µM G36 or G1. The software uEye Cockpit 2.0 (IDS Imaging Development Systems, Obersulm, Germany) was used to obtain pictures of the wells at t0–t3, with t0 after treatment and the others in a 24 h interval. The cells were analyzed with ImageJ (Wayne Rasband, National Institutes of Health, Bethesda, MD, USA).

4.6. Gap Closure

One hundred forty thousand cells per well of A431 and CAL-39 were seeded in a 24-well plate (Greiner Bio-One) separated by an insert. After incubation for 24 h, the cells were washed with Dulbecco's Phosphate Buffered Saline (DPBS; Pan Biotech) and treated with G1/G36. Pictures of the cells were taken using uEye Cockpit 2.0 (IDS Imaging Development Systems, Obersulm, Germany) at t0 and every 10 h until the gap was closed. The gap closure was analyzed by using Adobe Photoshop CS5 12.0 (Adobe Inc., San José, CA, USA), ImageJ 1.52a (Wayne Rasband), and Microsoft Excel (Microsoft).

4.7. Colony Formation

One thousand cells of A431 and CAL-39 per well were seeded in a 6-well plate (Greiner Bio-One). After 24 h, the cells were treated with G1 or G36 and incubated for five to seven days until the colonies emerged. The plates were stained with crystal violet (Sigma-Aldrich) and scanned using Epson Perfection V850 pro (Epson, Suwa, Japan) and the Epson Scan 2 software (Epson). Colonies were analyzed with ImageJ 1.52a (Wayne Rasband).

4.8. Tumor Sphere Formation

One thousand cells per well were seeded in an ultralow attachment 96-well plate with flat bottom (Corning Life Sciences). After 24 h of incubation, the cells were treated with G1/G36. Using the Celigo Cyntellect cell imaging cytometer (Nexcelom Bioscience LLC, Lawrence, KS, USA) and the Celigo 2.1.0.96 software (Celigo Inc., San Mateo, CA, USA), the plates were photographed every 4 days, starting after treatment with t0 until t5. The tumor spheres were analyzed using ImageJ 1.52a (Wayne Rasband).

4.9. Tissue Microarray

The vulvar carcinoma tissue microarray (US Biomax, Derwood, MD, USA) was deparaffinized, unmasked, and incubated with rabbit antihuman GPER1 antibody at 1:200 dilution (Thermo Fisher Scientific) overnight in a wet chamber. The next day, after washing, it was incubated with Alexa Fluor 488 antirabbit IgG antibody solution (Invitrogen, Waltham, MA, USA) at 1:10 dilution. Nuclei were stained using 4′,6-diamidino-2-phenylindole dihydrochloride at 1:1000 dilution (DAPI; Novus Biologicals, Littleton, CO, USA). The tissue microarray was embedded with fluorescence mounting medium (Dako North America, Carpinteria, CA, USA). Pictures were taken using the software cellSens Dimension 3.2 (Olympus Life Science Solutions, Tokyo, Japan) and an Olympus IX83microscope (Olympus Life Science Solutions).

4.10. Immune Cytology

Using an 8-well plate on glass (Sarstedt, Nümbrecht, Germany), 4000 and 8000 cells of A431 and 12,000 and 16,000 cells of CAL-39 were seeded and incubated for 24 h. After incubation with formaldehyde (3.7%; Sigma-Aldrich) for 15–30 min, the cells were washed twice with fluorescence staining solution/DPBS (Pan Biotech), 2% BSA (Carl Roth, Karlsruhe, Germany) and 0.25% Triton X-100 (Sigma-Aldrich). The cells were incubated overnight with the rabbit antihuman GPER1 antibody at 1:100 dilution (Thermo Fisher Scientific) in a wet chamber. After washing, fluorescence staining solution containing DAPI at 1:1000 dilution (Novus Biologicals) and Alexa Fluor 488 antirabbit IgG antibody solution (Invitrogen) at 1:10 dilution were added. After 30 min incubation, the cells were washed twice with fluorescence staining solution and once with DPBS. Fluorescence mounting medium (Dako North America) was added and covered with a cover slip (Menzel-Gläser, Braunschweig, Germany) and allowed to dry. The stained cells were photographed using an Olympus IX83 microscope (Olympus Life Science Solutions) and cellSens Dimension 3.2 software (Olympus Life Science Solutions).

4.11. Western Blot

Initially, 250,000 cells of A431 and 500,000 cells of CAL-39 were seeded and incubated for 24 h. The cells were then treated with G36 and G1 and incubated for 72 h. After washing the cells with DPBS (Pan Biotech), they were incubated 15–20 min with PBS-EDTA (100 mL DPBS, 200 mg EDTA tetra sodium salt (Affymetrix, Cleveland, OH, USA), detached, and centrifuged at 1300× g rcf for 5 min. Then, they were lysed in lytic buffer (Sigma-Aldrich, St. Louis, USA), 100 µL phosphatase inhibitor (Sigma-Aldrich, St. Louis, MO, USA), and 100 µL protease inhibitor (Sigma-Aldrich, St. Louis, MO, USA). Protein concentration was measured by the Bradford method using a multidetection microplate reader (BioTek Instruments) at 590 nm and GEN5 1.08 software (BioTek Instruments). Protein was diluted 1:1 with Laemmli buffer (Fresenius Kabi, Bad Homburg vor der Höhe,

Germany), 5 mL LDS sample buffer (Thermo Fisher Scientific), and 1 mL NuPage sample-seducing agent (Thermo Fisher Scientific). The proteins were fractioned using SDS-PAGE and Western blot apparatus (Bio-Rad Laboratories, Hercules, CA, USA). The membrane was incubated overnight at 4 °C with rabbit antihuman GAPDH at 1:1000 dilution (Cell Signaling Technology, Danvers, MA, USA), rabbit antihuman GPER1 at 1:2000 dilution (Thermo Fisher Scientific), rabbit antihuman ERα at 1:200 dilution from a 100 μg/mL stock solution (Santa Cruz Technology, Dallas, TX, USA), or rabbit antihuman ERβ at 1:200 dilution from a 100 μg/mL stock solution (Santa Cruz Technology). Before incubating with antirabbit IgG at 1:10,000 or 1:40,000 (GAPDH) dilution (Dianova, Hamburg, Germany) for 1 h, the membrane was washed three times for 10 min with TBST (0.1% Tween-20 (AppliChem, Darmstadt, Germany) in TBS). Chemiluminescence solution (400 μL peroxide solution (Westar Supernova, Cyanagen, Italy) and 400 μL Luminol amplification solution (Westar Supernova)) was applied to the membrane for 1 min. The Western blots were detected using a C-Digit blot scanner (Li-Cor Biosciences, Lincoln, NE, USA) and analyzed with Image Studio 4.0 software (Li-Cor Biosciences).

4.12. Statistical Analysis

The experiments were performed on at least three technical and biological replicates. Data was statistically analyzed using one-way ANOVA or rarely unpaired, two-tailed, parametric t-tests by GraphPad Prism 8.0.1 (GraphPad Software, San Diego, CA, USA). In the figures presented, the significances are represented by * ($p < 0.05$), ** ($p < 0.01$), *** ($p < 0.001$), and **** ($p < 0.0001$).

5. Conclusions

Based on the findings of this study, the activation of GPER1 in vulvar carcinoma appears to have a tumor-suppressive effect. This is supported by the decrease in migration, proliferation, colony formation, and tumor sphere formation, together with the presence of morphological signs of necrosis and the reduction in cell viability after G1 treatment in both vulvar carcinoma cell lines. In the CAL-39 cell line compared to the A431 cell line, these effects were observed at lower doses in the Resazurin, BrdU, colony formation, and sphere formation assays. Changes in expression of estrogen receptors do not seem to be the underlying cause of these observations. Additionally, there may be a correlation between GPER1 expression and malignancy grade of vulvar carcinoma. Further evidence is needed to better understand the function of GPER1 in vulvar carcinoma and to potentially utilize GPER1 activation in medical treatment.

Supplementary Materials: The following supporting information can be downloaded at: https://www.mdpi.com/article/10.3390/ijms241813705/s1.

Author Contributions: Conceptualization, C.G.; investigation, J.L. and L.H.; writing—original draft preparation, J.L. and C.G.; writing—review and editing, G.B. and J.G.; project administration, C.G. All authors have read and agreed to the published version of the manuscript.

Funding: This research received no external funding.

Institutional Review Board Statement: Not applicable.

Informed Consent Statement: Not applicable.

Data Availability Statement: The datasets used and/or analyzed during the current study are available from the corresponding author on reasonable request.

Acknowledgments: The authors thank Sonja Blume and Matthias Läsche for excellent technical assistance. We thank Anja Uhmann, Institute for Human Genetics, University Medical Center Göttingen, for her help in performing the BrdU assay.

Conflicts of Interest: The authors declare no conflict of interest.

References

1. Goldnau, C.; Köninger, A.; Kimmig, R. Das äußere Genitale der Frau: Pathologische Befunde und initiale Behandlungsschritte. *Der Urol.* **2010**, *49*, 1496–1502. [CrossRef]
2. Hacker, N.F.; Eifel, P.J.; van der Velden, J. Cancer of the vulva. *Int. J. Gynaecol. Obstet. Off. Organ Int. Fed. Gynaecol. Obstet.* **2015**, *131* (Suppl. 2), S76–S83. [CrossRef]
3. Merlo, S. Modern treatment of vulvar cancer. *Radiol. Oncol.* **2020**, *54*, 371–376. [CrossRef] [PubMed]
4. Nugent, E.K.; Brooks, R.A.; Barr, C.D.; Case, A.S.; Mutch, D.G.; Massad, L.S. Clinical and Pathologic Features of Vulvar Intraepithelial Neoplasia in Premenopausal and Postmenopausal Women. *J. Low. Genit. Tract Dis.* **2011**, *15*, 15–19. [CrossRef] [PubMed]
5. Guidozzi, F. Estrogen therapy in gynecological cancer survivors. *Climacteric* **2013**, *16*, 611–617. [CrossRef] [PubMed]
6. van der Velden, J.; van Lindert, A.C.; Gimbrere, C.H.; Oosting, H.; Heintz, A.M. Epidemiologic Data on Vulvar Cancer: Comparison of Hospital with Population-Based Data. *Gynecol. Oncol.* **1996**, *62*, 379–383. [CrossRef] [PubMed]
7. De Angelis, R.; Sant, M.; Coleman, M.P.; Francisci, S.; Baili, P.; Pierannunzio, D.; Trama, A.; Visser, O.; Brenner, H.; Ardanaz, E.; et al. Cancer survival in Europe 1999-2007 by country and age: Results of EUROCARE-5-A population-based study. *Lancet Oncol.* **2014**, *15*, 23–34. [CrossRef]
8. Preti, M.; Bucchi, L.; Micheletti, L.; Privitera, S.; Corazza, M.; Cosma, S.; Gallio, N.; Borghi, A.; Bevilacqua, F.; Benedetto, C. Four-decade trends in lymph node status of patients with vulvar squamous cell carcinoma in northern Italy. *Sci. Rep.* **2021**, *11*, 5661. [CrossRef]
9. Bologa, C.G.; Revankar, C.M.; Young, S.M.; Edwards, B.S.; Arterburn, J.B.; Kiselyov, A.S.; Parker, M.A.; Tkachenko, S.E.; Savchuck, N.P.; Sklar, L.A.; et al. Virtual and biomolecular screening converge on a selective agonist for GPR30. *Nat. Chem. Biol.* **2006**, *2*, 207–212. [CrossRef]
10. Hernandez-Silva, C.D.; Riera-Leal, A.; Ortiz-Lazareno, P.C.; Jave-Suárez, L.F.; Ramírez de Arellano, A.; Lopez-Pulido, E.I.; Macías-Barragan, J.G.; Montoya-Buelna, M.; Dávila-Rodríguez, J.R.; Chabay, P.; et al. GPER Overexpression in Cervical Cancer Versus Premalignant Lesions: Its Activation Induces Different Forms of Cell Death. *Anti-Cancer Agents Med. Chem.* **2019**, *19*, 783–791. [CrossRef]
11. Ino, Y.; Akimoto, T.; Takasawa, A.; Takasawa, K.; Aoyama, T.; Ueda, A.; Ota, M.; Magara, K.; Tagami, Y.; Murata, M.; et al. Elevated expression of G protein-coupled receptor 30 (GPR30) is associated with poor prognosis in patients with uterine cervical adenocarcinoma. *Histol. Histopathol.* **2020**, *35*, 351–359. [CrossRef] [PubMed]
12. Lan, J.; Gao, X.-H.; Kaul, R. Estrogen receptor subtype agonist activation in human cutaneous squamous cell carcinoma cells modulates expression of CD55 and Cyclin D1. *EXCLI J.* **2019**, *18*, 606–618. [CrossRef] [PubMed]
13. Bai, L.-Y.; Weng, J.-R.; Hu, J.-L.; Wang, D.; Sargeant, A.M.; Chiu, C.-F. G15, a GPR30 antagonist, induces apoptosis and autophagy in human oral squamous carcinoma cells. *Chem. Interactions* **2013**, *206*, 375–384. [CrossRef]
14. Hernández-Silva, C.D.; Villegas-Pineda, J.C.; Pereira-Suárez, A.L. Expression and Role of the G Protein-Coupled Estrogen Receptor (GPR30/GPER) in the Development and Immune Response in Female Reproductive Cancers. *Front. Endocrinol.* **2020**, *11*, 544. [CrossRef] [PubMed]
15. Jacenik, D.; Cygankiewicz, A.I.; Krajewska, W.M. The G protein-coupled estrogen receptor as a modulator of neoplastic transformation. *Mol. Cell. Endocrinol.* **2016**, *429*, 10–18. [CrossRef] [PubMed]
16. Girgert, R.; Emons, G.; Gründker, C. Estrogen Signaling in ERα-Negative Breast Cancer: ERβ and GPER. *Front. Endocrinol.* **2019**, *9*, 781. [CrossRef] [PubMed]
17. Dennis, M.K.; Field, A.S.; Burai, R.; Ramesh, C.; Petrie, W.K.; Bologa, C.G.; Oprea, T.I.; Yamaguchi, Y.; Hayashi, S.-I.; Sklar, L.A.; et al. Identification of a GPER/GPR30 antagonist with improved estrogen receptor counterselectivity. *J. Steroid Biochem. Mol. Biol.* **2011**, *127*, 358–366. [CrossRef]
18. Ku, T.K.S.; Crowe, D.L. Coactivator-mediated estrogen response in human squamous cell carcinoma lines. *J. Endocrinol.* **2007**, *193*, 147–155. [CrossRef]
19. Mancuso, M.; Gallo, D.; Leonardi, S.; Pierdomenico, M.; Pasquali, E.; De Stefano, I.; Rebessi, S.; Tanori, M.; Scambia, G.; Di Majo, V.; et al. Modulation of basal and squamous cell carcinoma by endogenous estrogen in mouse models of skin cancer. *Carcinog.* **2009**, *30*, 340–347. [CrossRef]
20. Han, N.; Heublein, S.; Jeschke, U.; Kuhn, C.; Hester, A.; Czogalla, B.; Mahner, S.; Rottmann, M.; Mayr, D.; Schmoeckel, E.; et al. The G-Protein-Coupled Estrogen Receptor (GPER) Regulates Trimethylation of Histone H3 at Lysine 4 and Represses Migration and Proliferation of Ovarian Cancer Cells In Vitro. *Cells* **2021**, *10*, 619. [CrossRef]
21. Girgert, R.; Emons, G.; Gründker, C. Inhibition of growth hormone receptor by Somavert reduces expression of GPER and prevents growth stimulation of triple-negative breast cancer by 17β-estradiol. *Oncol. Lett.* **2018**, *15*, 9559–9566. [CrossRef]
22. Liu, Y.; Du, F.-Y.; Chen, W.; Fu, P.-F.; Yao, M.-Y.; Zheng, S.-S. G15 sensitizes epithelial breast cancer cells to doxorubicin by preventing epithelial-mesenchymal transition through inhibition of GPR30. *Am. J. Transl. Res.* **2015**, *7*, 967–975.
23. Molina, L.; Bustamante, F.; Ortloff, A.; Ramos, I.; Ehrenfeld, P.; Figueroa, C.D. Continuous Exposure of Breast Cancer Cells to Tamoxifen Upregulates GPER-1 and Increases Cell Proliferation. *Front. Endocrinol.* **2020**, *11*, 563165. [CrossRef] [PubMed]
24. Schmitz, V.; Bauerschmitz, G.; Gallwas, J.; Gründker, C. Suppression of G Protein-coupled Estrogen Receptor 1 (GPER1) Enhances the Anti-invasive Efficacy of Selective ERβ Agonists. *Anticancer. Res.* **2022**, *42*, 5187–5194. [CrossRef] [PubMed]

25. Yang, S.-Y.V.; Lee, J.-W.; Kim, W.-S.; Jung, K.-L.; Lee, S.-J.; Lee, J.-H.; Bae, D.-S.; Kim, B.-G. Adenoid cystic carcinoma of the Bartholin's gland: Report of two cases and review of the literature. *Gynecol. Oncol.* **2006**, *100*, 422–425. [CrossRef] [PubMed]
26. Friese, K.; Kost, B.; Vattai, A.; Marmé, F.; Kuhn, C.; Mahner, S.; Dannecker, C.; Jeschke, U.; Heublein, S. The G protein-coupled estrogen receptor (GPER/GPR30) may serve as a prognostic marker in early-stage cervical cancer. *J. Cancer Res. Clin. Oncol.* **2018** *144*, 13–19. [CrossRef] [PubMed]
27. Ramírez-López, I.G.; De Arellano, A.R.; Jave-Suárez, L.F.; Hernández-Silva, C.D.; García-Chagollan, M.; Hernández-Bello, J. Lopez-Pulido, E.I.; Macias-Barragan, J.; Montoya-Buelna, M.; Muñoz-Valle, J.F.; et al. Interaction between 17β-estradiol, prolactin and human papillomavirus induce E6/E7 transcript and modulate the expression and localization of hormonal receptors. *Cancer Cell Int.* **2019**, *19*, 227. [CrossRef] [PubMed]
28. Ruckriegl, S.; Loris, J.; Wert, K.; Bauerschmitz, G.; Gallwas, J.; Gründker, C. Knockdown of G Protein-coupled Estrogen Receptor 1 (GPER1) Enhances Tumor-supportive Properties in Cervical Carcinoma Cells. *Cancer Genom. Proteom.* **2023**, *20*, 281–297 [CrossRef]
29. Akimoto, T.; Takasawa, A.; Takasawa, K.; Aoyama, T.; Murata, M.; Osanai, M.; Saito, T.; Sawada, N. Estrogen/GPR30 Signaling Contributes to the Malignant Potentials of ER-Negative Cervical Adenocarcinoma via Regulation of Claudin-1 Expression *Neoplasia* **2018**, *20*, 1083–1093. [CrossRef]
30. Zhu, C.-X.; Xiong, W.; Wang, M.-L.; Yang, J.; Shi, H.-J.; Chen, H.-Q.; Niu, G. Nuclear G protein-coupled oestrogen receptor (GPR30) predicts poor survival in patients with ovarian cancer. *J. Int. Med. Res.* **2018**, *46*, 723–731. [CrossRef]
31. Gerdes, J. Ki-67 and other proliferation markers useful for immunohistological diagnostic and prognostic evaluations in human malignancies. *Semin. Cancer Biol.* **1990**, *1*, 199–206.
32. Camp, R.L.; Charette, L.A.; Rimm, D.L. Validation of Tissue Microarray Technology in Breast Carcinoma. *Lab. Investig.* **2000** *80*, 1943–1949. [CrossRef]
33. Lv, X.; He, C.; Huang, C.; Hua, G.; Wang, Z.; Remmenga, S.W.; Rodabough, K.J.; Karpf, A.R.; Dong, J.; Davis, J.S.; et al. G-1 Inhibits Breast Cancer Cell Growth via Targeting Colchicine-Binding Site of Tubulin to Interfere with Microtubule Assembly. *Mol Cancer Ther.* **2017**, *16*, 1080–1091. [CrossRef] [PubMed]
34. Mori, T.; Ito, F.; Matsushima, H.; Takaoka, O.; Tanaka, Y.; Koshiba, A.; Kusuki, I.; Kitawaki, J. G protein–coupled estrogen receptor 1 agonist G-1 induces cell cycle arrest in the mitotic phase, leading to apoptosis in endometriosis. *Fertil. Steril.* **2015** *103*, 1228–1235.e1221. [CrossRef] [PubMed]
35. Wang, C.; Lv, X.; Jiang, C.; Davis, J.S. The putative G-protein coupled estrogen receptor agonist G-1 suppresses proliferation of ovarian and breast cancer cells in a GPER-independent manner. *Am. J. Transl. Res.* **2012**, *4*, 390–402. [PubMed]
36. Padet, L.; St-Amour, I.; Aubin, E.; Proulx, D.P.; Bazin, R.; Lemieux, R. Dose-Dependent inhibition of brdu detection in the cell proliferation elisa by culture medium proteins. *J. Immunoass. Immunochem.* **2009**, *30*, 348–357. [CrossRef]
37. Girgert, R.; Emons, G.; Gründker, C. Inactivation of GPR30 reduces growth of triple-negative breast cancer cells: Possible application in targeted therapy. *Breast Cancer Res. Treat.* **2012**, *134*, 199–205. [CrossRef]
38. Riss, T.L.; Moravec, R.A.; Niles, A.L.; Duellman, S.; Benink, H.A.; Worzella, T.J.; Minor, L. Cell Viability Assays. In *Assay Guidance Manual*; Eli Lilly & Company and the National Center for Advancing Translational Sciences: Bethesda, MD, USA, 2016.
39. Van Horssen, R.; Hagen, T.L.T. Crossing barriers: The new dimension of 2D cell migration assays. *J. Cell. Physiol.* **2011**, *226*, 288–290. [CrossRef]
40. Horibata, S.; Vo, T.V.; Subramanian, V.; Thompson, P.R.; Coonrod, S.A. Utilization of the Soft Agar Colony Formation Assay to Identify Inhibitors of Tumorigenicity in Breast Cancer Cells. *J. Vis. Exp.* **2015**, *99*, e52727. [CrossRef]
41. Lee, C.H.; Yu, C.C.; Wang, B.Y.; Chang, W.W. Tumorsphere as an effective in vitro platform for screening anti-cancer stem cell drugs. *Oncotarget* **2015**, *7*, 1215–1226. [CrossRef]

MDPI AG
Grosspeteranlage 5
4052 Basel
Switzerland
Tel.: +41 61 683 77 34

International Journal of Molecular Sciences Editorial Office
E-mail: ijms@mdpi.com
www.mdpi.com/journal/ijms

www.ingramcontent.com/pod-product-compliance
Lightning Source LLC
Chambersburg PA
CBHW041404210926
51850CB00013B/3170
* 9 7 8 3 7 2 5 8 2 4 9 2 2 *